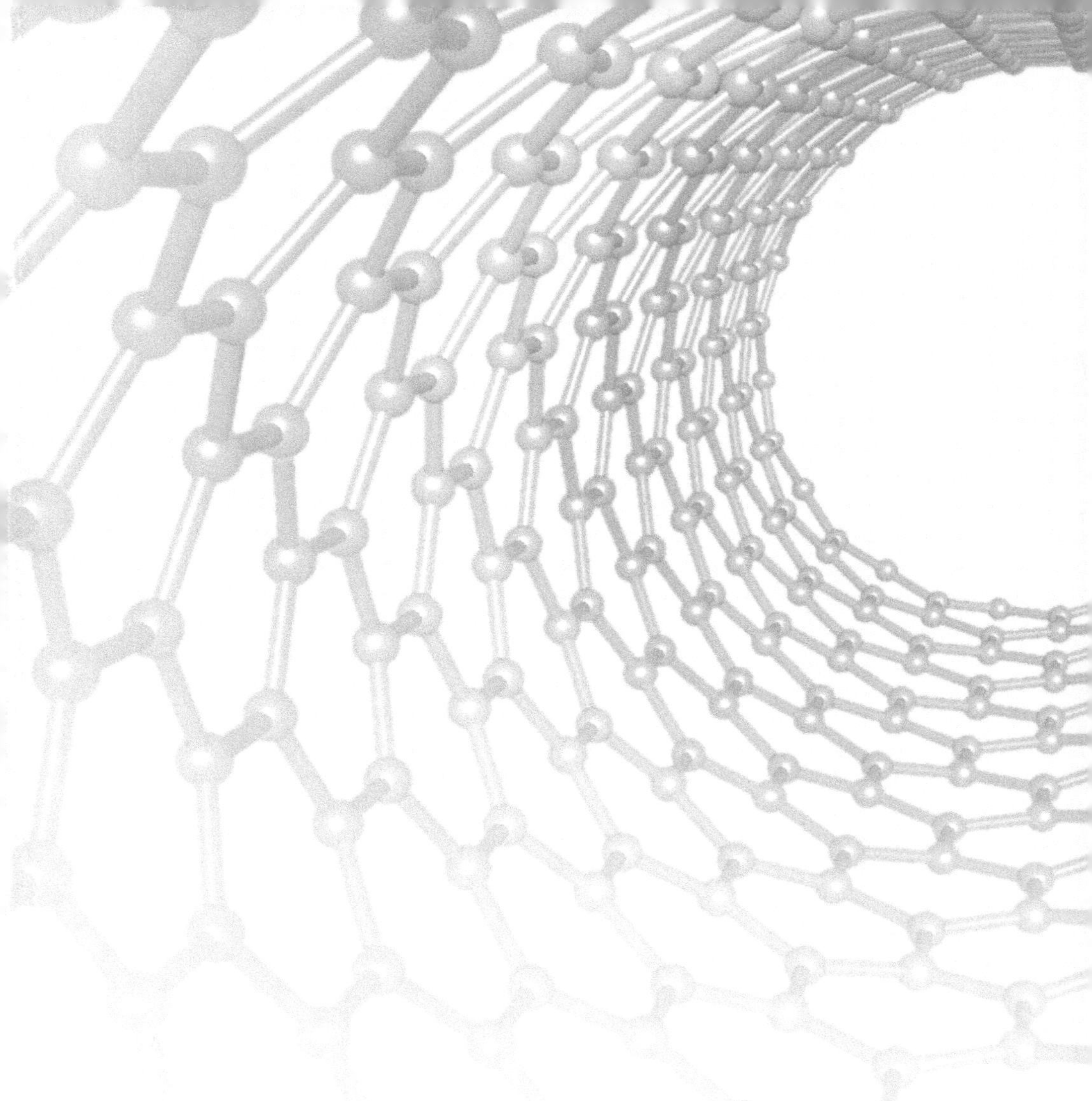

Nanocarbon Electronics

Nanocarbon Electronics

edited by

Changjian Zhou | Min Zhang | Cary Y. Yang

JENNY STANFORD
PUBLISHING

Published by

Jenny Stanford Publishing Pte. Ltd.
Level 34, Centennial Tower
3 Temasek Avenue
Singapore 039190

Email: editorial@jennystanford.com
Web: www.jennystanford.com

British Library Cataloguing-in-Publication Data
A catalogue record for this book is available from the British Library.

Nanocarbon Electronics

ISBN 978-981-4877-11-4 (Hardcover)
ISBN 978-1-003-04308-9 (eBook)

Contents

Preface

Nanocarbon, or a form of carbon in the nanoscale, has been around for many millennia. The discoveries of carbon nanotubes (CNTs), fullerenes, graphene, and other nanocarbons in the past few decades, together with the advancement of scientific knowledge and laboratory and analytical research tools, brought to light the enormous potential of this class of nanomaterials. Nanocarbons consist of a large family including zero-dimensional (0D) fullerene, carbon dots, and graphene quantum dots; one-dimensional (1D) CNT, carbon nanofibers, and carbon nanohorns; and two-dimensional (2D) graphene and graphene nanoribbons. Our focus here is on the 1D CNT and 2D graphene to review past and current studies on the fundamental properties, synthesis methods, and various electronic devices fabricated using either or both of these two nanocarbons. The superior properties of these nanocarbons have provided fertile grounds for numerous studies of their physics and chemistry, as well as their applicability to technologies such as integrated circuits (IC or chip), sensors, and energy conversion and storage. In this edited volume, we have attempted to address the suitability of CNT and graphene as the active materials in transistors and sensors, as on-chip interconnects, and in energy conversion and storage systems. We hope that the contents of this volume will stimulate further studies, leading to the eventual realization of these nanocarbons' full potential in electronics and related fields.

The book consists of eight chapters. Chapter 1 overviews the electronic properties of CNT and graphene while highlighting their most-studied potential applications in electronics. It also includes a summary of each chapter. More in-depth discussions on materials and device fabrication are given in Chapter 2, followed by applications in interconnects in Chapter 3 and transistors in Chapters 4 and 5. Chapter 6 covers potential applications in flexible sensors, while Chapter 7 presents applications for energy conversion and storage. The final chapter presents the state of the nanocarbon market, current challenges, and a projection of where the field of nanocarbon electronics might be heading, based on expected advances in its science and technology in the coming years.

A meaningful study of nanocarbons and their applications would not be possible without the groundbreaking works of pioneers over the past three decades, which are cited throughout this volume. We are grateful for this opportunity to share the knowledge gained from this study with the scientific community. Throughout the compilation of this volume, we were fortunate to have received the help and support of our many colleagues and students, whose advice and contributions at various stages of the manuscript preparation made the completion of this task possible. We are deeply grateful to all chapter contributors, especially Dr. Jessica Koehne, Dr. Morgan Anderson, Prof. Jun Li, Prof. Jeongwon Park, Patrick Wilhite, Dr. Anshul Vyas, and Dr. Yu Zheng for their valuable and timely efforts. To our publishers, Stanford Chong and Jenny Rompas, and their editor, Shivani Sharma, we express our heartfelt gratitude for their patience and assistance throughout the seemingly never-ending manuscript preparation process. Finally, we thank our respective institutions, South China University of technology and Peking University Shenzhen Graduate School, China, and Santa Clara University, California, USA, for their support and encouragement.

Changjian Zhou
Min Zhang
Cary Y. Yang
Summer 2020

Chapter 1

Overview of Nanocarbon Electronics

Changjian Zhou[a] and Cary Y. Yang[b]
[a]*School of Microelectronics, South China University of Technology, Guangzhou 510641, China*
[b]*Center for Nanostructures, Santa Clara University, Santa Clara, California 95053, USA*
zhoucj86@gmail.com

1.1 Introduction

As the most prominent representative of nanomaterials, the family of nanocarbons provides an excellent platform for discovering novel physical phenomenon, as well as inventing emerging devices with superior performances in various applications including integrated circuits (ICs), sensors, electrochemical catalysts, and energy storage [1–3]. In the IC industry, Moore's law has been used to predict the development trend in the last six decades [4]. As the technology of this industry enters into the sub-10 nm nodes, various nanomaterials and device structures have been proposed to tackle the emerging challenges such as gate controllability and leakage current in field-effect transistors (FETs) [5], delay and metal line electromigration in

Nanocarbon Electronics
Edited by Changjian Zhou, Min Zhang, and Cary Y. Yang

ISBN 978-981-4877-11-4 (Hardcover), 978-1-003-04308-9 (eBook)
www.jennystanford.com

interconnect structures [6], and on-chip and chip-package thermal management [7]. Since the discovery of carbon nanotube (CNT) in 1991 [8], various electronic devices, circuits, and even entire microprocessor chips have incorporated CNTs in them as both passive and active components [9, 10], further demonstrating the promising role of this nanocarbon in next-generation ICs. CNT-based FET with a gate length scaled down to 5 nm has been successfully fabricated with performance superior to its silicon counterparts with the same dimensions [11].

The advent of graphene [12] has invoked a strong interest in two-dimensional materials and devices [13], which in turn results in attracting continuously increasing number of researchers from various disciplines [14, 15]. The remarkable carrier mobility, mechanical flexibility, and unique optoelectronic properties of graphene make it an ideal material for high-frequency transistors, flexible sensors, and photodetectors [16–18]. As suggested in the latest International Roadmap for Devices and Systems (IRDS™) [19], the IC industry is now adopting an application-oriented development strategy that focuses on integrating more functionality and power efficiency of the overall system, and not simply pursuing the continuous feature size shrinkage in the future. Thus, both CNT and graphene with their high surface-to-volume ratio are suitable for integrated flexible sensor applications, which will undoubtedly add much value to silicon-based ICs when a mature integrating strategy is developed [20]. Such applications are key to the development of electronic devices for the fields of artificial intelligence and Internet of things [21, 22].

While nanocarbons consist of a large family of materials including zero-dimensional (0D) fullerene, carbon dots, and graphene quantum dots; one-dimensional (1D) CNT, carbon nanofibers, and carbon nanohorns; and two-dimensional (2D) graphene and graphene nanoribbons, we mainly focus on CNT and graphene in this book to examine past and current studies on the fundamental properties, synthesis methods, and various electronic devices fabricated using either or both of these two nanocarbons. In this chapter, we present a brief overview of these two nanocarbons and the typical devices developed based on them. The organization of the chapters is then described followed by a summary of each.

1.2 Nanocarbon Materials Fundamentals

Carbon atom has four electrons in its outermost shell and its electronic configuration in the ground state is $1s^2 2s^2 2p^2$. The energy difference between 2s and 2p orbitals is negligible compared to the ionization energy of carbon atom, allowing wave functions of s and p orbitals to mix with each other, generally known as sp^n orbital hybridization [23]. In sp^n hybridization, (n+1) covalent bonds are responsible for the local structure that results in various allotropes of carbon. For example, sp^3 hybridization gives rise to four orbitals which overlap with neighboring carbon hybridized orbitals to form the strong bonds in diamond. In sp^2 hybridization, each carbon atom forms covalent bonds with three neighboring atoms resulting in a planar structure as in graphene and graphite. A weak π-bond among neighboring atoms exists for electrons in the $2p_z$ orbitals. For both graphene and CNT, the three strong in-plane covalent bonds are responsible for the high mechanical strength and high thermal conductivity, while the π electrons are delocalized over the graphitic plane and responsible for the high carrier mobility and other unique electronic transport properties [24, 25].

Graphene is made up of sp^2 hybridized C–C atoms in a planar honeycomb structure. The graphene crystal and its unit lattice vectors ($\mathbf{a_1}, \mathbf{a_2}$) are shown in Fig. 1.1(a). The honeycomb structure is not a Bravais lattice, but can be defined by a hexagonal lattice with a two-atom basis, with its unit cell defined by ($\mathbf{a_1}, \mathbf{a_2}$) as shown in Fig. 1.1(a). The lattice constant $a = |\mathbf{a_1}| = |\mathbf{a_2}|$ is related to the C–C bond length or nearest-neighbor distance d_{CC} = 1.42 Å by $a = \sqrt{3}d_{CC}$, and $\angle(\mathbf{a_1}, \mathbf{a_2}) = 60°$. Graphene can exist in the form of a monolayer or as multilayer graphene (MLG). Energy band structure for either configuration can be computed based on the graphene structure shown in Fig. 1.1(a). The full energy band structure of a graphene monolayer is given in Fig. 1.1(c), showing the valence and conduction bands. For a material to be classified as conductor, electrons must populate the conduction band. In graphene, the valence and conduction bands meet at K and K′ points, which allow electrons to easily populate the conduction band at finite temperatures and participate in conduction. The point in reciprocal lattice where valence and conduction bands meet is called the Dirac point. In graphene, it happens to coincide with a K or K′ point. As there is no bandgap between conduction and valence bands, monolayer graphene is considered a zero-bandgap material.

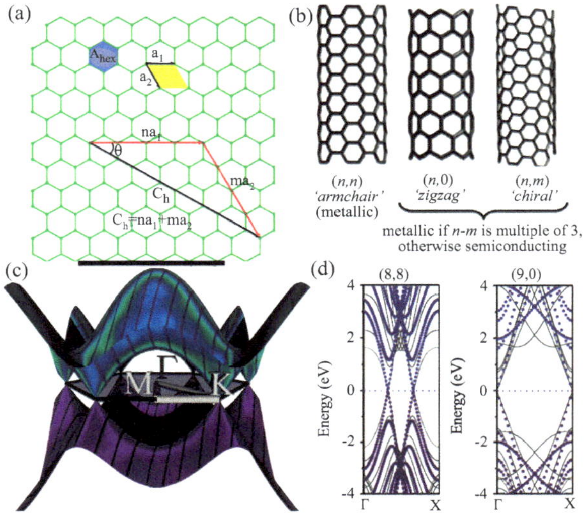

Figure 1.1 (a) Crystal structure of graphene. A_{hex} is the area of a hexagon (light blue), which equals to that of the unit cell containing two carbon atoms (yellow). (b) Schematics of SWCNT with different chiralities (n, m). n and m determine the electronic and optical properties of the SWCNT. (c) Electronic band structure of graphene showing the Γ, K, and M points in the Brillouin zone (BZ). The cutting lines on valence and conduction bands correspond to the CNT's allowed states. (d) Electronic band structure of an (8, 8) armchair CNT and a (9, 0) zigzag CNT obtained from zone-folding (solid circles) and ab initio calculations (solid lines). Figures (a) and (b) reprinted with permission from Ref. [31], Copyright 2015, by IOP Publishing and Figs. (c) and (d) reprinted with permission from Ref. [32], Copyright 2019, by the American Physical Society.

The structure of CNT is closely related to that of graphene, as shown in Figs. 1.1(a) and (b). Single-walled CNT (SWCNT) consists of a single shell of carbon atoms formed by rolling a graphene sheet into a cylinder, and multi-walled CNT (MWCNT) is made up of concentric shells of graphene with an inter-shell spacing of 3.354 Å. The direction of such rolling, known as chirality, determines the electronic and optical properties of the resulting CNT. Identification of SWCNT's chirality has been carried out using optical absorption spectroscopy [26]. The orientation and atomic arrangements of a single CNT shell can be specified using three vectors: chiral vector $\mathbf{C_h}$, translational vector $\mathbf{T}$, and symmetry vector $\mathbf{R}$ [27]. The first two of these vectors are used in specifying CNT unit cells in real and reciprocal spaces. The chiral angle θ is defined and labeled as the angle between vectors $\mathbf{C_h}$ and $\mathbf{a_1}$ in Fig. 1.1(a). In particular, zigzag and armchair SWCNTs shown in Fig. 1.1(b) correspond to $\theta = 0°$ with chiral indices or chirality (n, 0), and $\theta = 30°$ with (n, n), respectively. Note from Fig. 1.1(b) that the names 'armchair' and 'zigzag' each reflects the shape of the chiral vector along which the graphene is rolled to form a SWCNT. The translational vector $\mathbf{T}$ is defined as the unit lattice vector in a carbon nanotube wall, along the nanotube axis and normal to the chiral vector.

The SWCNT electronic structure, similar to its atomic structure, can be derived from that of graphene by neglecting the cylindrical curvature of the CNT wall. One can project the cutting lines onto the reciprocal lattice of graphene to derive the 1D CNT band structure from that of graphene, using a scheme known as zone folding (ZF) [28]. The 1D energy sub-bands of CNT are constructed by "cutting" into the 2D graphene bands in its Brillouin zone (BZ) with the set of cutting lines. An example of ZF for an armchair and zigzag CNT is shown in Fig. 1.1(d). In either case, the cutting line crosses the Fermi level at K and K′ points of the graphene BZ, thus resulting in metallic behavior for CNT, since the valence and conduction bands of graphene intersect at these points. The ZF method thus provides a good approximation for obtaining electronic band structure of CNT from that of graphene for CNTs having large diameters, as curvature effects can be neglected. According to the chiral indices (n, m), the condition (n–m) = 3i (integer i) infers that for all the possible values of $\mathbf{C_h}$, 1/3 of SWCNTs are metallic and the remaining 2/3 are semiconducting. As we can see from the following chapters, this statistical distribution of the SWCNTs has profound consequences on the performance of electron devices [29, 30].

Furthermore, CNT can be broadly classified into two types, SWCNT and MWCNT. SWCNT can be further classified into metallic and semiconducting types, depending on its chiral vector as discussed above. MWCNT can sometimes appear as consisting of concentric graphitic cylinders with a hollow interior or a stacked-cup-shaped structure known as carbon nanofiber (CNF) [33]. As far as the growth of nanocarbons goes, the synthesis of fullerene serves as the foundation (although we do not specifically address fullerene in this volume), because the methods used to synthesize CNTs are exactly the same as those for fullerene [34, 35]. The first experimental evidence of MWCNT dates back to 1952 [36], when it was reported that the microscopic images of carbon filaments showed clear inner cavity, although the graphitic walls of the carbon structure were not clearly identified, probably due to the resolution limit of the transmission electron microscope (TEM) used. The rediscovery of MWCNT by Iijima in 1991 [8], who observed the clear high-resolution TEM images of a MWCNT, has spurred enormous research interests in CNTs since. Figure 1.2(a) shows the first experimental evidence of MWCNT with 50 nm diameter reported in 1952 [37]. The TEM images of MWCNT reported by Iijima [8] is shown in Fig. 1.2(b), in which the graphitic walls are clearly visible [8]. Soon after the rediscovery of MWCNT, two groups including Iijima and Bethune from IBM synthesized SWCNT independently using the same arc-discharge growth method but with metal catalysts additives [38, 39]. Figures 1.2(c) and (d) clearly reveal the single graphitic shell in the SWCNTs. The mass production of CNTs was realized with the arc-discharge and laser ablation methods [40–42]. The integration of CNT with solid substrate for potential applications in electron devices has been achieved with the development of the chemical vapor deposition (CVD) method.

While the theoretical properties of graphene were well documented even before the advent of CNT, the isolation of monolayer graphene and related experimental research have a short history dating back to 2004, when Geim and Novoselov adopted an ingenious yet simple exfoliation method to obtain the first monolayer graphene [12]. Prior to this discovery, monolayer graphene was thought to be thermodynamically unstable and tend to form fullerene, CNT, or other nanocarbons to minimize the total free energy [43, 44]. Indeed, the exfoliated graphene was transferred onto a SiO_2/Si substrate, and turned out to be not perfectly flat but featured multiple ripples to stabilize the 2D structure.

The discovery of graphene presented physicists and materials scientists with fertile ground for theoretical studies, such as its unique charge carrier transport and optoelectronic properties [45–47]. At the same time, graphene-based electronics received great attention soon after its discovery, and leading companies such as

IBM and Intel have invested in studying the potential applications of graphene in future electron devices [48, 49]. Its high carrier mobility, room-temperature ballistic transport, and high saturated carrier velocity make graphene an ideal material for high-frequency FET applications [49, 50]. While the low on/off current ratio of early demonstrated transistors inhibited its application for logic devices, it opened up new research efforts to engineer its bandgap [51]. Similar to CNT, large-scale synthesis of graphene has been achieved with CVD methods, which paved the way toward wafer-scale integration with current electronic devices.

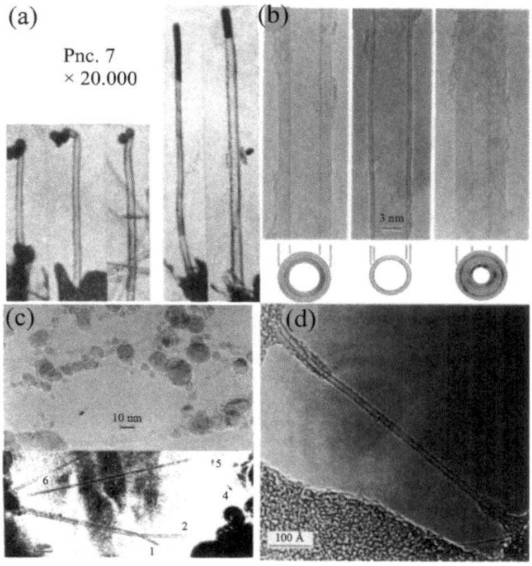

Figure 1.2 (a) First TEM images of MWCNTs [37]. (b) TEM images of several MWCNTs with different number of walls. Parallel dark lines correspond to the (002) lattice images of graphite. A cross section of each tubule is illustrated. (c) TEM images showing bundles of SWCNTs with the lower panel showing several individual SWCNTs. The SWCNTs labelled 1 and 2 are 0.75 nm and 1.37 nm in diameter, respectively. (d) TEM image of a single SWCNT. Figure (b) reprinted from Ref. [8], Copyright 1991; Fig. (c) from Ref. [38], Copyright 1993; and Fig. (d) from Ref. [39], Copyright 1993, by permission from Springer Nature.

1.3 Nanocarbon-Based Electronic Devices

Following Moore's law, reducing transistor dimensions results in lower operating voltages, increased switching speeds, and less power consumption, which in turn yield orders of magnitude increase in switching speed or clock frequency of microprocessors from the initial generation of Intel 4004 (750 kHz) to the current tenth generation Core i7, consisting of four data processing cores each with clock speed 4.8 GHz, built using Intel's 10 nm process technology [52]. For each new technology node, device structures and/or new materials are introduced to tackle the problems such as short-channel effects partially mitigated using FinFET structure, decreasing leakage current with high-k dielectrics, and countering mobility decrease with strained silicon and silicon–germanium channel materials [53–55]. The current sub-10 nm nodes under development would require significant revolutionary revision to the existing silicon-based CMOS fabrication lines, which is why any emerging novel materials including nanocarbons and two-dimensional materials would attract much research investment to assess its potential applications in ICs.

The fabrication of ICs can be roughly classified into front-end-of-the-line (FEOL) and back-end-of-the-line (BEOL) processes. Billions of transistors are fabricated on a single chip with FEOL processes using silicon-based channel materials, and the lines and vias used to connect these billions of FETs on the same chip are fabricated with BEOL processes using metals such as Cu and W as well as low-k intermetallic dielectrics. In this section, these two kinds of electronic devices, active FETs and passive interconnects, fabricated using nanocarbon materials as alternatives to Si and Cu (or W), respectively, are reviewed.

In 1998, Tans reported the first FET using semiconducting CNT as the channel material [56]. A Si global back-gate is adopted to tune the band bending of the SWCNT through the SiO_2 gate dielectric (Fig. 1.3a), and a relatively high on/off current ratio of 10^6 is obtained, demonstrating that a single semiconducting SWCNT can be applied as a switching transistor [56]. While the channel length of the transistor is around 400 nm, the on current is limited to ~1 μA due to the large contact resistance between CNT and the source/drain metal electrodes. In the early reports on SWCNT-based field-effect transistors (CNFETs), typical p-channel characteristics were recorded, probably due to the oxygen doping and water adsorbed on the CNT surface. The Fermi-level pinning of the SWCNT with contact metal electrodes having a high work function facilitates hole transport across the interface but blocks electron transmission, resulting in the reported p-channel transistor characteristics. Based on p-channel transistors, CNT circuits for digital operations including inverter, memory, and ring oscillator were demonstrated [57].

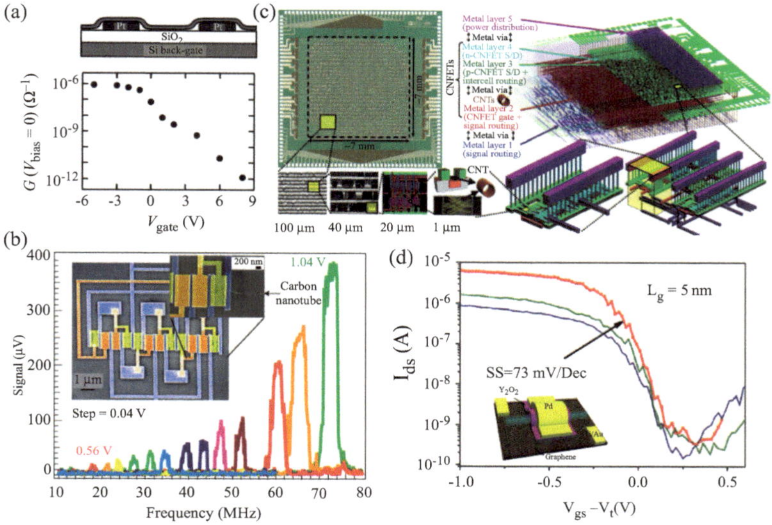

Figure 1.3 (a) Schematic cross-sectional device structure of CNFET and the experimental transconductance versus gate voltage curve showing the p-channel characteristics. (b) Ring-oscillator circuit based on a single SWCNT. The graph shows how the ringing frequency changes as the supply voltage is increased from 0.56 V to 1.04 V in 0.04 V steps. Inset shows a scanning electron microscope (SEM) image of the ring oscillator [67]. (c) Image of a fabricated CNT microprocessor and the three-dimensional schematic of the physical layout of the CNT chip. (d) I_{ds}–V_g curve of the 5 nm-gate CNFET, with inset showing the device structure [11]. Figure (a) reprinted from Ref. [56], Copyright 1998; Fig. (b) reprinted from Ref. [67], Copyright 2007; and Fig. (c) from Ref. [68], Copyright 2019, by permission from Springer Nature.

For future low-power and high-performance large-scale IC applications, the complementary metal oxide semiconductor (CMOS) inverter as the most fundamental building block would require both p- and n-channel transistors. Various methods have been proposed to realize n-CNFET, such as polymer functionalization of CNT [58], potassium doping of the CNT channel or near the contact interface

[59], annealing to remove adsorbed oxygen [60], and adopting low-work-function metal electrodes such as Ca and Sc. A CMOS inverter was first demonstrated using a p-CNFET connected to an n-CNFET [61]. At the early stage of CNT transistor development, the channel of a typical CNFET usually consists of one or a very limited number of semiconducting CNTs. As the transistor performance is directly related to the CNT chirality, the performance of the CNT transistors and the inverters built from them varied widely, thus a relatively more complex digital IC, either memory or logic, seems too far from reality at that time. Nevertheless, other devices including p-n junction, bipolar junction transistor, tunneling FET, and optoelectronic devices have been demonstrated utilizing the unique electronic properties of SWCNTs [62–66].

For traditional silicon-based FETs, hole exhibits a much lower mobility than electron, which poses a great challenge for IC design with both p-type MOSFET and n-type MOSFET, as a more complex layout design strategy has to be adopted to compensate for this asymmetry. In this regard, the intrinsic electronic transport property of CNT derived from its symmetrical band structure suggests an equal effective mass for electron and hole, which is beneficial for CMOS IC design. To build a CMOS inverter and even a more complex circuit, it is possible to utilize the intrinsic ambipolar transport property of a single CNT. A ring-oscillator fabricated with a single SWCNT featuring an oscillating frequency of 72 MHz was reported, in which the CMOS inverter consisted of two CNT transistors with different gate metals to tune the threshold voltage [Fig. 1.3(b)] [67]. However, the mature doping technology in silicon and other bulk materials cannot be applied to precisely control the electrical type of semiconducting nanocarbons without destroying the crystalline structure. A reliable method to control the CNT electrical type typically involves careful choice of work function combinations for gate electrodes, source/drain electrodes, and surrounding dielectrics. The underlying mechanisms includes threshold voltage shift, Schottky barrier formation between the source/drain contact and the CNT that facilitates a particular carrier transport, and the dielectric-induced unintentional doping to the CNT [64, 69, 70]. By adopting Pd and Sc as the source/drain electrodes, p- and n-CNFETs were fabricated on the same SWCNT, which reveals that the high work function Pd and low work function Sc can be utilized to allow only hole and electron transport through the respective Schottky barriers [71]. Air-stable n-CNFETs can also be achieved by passivation of the CNT channel with Si_3N_4. Combining all these techniques, various high-performance CNFETs have been demonstrated without relying on doping techniques. On the other hand, it is also possible to utilize only p-CNFETs to build a complex circuit. One recent successful example is a CNT-based microprocessor which consists of 178 p-CNFETs [72]. However, special design method has to be adopted to allow the imperfect CNT placement and to burn out the metallic CNTs for a functional system, which probably precludes its competing with existing CMOS technology. Very recently, the same group reported a microprocessor fabricated using more than 14,000 p- and n-CNFETs [Fig. 1.3(c)] [68]. As far as a single device goes, 5-nm-gate-length complementary CNFETs exhibiting better performance than their silicon counterparts have been reported [Fig. 1.3(d)] [11].

To satisfy the requirements of practical applications in ICs, the CNT quality and uniformity must be substantially improved to meet the electronic grade. Short of meeting such challenge, CNTs can only be functionalized as ingredients in composites, thermal interface materials, or thin films consisting of CNT networks that could find applications in thin-film transistors [73]. Over the last two decades, the performance of a single transistor developed with a single CNT has been improved to the level comparable with state-of-the-art silicon FETs, in terms of device driving current, on/off ratio, subthreshold swing, and device scaling. To make the most advanced chip, billions of FETs must be fabricated on the same IC with controllable device-to-device performance deviations, which poses a great challenge for CNT growth process. For SWCNTs, statistically one-third is metallic which should be removed for FET applications, and even one percent of metallic SWCNTs would be detrimental to the overall performance of the CNFETs, such as decreasing the on/off ratio to ~1. On the other hand, if the target application is interconnect, the percentage of metallic CNTs should be much higher, while a small percentage of semiconducting CNT is still acceptable for interconnect applications. Thus the main hurdle toward CNT-based ICs is the purification and precise placement of the CNTs to achieve pure semiconducting SWCNTs with a single chirality and also a reasonable packing density on the substrate for device fabrications. After decades of refining CNT growth methods, it is possible now to grow purely SWCNTs with predominant semiconducting ones [74]. However, uniform transistor performance in an IC requires all CNTs having the same chirality. Currently, the commercial CNT solutions containing 99.9% semiconducting SWCNTs are available, but the chirality control is yet to be demonstrated [75].

For interconnect applications, metallic CNTs are preferred. The challenges are not so much selectivity of metallic over semiconducting CNTs, but packing density, alignment along a specific direction for vertical vias or horizontal lines, and CNT contact resistance. These challenges and attempts to mitigate them are discussed in depth in Chapter 3. MWCNTs have been studied extensively over the past two decades as a potential on-chip interconnect material, pioneered by Graham [76] who first reported findings on a 30-nm-wide CNT via. Subsequently, various academic and industrial research groups have focused on different aspects of realizing CNT vias. Nihei [77] reported a process to integrate CNT vias on-chip, compatible with BEOL thermal budget of 400~450 °C. Their most recent report demonstrated a CNT via with 70 nm width and aspect ratio of ~1, yielding a resistance of 11 kΩ [78]. Vollebregt [79] also focused on lowering the CNT growth temperature to as low as 350 °C. However, the feature sizes of interconnects reported were in the 1~3 μm range, much larger than those in advanced technology nodes. Chiodarelli [80] demonstrated an increase in CNT packing density in a via by employing an AlCu alloy underlayer, which enhanced Fe catalyst dewetting. This approach resulted in more CNT shells as conducting paths and could potentially yield lower via resistance. However, their low-temperature CNT growth resulted in via resistance in the 10 kΩ range. Growing CNTs at low temperature introduces defects which in turn reduces electron mean free path and increases resistance. Therefore, for BEOL-compatible process CNT via resistance remains significantly higher than its Cu and W counterparts. Nevertheless, CNTs exhibit high mechanical strength in their structure due to sp^2 C–C bonding which makes them resistant to electromigration. Current-carrying capacities reported for CNT vias generally range from 100 to 1000 MA/cm^2 [81]. Such attractive reliability makes CNT a contender to replace Cu and W, especially in vias, if the CNT via resistance can be reduced to approach its Cu and W counterparts. To achieve better CNT via performance is indeed a daunting challenge, which at the very least requires lowering contact resistance and via resistance by optimizing CNT growth and contact metallization processes.

For graphene-based electronic devices, the pioneering work by the Manchester group reported field effect in single- and few-layer graphene [13], which suggests applications for FETs. A back-gated structure was used in their work to demonstrate the carrier density tunability by the silicon global gate. While that was useful for prototype demonstration, a practical application would require a local top-gate structure. The first top-gate graphene FETs were demonstrated using an evaporated 20 nm SiO_2 as the gate dielectric [82]. Subsequent studies used different substrates and gate dielectrics to improve the device performance. For example, using diamond-like carbon and 2D h-BN as substrates was shown to result in higher carrier mobility in graphene by suppressing phonon scattering [83, 84]. Despite its high carrier mobility, high saturation current, and atomic-scale thickness that provide graphene FETs with potential advantages over nanoscale silicon FETs in overcoming short-channel effects, the fact that graphene FETs cannot be effectively switched off precludes any digital applications. Unlike semiconducting SWCNTs, graphene has a zero bandgap, and the FETs built on pristine graphene exhibit a very low on/off current ratio (<100) at room temperature, which is at least two orders of magnitude lower than the minimum CMOS ICs' requirement.

After the first demonstration of graphene FETs, various ways have been proposed to open a bandgap in graphene, such as forming a graphene nanoribbon [85], field-biasing bilayer graphene [51], applying strain [86], and functionalization of graphene [87]. Theoretically, the bandgap of graphene nanoribbon is inversely proportional to its width, similar to the bandgap-diameter relationship in SWCNTs. To achieve a sizable bandgap, a narrow graphene nanoribbon with width in the sub-10 nm would be required, which poses great challenges in fabrication. Xia [49] demonstrated biasing a bilayer graphene to achieve a bandgap of ~130 meV when a vertical electric field of 2.2 V/nm is applied through the graphene, resulting in an on/off ratio of ~100 at room temperature. While it is also possible to apply strain to induce a bandgap, the strain is too large for practical applications. The experimental data reported thus far suggested that bandgap opening was at the expense of lowering the mobility [88]. Forming graphene nanoribbon seems to be the most promising method, but the process has to be controlled at atomic precision, which is beyond current experimental capabilities.

Although graphene FETs cannot be effectively switched off, its high carrier mobility and high saturation velocity are appealing to applications in radio frequency (RF) devices. A top-gated graphene FET with high transconductance and current saturation characteristics has also been reported [89]. Another study employed a large-area CVD-grown graphene to obtain a cutoff frequency of 155 GHz for a 40 nm channel length top-gated graphene FET [90]. Various RF circuits, including RF mixers, low-noise amplifiers, resonators, and receivers have been proposed and showed promising performances [91–93]. RF receivers integrating on-chip capacitors and inductors with graphene FETs that could wirelessly transmit signals with a carrier frequency of 4.3 GHz were demonstrated [91].

Another important application of graphene not requiring high on/off current ratio is on-chip interconnects, in which graphene is a promising candidate for replacing the horizontal metal lines due to its 2D geometry, large current-carrying capability, and potentially lower resistances compared to the existing Cu and W metal interconnects. Report on graphene as horizontal interconnect showed that a fabricated 8 nm–wide MLG line exhibited a resistivity of 3.2 μΩ·cm, which is comparable to that of bulk Cu and less than that for a Cu line with the same width [6, 94]. Generally, for all applications of graphene to electronic devices, whether it be FET or interconnect, a post-growth transfer process is usually required. Research efforts toward ultraclean graphene transfer process to avoid polymer and solution contaminants can eventually lead to uniform device performance.

1.4 Chapter Organization and Summaries

Figure 1.4 presents a brief chronology of major breakthroughs in the development of CNT and graphene and their applications to electronic devices. While these novel materials have given rise to immense opportunities for wide-ranging applications, especially in electronics, one must keep in mind the daunting challenges that currently preclude mass production of nanocarbon-based electronic devices. And their supplanting Si and Cu in the most advanced ICs remains an elusive target in the future. Nevertheless, the superior physical and chemical properties of CNT and graphene are simply too good for scientists and technologists to ignore as we are now entering the post-Moore's law era. To continue the exploration and development of these wonderful nanomaterials for electronic applications will require multi-disciplinary efforts and collaboration as well as strong support from industry. It is our fervent hope that this volume can contribute to stimulating further work in this field that will lead to the eventual realization of nanocarbon-based electronic devices that impact our daily lives.

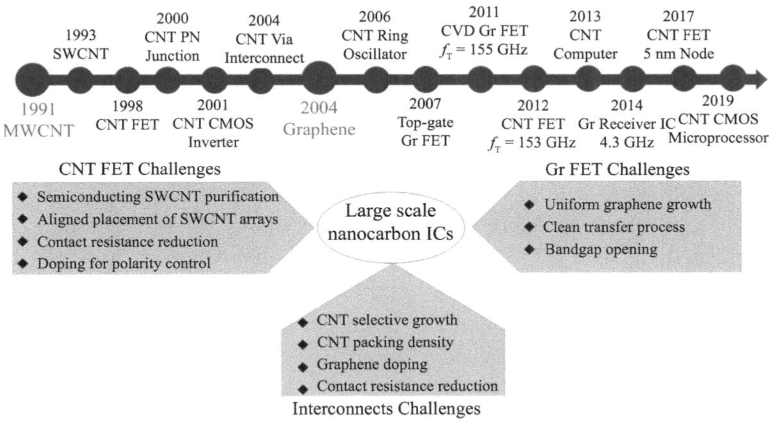

Figure 1.4 Brief chronology of major breakthroughs in the development of graphene- and CNT-based electronic devices, and the main challenges for integrating them into nanocarbon-based circuits.

The organization of this volume and brief summaries of Chapters 2 through 8 are presented as follows. The overview presented in Chapter 1 is intended to familiarize readers with the basic properties of graphene and CNT and their most-studied potential applications in electronics. The more in-depth discussions on materials and device fabrication are given in Chapter 2, followed by applications in interconnects in Chapter 3 and transistors in Chapters 4 and 5. Chapter 6 covers potential applications in flexible sensors, while Chapter 7 presents applications for energy conversion and storage. The volume concludes in Chapter 8 with a projection of where the field of nanocarbon electronics might be heading.

In Chapter 2, three CNT growth methods, arc-discharge, laser ablation, and chemical vapor deposition are introduced and the growth mechanisms and current status in terms of quality and growth yield are discussed. For graphene, the advantages and disadvantages of exfoliation, CVD, and epitaxial growth are presented. In particular, keeping in mind electronic device applications, the potential of integrating each growth method into the current Si-based process technology is assessed.

The opportunities and challenges of using CNTs for on-chip vertical vias and horizontal lines interconnects are discussed in Chapter 3 in the context of the end of current IC technology roadmap. While the reliability of CNT vias is widely demonstrated to be far superior to those of current IC interconnect materials, Cu and W, the via resistance remains a daunting challenge, due largely to high contact resistance. An all-carbon 3D interconnect structure consisting of CNT grown directly on graphene is presented as a possible way to mitigate this contact resistance problem.

Chapter 4 presents the current state in fabricating and characterizing CNT FETs, as well as their usage in ICs, display electronics, and sensors. The main challenges to overcome in materials preparation are the removal of metallic CNTs and CNT diameter variations. In device fabrication, the problems include alignment and density variation of CNT channel, metal-CNT contacts, stable channel doping techniques, large-area fabrication for CNT arrays, reliable high-κ dielectric, and CNT-dielectric interface. For CNFET circuits, critical issues involve mismatched performance between p- and n-channel devices, sensitivity to device variations, and process compatibility of both device types.

The chronology of progress on graphene FETs is presented in Chapter 5. The latest IC roadmap strongly recommended intensified research into graphene. Challenges include bandgap engineering, carrier mobility, and design of graphene structure in transistor as they all impact its performance. Research on integration of graphene-based circuits is still in its infancy. Ambipolar conduction of graphene creates new opportunities for realizing RF circuit functions with fewer components.

The potential use of nanocarbons in the fast-emerging field of flexible wearable electronics is presented in Chapter 6. Fabrication considerations and applications of flexible nanocarbon devices are discussed. Flexible electronics offer applications ranging from wearable sensors for continuous health monitoring to on-demand fabrication of complex circuits. The ability to rapidly fabricate using 3D printing methodologies drastically reduces cost and allows for a high level of device tunability. Integration of nanocarbons in general, and CNTs and graphene in particular, can further expand the capabilities of flexible electronics by enabling direct manipulation of material properties for a variety of functions.

Chapter 7 describes the fabrication of carbon nanofibers using electrospinning of polymeric precursors followed by a thermal carbonization process. The unique properties of electrospinning allow carbon nanofibers to be prepared in controlled structures (solid, porous, or hollow) and with desired physical properties through addition of additives into the polymeric cursors during electrospinning. The produced carbon nanofibers form self-standing membranes with interconnected web-like internal structures with high porosities (over 90%). These properties allow deposition of proper active energy materials onto the internal surface inside the membrane to form unique hybrid materials by further electrophoretic deposition or electrochemical deposition. Specific electrospun carbon nanofibers are illustrated in energy conversion and storage applications including dye-sensitized solar cells, polymer electrolyte membrane fuel cells, and lithium-ion batteries.

The book concludes with a brief narrative on the prospects of nanocarbon-based electronics. A roadmap is presented to target the more promising applications of CNTs and graphene in the coming decade. The purpose of such roadmap is not necessarily a prediction of the future, but rather is intended to stimulate further work on the science of nanocarbons and the transfer of new scientific knowledge to applications in electronics.

References

1. W. J. Yu, S. Y. Lee, S. H. Chae, et al. (2011), Small hysteresis nanocarbon-based integrated circuits on flexible and transparent plastic substrate, *Nano Letters*, **11**, 1344–1350.
2. Y. Y. Liang, Y. G. Li, H. L. Wang, et al. (2013), Strongly coupled inorganic/nanocarbon hybrid materials for advanced electrocatalysis, *Journal of the American Chemical Society*, **135**, 2013–2036.

3. D. Wei, P. Hiralal, H. Wang, et al. (2013), Hierarchically structured nanocarbon electrodes for flexible solid lithium batteries, *Nano Energy*, **2**, 1054–1062.
4. G. E. Moore (2000), *Cramming More Components onto Integrated Circuits*, Morgan Kaufmann Publishers, 56–59.
5. H. Liu, D. K. Mohata, A. Nidhi, et al. (2012), Exploration of vertical MOSFET and tunnel FET device architecture for Sub 10nm node applications, *2012 70th Annual Device Research Conference* (DRC), 233–234.
6. D. Kondo, H. Nakano, B. Zhou, et al. (2014), Sub-10-nm-wide intercalated multi-layer graphene interconnects with low resistivity, *IEEE International Interconnect Technology Conference*, 189–192.
7. A. D. Franklin, M. Luisier, S-J. Han, et al. (2012), Sub-10 nm carbon nanotube transistor, *Nano Letters*, **12**, 758–762.
8. Sumio Iijima (1991), Helical microtubules of graphitic carbon, *Nature*, **354**, 56–58.
9. L. Ding, Z. Zhang, S. Liang, et al. (2012), CMOS-based carbon nanotube pass-transistor logic integrated circuits, *Nature Communications*, **3**, 677.
10. M. Loong, P. Tan, G. Lentaris, et al. (2012), Device and circuit-level performance of carbon nanotube field-effect transistor with benchmarking against a nano-MOSFET, *Nanoscale Research Letters*, **7**, 467–467.
11. C. Qiu, Z. Zhang, M. Xiao, et al. (2017), Scaling carbon nanotube complementary transistors to 5-nm gate lengths, *Science*, **355**, 271.
12. K. S. Novoselov, A. K. Geim, S. V. Morozov, et al. (2004), Electric field effect in atomically thin carbon films, *Science*, **306**, 666.
13. A. Geim and K. S. Novoselov (2007), The rise of graphene, *Nature Materials*, **6**, 183–191.
14. K. S. Novoselov, A. K. Geim, S. V. Morozov, et al. (2005), Two-dimensional gas of massless Dirac fermions in graphene, *Nature*, **438**, 197–200.
15. A. H. Castro Neto, F. Guinea, N. M. R. Peres, et al. (2009), The electronic properties of graphene, *Reviews of Modern Physics*, **81**, 109–162.
16. F. Schwierz (2010), Graphene transistors, *Nature Nanotechnology*, **5**, 487–496.
17. Y. Y. Shao, J. Wang, H. Wu, et al. (2010), Graphene based electrochemical sensors and biosensors: a review, *Electroanalysis*, **22**, 1027–1036.
18. F. N. Xia, T. Mueller, Y. M. Lin, et al. (2009), Ultrafast graphene photodetector, *Nature Nanotechnology*, **4**, 839–843.

19. P. Gargini and IEEE (2017), Roadmap evolution: from NTRS to ITRS, from ITRS 2.0 to IRDS, *IEEE 2017 Fifth Berkeley Symposium on Energy Efficient Electronic Systems & Steep Transistors Workshop*.

20. Z. Zhu (2017), An overview of carbon nanotubes and graphene for biosensing applications, *Nano-Micro Letters*, **9**, 25.

21. Z. Ma, A. Wei, J. Ma, et al. (2018), Lightweight, compressible and electrically conductive polyurethane sponges coated with synergistic multiwalled carbon nanotubes and graphene for piezoresistive sensors, *Nanoscale*, **10**, 7116–7126.

22. L. Chen, M. Weng, P. Zhou, et al. (2017), Multi-responsive actuators based on a graphene oxide composite: intelligent robot and bioinspired applications, *Nanoscale*, **9**, 9825-9833.

23. B. Kirtman, D. M. Chipman and W. E. Palke (1977), Orbital hybridization, *Journal of the American Chemical Society*, **99**, 1305–1307.

24. A. V. Eletskii (2009), Transport properties of carbon nanotubes, *Physics-Uspekhi*, **52**, 209–224.

25. K. Wakabayashi, Y. Takane, M. Yamamoto, et al. (2009), Electronic transport properties of graphene nanoribbons, *New Journal of Physics*, **11**, 095016.

26. M. S. Dresselhaus, G. Dresselhaus, R. Saito, et al. (2005), Raman spectroscopy of carbon nanotubes, *Physics Reports*, **409**, 47–99.

27. M. S. Dresselhaus, G. Dresselhaus and R. Saito (1995), Physics of carbon nanotubes, *Carbon*, **33**, 883–891.

28. T. Boykin and G. Klimeck (2005), Practical application of zone-folding concepts in tight-binding calculations, *Physical Review B*, **71**, 115215.

29. M. Kaempgen, C. K. Chan, J. Ma, et al. (2009), Printable thin film supercapacitors using single-walled carbon nanotubes, *Nano Letters*, **9**, 1872–1876.

30. A. Vijayaraghavan, F. Hennrich, N. Stürzl, et al. (2010), Toward single-chirality carbon nanotube device arrays, *ACS Nano*, **4**, 2748–2754.

31. A. J. Page, F. Ding, S. Irle, et al. (2015), Insights into carbon nanotube and graphene formation mechanisms from molecular simulations: a review, *Reports on Progress in Physics*, **78**, 036501.

32. V. Meunier, A. G. S. Filho, E. B. Barros, et al. (2016), Physical properties of low-dimensional sp^2-based carbon nanostructures, *Reviews of Modern Physics*, **88**, 025005.

33. P. Serp, M. Corrias, and P. Kalck (2003), Carbon nanotubes and nanofibers in catalysis, *Applied Catalysis A: General*, **253**, 337–358.

34. A. Nimibofa, A. Ebelegi, C. Abasi, et al. (2018), Fullerenes: synthesis and applications, *Journal of Materials Science Research*, **7**, 22–36.

35. P. Hebgen, A. Goel, J. B. Howard, et al. (2000), Synthesis of fullerenes and fullerenic nanostructures in a low-pressure benzene/oxygen diffusion flame, *Proceedings of the Combustion Institute*, **28**, 1397–1404.

36. M. Monthioux and V. Kuznetsov (2006), Who should be given the credit for the discovery of carbon nanotubes? *Carbon*, **44**, 1621–1623.

37. L. V. Radushkevich and V. M. Luk'yanovich (1952), The structure of carbon forming in thermal decomposition of carbon monoxide on an iron catalyst, *Russian Journal of Physical Chemistry A*, **26**, 88–95.

38. S. Iijima and T. Ichihashi (1993), Single-shell carbon nanotubes of 1-nm diameter, *Nature*, **363**, 603–605.

39. D. S. Bethune, C. H. Kiang, M. S. de Vries, et al. (1993), Cobalt-catalysed growth of carbon nanotubes with single-atomic-layer walls, *Nature*, **363**, 605–607.

40. R. Purohit, K. Purohit, S. Rana, et al. (2014), Carbon nanotubes and their growth methods, *Procedia Materials Science*, **6**, 716–728.

41. Y. Ando and X. Zhao (2006), Synthesis of carbon nanotubes by arc-discharge method, *New Diamond and Frontier Carbon Technology*, **16**, 123–137.

42. J. Chrzanowska, J. Hoffman, A. Małolepszy, et al. (2015), Synthesis of carbon nanotubes by the laser ablation method: effect of laser wavelength, *Physica Status Solidi B*, **252**, 1860–1867.

43. N. D. Mermin (1968), Crystalline order in two dimensions, *Physical Review*, **176**, 250–254.

44. K. S. Novoselov, D. Jiang, F. Schedin, et al. (2005), Two-dimensional atomic crystals, *Proceedings of the National Academy of Sciences of the United States of America*, **102**, 10451–10453.

45. D. S. L. Abergel, V. Apalkov, J. Berashevich, et al. (2010), Properties of graphene: a theoretical perspective, *Advances in Physics*, **59**, 261–482.

46. I. Yudhistira and S. Adam (2015), Theory for electron transport in graphene, *Synthetic Metals*, **210**, 2–8.

47. F. Bonaccorso, Z. Sun, T. Hasan, et al. (2010), Graphene photonics and optoelectronics, *Nature Photonics*, **4**, 611.

48. A. A. Lebedev and P. Bulat (2014), Overview of the graphene production technologies and instrument application, *Life Science Journal*, **11**, 404–409.

49. F. Xia, D. B. Farmer, Y-M. Lin, et al. (2010), Graphene field-effect transistors with high on/off current ratio and large transport band gap at room temperature, *Nano Letters*, **10**, 715–718.

50. Y. Wu, K. A. Jenkins, A. Valdes-Garcia, et al. (2012), State-of-the-art graphene high-frequency electronics, *Nano Letters*, **12**, 3062–3067.

51. E. Sano and T. Otsuji (2009), Bandgap engineering of bilayer graphene for field-effect transistor channels, *Japanese Journal of Applied Physics*, **48**, 091605.

52. J. Burek (2019), Intel 'Ice Lake' benchmarked: how 10nm CPUs could bring major GPU grunt to new laptops (https://www.pcmag.com/article/369884/intel-ice-lake-benchmarked-how-10nm-cpus-could-bring-majo).

53. D. Guo, H. Shang, K. Seo, et al. (2014), 10nm FINFET technology for low power and high performance applications, *2014 12th IEEE International Conference on Solid-State and Integrated Circuit Technology (ICSICT)*, 1–4.

54. Y-S. Lin, P-H. Cheng, K-W. Huang, et al. (2018), Atomic layer deposition of sub-10 nm high-K gate dielectrics on top-gated MoS_2 transistors without surface functionalization, *Applied Surface Science*, **443**, 421–428.

55. I. Aberg, O. O. Olubuyide, C. N. Chleirigh, et al. (2004), Electron and hole mobility enhancements in sub-10 nm-thick strained silicon directly on insulator fabricated by a bond and etch-back technique, *Digest of Technical Papers, 2004 Symposium on VLSI Technology, 2004*, 52–53.

56. S. J. Tans, A. R. M. Verschueren, and C. Dekker (1998), Room-temperature transistor based on a single carbon nanotube, *Nature*, **393**, 49–52.

57. A. Bachtold, P. Hadley, T. Nakanishi, et al. (2001), Logic circuits with carbon nanotube transistors, *Science*, **294**, 1317.

58. M. Shim, A. Javey, N. W. S. Kam, et al. (2001), Polymer functionalization for air-stable n-type carbon nanotube field-effect transistors, *Journal of the American Chemical Society*, **123**, 11512–11513.

59. A. Javey, R. Tu, D. B. Farmer, et al. (2005), High-performance n-type carbon nanotube field-effect transistors with chemically doped contacts, *Nano Letters*, **5**, 345–348.

60. V. Derycke, R. Martel, J. Appenzeller, et al. (2002), Controlling doping and carrier injection in carbon nanotube transistors, *Applied Physics Letters*, **80**, 2773–2775.

61. X. Liu, C. Lee, C. Zhou, et al. (2001), Carbon nanotube field-effect inverters, *Applied Physics Letters*, **79**, 3329–3331.
62. J. U. Lee (2007), Carbon nanotube p-n diodes, *LEOS 2007—IEEE Lasers and Electro-Optics Society Annual Meeting Conference Proceedings*, 531–532.
63. W. Zhang, Q. Zhang, Y. Chai, et al. (2007), Gate voltage dependent characteristics of p-n diodes and bipolar transistors based on multiwall CNx/carbon nanotube intramolecular junctions, *Nanotechnology*, **18**, 395205.
64. L. Yu-Ming, J. Appenzeller, J. Knoch, et al. (2005), High-performance carbon nanotube field-effect transistor with tunable polarities, *IEEE Transactions on Nanotechnology*, **4**, 481–489.
65. N. V. Karimi and Y. Pourasad (2015), Tunneling carbon nanotube field effect transistor with asymmetric graded double halo doping in channel: asym-GDH-T-CNTFET, *Procedia Materials Science*, **11**, 287–292.
66. S. Wang, Z. Zhang, and L-M. Peng (2011), Doping-free carbon nanotube optoelectronic devices, *Chinese Science Bulletin*, **57**, 149–156.
67. P. Avouris, Z. Chen, and V. Perebeinos (2007), Carbon-based electronics, *Nature Nanotechnology*, **2**, 605–615.
68. G. Hills, C. Lau, A. Wright, et al. (2019), Modern microprocessor built from complementary carbon nanotube transistors, *Nature*, **572**, 595–602.
69. A. Lin, N. Patil, K. Ryu, et al. (2009), Threshold voltage and on–off ratio tuning for multiple-tube carbon nanotube FETs, *IEEE Transactions on Nanotechnology*, **8**, 4–9.
70. A. L. Falk, K-C. Chiu, D. B. Farmer, et al. (2017), Coherent plasmon and phonon-plasmon resonances in carbon nanotubes, *Physical Review Letters*, **118**, 257401.
71. Z. Zhang, X. Liang, S. Wang, et al. (2007), Doping-free fabrication of carbon nanotube based ballistic CMOS devices and circuits, *Nano Letters*, **7**, 3603–3607.
72. M. M. Shulaker, G. Hills, N. Patil, et al. (2013), Carbon nanotube computer, *Nature*, **501**, 526–530.
73. V. K. Sangwan, R. P. Ortiz, J. M. P. Alaboson, et al. (2012), Fundamental performance limits of carbon nanotube thin-film transistors achieved using hybrid molecular dielectrics, *ACS Nano*, **6**, 7480–7488.

74. J. Liu, C. Wang, X. Tu, et al. (2012), Chirality-controlled synthesis of single-wall carbon nanotubes using vapour-phase epitaxy, *Nature Communications*, **3**, 1199.

75. J. Lefebvre, J. Ding, Z. Li, et al. (2017), High-purity semiconducting single-walled carbon nanotubes: a key enabling material in emerging electronics, *Accounts of Chemical Research*, **50**, 2479–2486.

76. A. P. Graham, G. Duesberg, W. Hoenlein, et al. (2005), How do carbon nanotubes fit into the semiconductor roadmap? *Applied Physics A*, **80**, 1141–1151.

77. M. Nihei, A. Kawabata, D. Kondo, et al. (2005), Electrical properties of carbon nanotube bundles for future via interconnects, *Japanese Journal of Applied Physics*, **44**, 1626–1628.

78. M. Katagiri, M. Wada, B. Ito, et al. (2012), Fabrication and characterization of planarized carbon nanotube via interconnects, *Japanese Journal of Applied Physics*, **51**, 05ED02.

79. S. Vollebregt, F. D. Tichelaar, H. Schellevis, et al. (2014), Carbon nanotube vertical interconnects fabricated at temperatures as low as 350°C, *Carbon*, **71**, 249–256.

80. N. Chiodarelli, K. Kellens, D. Cott, et al. (2010), Integration of vertical carbon nanotube bundles for interconnects, *ECS Transactions*, **19**, 215th ECS Meeting, May 24–29, 2009, San Francisco, CA, No. 24, 2009.

81. M. Sato, T. Hyakushima, A. Kawabata, et al. (2010), High-current reliability of carbon nanotube via interconnects, *Japanese Journal of Applied Physics*, **49**, 105102.

82. M. C. Lemme, T. J. Echtermeyer, M. Baus, et al. (2007), A graphene field-effect device, *IEEE Electron Device Letters*, **28**, 282–284.

83. F. Zhao, A. Vrajitoarea, Q. Jiang, et al. (2015), Graphene-nanodiamond heterostructures and their application to high current devices, *Scientific Reports*, **5**, 13771–13771.

84. C. R. Dean, A. F. Young, I. Meric, et al. (2010), Boron nitride substrates for high-quality graphene electronics, *Nature Nanotechnology*, **5**, 722–726.

85. L. Villamagua, M. Carini, A. Stashans, et al. (2016), Band gap engineering of graphene through quantum confinement and edge distortions, *Ricerche di Matematica*, **65**, 579–584.

86. G. Gui, J. Li, and J. Zhong (2008), Band structure engineering of graphene by strain: first-principles calculations, *Physical Review B*, **78**, 075435.

87. H. Zhang, E. Bekyarova, J-W. Huang, et al. (2011), Aryl functionalization as a route to band gap engineering in single layer graphene devices, *Nano Letters*, **11**, 4047–4051.

88. J. Wang, R. Zhao, M. Yang, et al. (2013), Inverse relationship between carrier mobility and bandgap in graphene, *Journal of Chemical Physics*, **138**, 084701.

89. I. Meric, M. Y. Han, A. F. Young, et al. (2008), Current saturation in zero-bandgap, top-gated graphene field-effect transistors, *Nature Nanotechnology*, **3**, 654–659.

90. Y. Wu, Y-M. Lin, A. A. Bol, et al. (2011), High-frequency, scaled graphene transistors on diamond-like carbon, *Nature*, **472**, 74–78.

91. S-J. Han, A. V. Garcia, S. Oida, et al. (2014), Graphene radio frequency receiver integrated circuit, *Nature Communications*, **5**, 3086.

92. L. Liao and X. Duan (2012), Graphene for radio frequency electronics, *Materials Today*, **15**, 328–338.

93. J. S. Moon, M. Antcliffe, H. C. Seo, et al. (2011), Graphene transistors for RF applications: opportunities and challenges, *2011 International Semiconductor Device Research Symposium (ISDRS)*, DOI: 10.1109/ISDRS.2011.6135155.

94. A. Pyzyna, R. Bruce, M. Lofaro, et al. (2015), Resistivity of copper interconnects beyond the 7 nm node, *2015 Symposium on VLSI Technology (VLSI Technology)*, T120–T121.

Chapter 2

Nanocarbon Growth Methods and Device Integration

Changjian Zhou,[a] Wei Du,[a] and Cary Y. Yang[b]
[a]*School of Microelectronics, South China University of Technology, Guangzhou 510641, China*
[b]*Center for Nanostructures, Santa Clara University, Santa Clara, California 95053, USA*
zhoucj86@gmail.com

2.1 Introduction

Nanocarbon materials form a large family, with some well-known members such as fullerene [1], carbon nanotube (CNT) [2], and graphene [3] standing out for their unique physical and/or chemical properties. For example, the large surface-to-volume ratio of both CNT and graphene enables the development of high-sensitivity sensors by taking advantage of the absorption property of the nanocarbon materials [4–6]. For the same reason, electrodes built from nanocarbon materials are promising candidates for catalysis [7, 8], energy storage [9, 10], supercapacitors [11, 12], and batteries [13, 14]. The long electron mean free path [15, 16], atomic surface

Nanocarbon Electronics
Edited by Changjian Zhou, Min Zhang, and Cary Y. Yang

ISBN 978-981-4877-11-4 (Hardcover), 978-1-003-04308-9 (eBook)
www.jennystanford.com

smoothness, and the high charge carrier mobility in CNT and graphene have attracted many research efforts from both academia and industry that aim to develop the next-generation electron devices [17, 18]. These electron devices include the field-effect transistor (FET), which is the most fundamental building block for integrated circuits (ICs), and interconnect structures that electrically connect the numerous FETs inside a single IC [19, 20]. As the current mainstream silicon technology encounters several problems such as increased leakage current, increased power density, reduced gate controllability [21], increased interconnect delay [22], and interconnect reliability [23], devices based on nanocarbon materials are becoming potential candidates in future advanced technology nodes.

It is a prerequisite that a mature fabrication method is developed, and the resulting synthesized material is of uniform high quality for any targeted practical application. There are many requirements before an emerging technology is adopted for real products. For nanomaterials, uniformity and quality of the materials, scalability of fabrication method, and integration strategy are the main considerations for their applications in electron devices. As silicon technology has dominated over the last six decades, any emerging technology should be strictly accessed on its suitability for large-scale production before the industry can adopt it to replace or just augment the existing technology [24].

The focus of this chapter is on growth methods of nanocarbon materials. In particular, we discuss the growth of CNT and graphene for development toward their eventual integration into IC manufacturing. In general, there are two different cases in terms of integrating nanocarbon with current silicon technology. If the nanomaterials are used to augment the silicon-based technology, the growth parameters, including growth temperature and catalyst, should be compatible with the mainstream CMOS technology. A typical example is nanocarbon interconnect whose integration involves the incorporation of CNT or graphene after the silicon transistor fabrication, and the growth temperature should not exceed ~450°C to avoid damaging the existing silicon devices [25]. On the other hand, if the nanocarbon materials are used to build all the required functional parts such as sensors and actuators, the requirements are not as stringent as for silicon chip integration. The suitability of a growth method should be assessed according to the specific application being considered.

In this chapter, we briefly introduce the growth mechanisms and common parameters for each growth method, and comment on its suitability for device integration purposes. In the ensuing sections, three kinds of nanocarbon materials will be discussed, CNT, graphene, and a novel structure containing both CNT and graphene, which is referred to hereafter as CNT–graphene heterostructure.

2.2 CNT Growth Method and Its Integration

2.2.1 Arc-Discharge Method

CNT was firstly grown by the arc-discharge method, as reported by Iijima who discovered the multi-walled CNT (MWCNT) when he studied the synthetization of fullerenes [26]. Figure 2.1 shows the transmission electron microscope (TEM) images of the grown MWCNTs, and the number of walls of 5, 2, and 7 can be clearly identified for the three MWCNTs. A typical apparatus used for arc-discharge synthesizing CNTs is shown in Fig. 2.2a [27]. Two graphitic electrodes are configured vertically with a tip-to-tip distance in the range of 1–2 mm. The chamber is firstly pumped to a high vacuum state and then flowed with rarefied gas such as helium before the growth initiates. During the growth stage, a DC voltage is applied between the two graphite electrodes, which is high enough to ignite the arc discharge between the opposing electrode tips. The deposition of carbon nanomaterials including fullerenes and MWCNTs can be found on the cathode electrode tip, as can be seen from Fig. 2.2b [28], while the anode electrode provides the carbon source for the nanocarbon material deposition. The CNTs synthesized using pure graphitic electrode are MWCNTs with diameters from 4 to 30 nm and length of 1 μm in an early report [26]. Furthermore, the growth of MWCNTs can be optimized by changing electrode separation, the nature of current (DC or AC), the reaction gas ambient, and pressure [29–31]. Ebbesen carried out the growth experiments under different inert gas and pressure, and they found that under an optimized condition of helium gas pressure of 500 Torr, a high yield of ~75% MWCNTs can be achieved [30]. The lengths of the

MWCNTs are also improved to be around several mm. It is also found that the growth of MWCNTs mainly happens at the center part of the electrode where the plasma is more stable, suggesting the high temperature induced by the arc plasma is essential in synthesizing the high-quality MWCNTs.

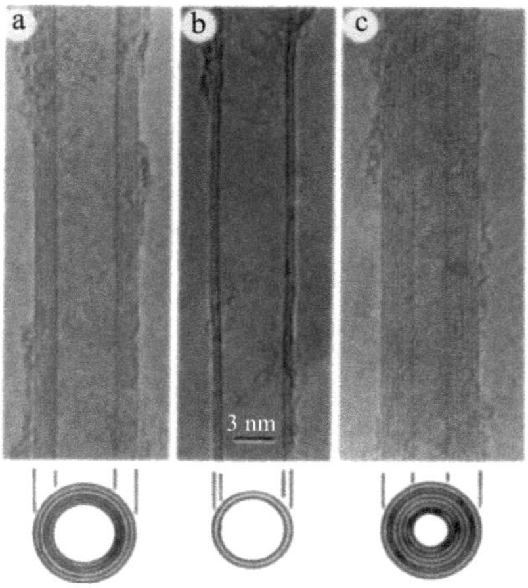

Figure 2.1 TEM images of the MWCNTs grown with the arc-discharge method. A cross section of each tubule is illustrated. Reprinted from Ref. [26], Copyright 1991, by permission from Springer Nature.

The synthesis of single-walled CNTs (SWCNTs) was reported by the Iijima group and an IBM group independently 2 years after the discovery of MWCNTs [32, 33], as shown in Fig. 2.3. The main difference between MWCNT synthesis and SWCNT synthesis is the catalyst introduction into the graphite electrode for SWCNT growth. Iron, cobalt, and nickel are commonly used catalysts. Another difference is that the grown SWCNT covers the entire chamber, while the grown MWCNT can be found only at the cathode electrode tip [27]. The synthesized SWCNTs usually contain impurities, including catalyst nanoparticles and amorphous carbon, which degrade the properties of the SWCNT for practical applications. To reduce the amorphous carbon byproduct, Yoshinori Ando optimized the arc-discharge method by using Fe-containing graphite rod and conducting the growth in an H_2/Ar mixture gas environment [34]. The macroscopic net of SWCNT up to several tens of cm can be achieved with the optimized method.

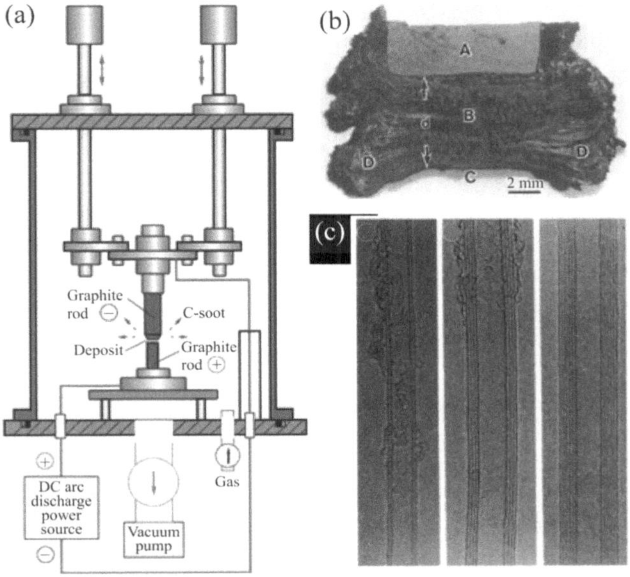

Figure 2.2 (a) Schematic diagram of CNT synthesizing apparatus by the arc-discharge method. (b) Optical image of cathode deposit: A is the cathode tip, B contains MWCNTs, C is the top of cathode facing the anode, and D is the hard carbon materials. (c) TEM images of MWCNTs with different number of walls prepared under 500 Torr helium gas. Figure (a) reprinted from Ref. [27], Copyright 2004, with permission from Elsevier; Fig. (b) reprinted from Ref. [28], Copyright 1993, with permission from The Japan Society of Applied Physics, IOP Publishing; Fig. (c) reprinted from Ref. [30], Copyright 1992, by permission from Springer Nature.

In terms of growth temperature, it is noted that the average temperature of the ignited plasma between the two electrodes is ~4000 K [31]. Such a high temperature results in high-quality CNT synthesis, while at the same time prohibits the direct integration of CNT on any silicon-based electron devices. The mass production of MWCNT and SWCNT can be obtained with proper apparatus and growth parameters [35–37]. The synthesized CNTs collected from the graphite electrode or the chamber soot should be post-processed to remove the byproducts, including fullerenes and metal particles for practical applications. The as-collected CNTs can be readily used as electrodes for supercapacitors [38]. For high-performance transistor applications, the purity and the uniformity of the as-collected

CNTs cannot be directly applied for several reasons. SWCNTs are preferable because most MWCNTs are conducting, which cannot be turned off if they are used in conjunction with SWCNTs. What is more, even the collected material only contains SWCNTs, the metallic SWCNTs will also results in poor transistor properties. Thus, the as-prepared CNTs cannot be directly grown on silicon substrates for transistors or interconnects. However, the synthesized materials can be purified and sorted by adopting techniques, including chemical functionalizing, electrophoretic separation, electrical breakdown, ultracentrifugation to isolate the semiconducting SWCNTs for thin-film transistor (TFT) applications [39, 40].

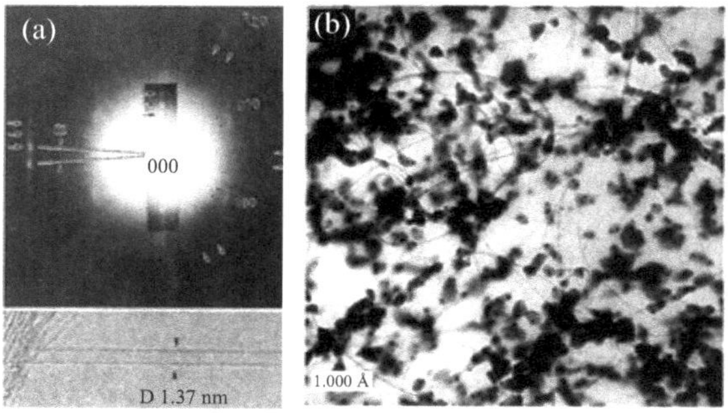

Figure 2.3 (a) Electron diffraction pattern and TEM image taken from an SWCNT of diameter 1.37 nm. (b) TEM picture of web-like material showing strands of thread-like fibers and cobalt clusters embedded in carbon soot particles. Figures (a) and (b) reprinted from Refs. [32] and [33], Copyright 1993, by permission from Springer Nature.

2.2.2 Laser Ablation Method

As the name indicates, the laser ablation method uses a high-power laser to vaporize the target, which is similar to the graphite rod used in the arc-discharge method [41]. The same method was developed to synthesize fullerene by Smalley's group [42] and further applied to synthesize SWCNTs using a laser-vaporized carbon–nickel–cobalt mixture at ~1200°C. Figure 2.4 shows a typical setup of the laser ablation method [27]. For SWCNT synthesizing, the target is usually added with metal catalyst, including Fe, Co, Ni, Pt, Y, and their mixtures

[43–45]. The quartz tube can be heated to a high temperature, and a window is installed to allow the laser beam illuminating from one side of the quartz tube. On the other side of the tube, a water-cooled collector is installed to collect the grown CNTs. When the high-power laser beam is focused on the graphite rod, the evaporated carbon-containing gas adsorbs on the metal catalysts and forms the SWCNT. Because the metal catalyst and carbon have a quite different vapor pressure, the ratio between the metal catalyst and carbon changes with time during the growth stage. Usually this results in SWCNTs with different structures such as different diameters and chirality. On the other hand, if the metal and carbon are evaporated separately, large-scale SWCNTs with quite uniform diameters could be obtained [46]. In this way, the two targets containing carbon and metal catalysts are installed in the same chamber, while the laser beam could be tuned to focus on each target separately. During the CNT growth, the chamber is flowed with inert gas such as argon, nitrogen, or helium gas. The flowed gas also serves as the carrier gas to transport the evaporated high-temperature gas phase carbon and metal catalyst from the focused targets to the low-temperature collector. The growth of SWCNT initiates as the evaporated carbon gas interacts with the metal catalysts particles. The chamber is also heated to slow down the condensation process, which facilitates the interaction of metal catalyst and the carbon gas. Once the microtubule nanotube structure forms on the collector, the laser irradiation provides new carbon feedstock for continuous CNT growth until the metal catalyst is saturated with carbon layer.

The growth parameters include the laser type and configurations [47], target [48], catalyst type [43], catalyst-to-carbon ratio, carrier gas [49], and chamber pressure and temperature [50, 51]. Both the continuous laser (e.g., CO_2 laser) and pulsed laser (e.g., Nd:YAG) are adopted to synthesize SWCNTs [43, 52]. For a pulsed Nd:YAG laser [52], the typical parameters of the laser configurations include pulse width (5–7 ns), wavelength (532 nm), laser power density (4 GW/cm^2), repetition rate, and focused beam size (1 mm). As fullerene is usually formed in the high-temperature zone at the initial stage of the CNT growth, it is reckoned that the role of metal catalysts is to open the cage structure of fullerene for following CNT structure formation. The catalyst-to-carbon ratio should be well controlled to produce the optimal yield, while at the same time to avoid the contamination of the synthesized SWCNTs [53]. A typical value of the catalyst mixture ratio is Ni/Co (0.6/0.6 at.%) [52] and Ni/Y (4/1 and 2/0.5 at.%) [54]. For the carrier gas, inert gas Ar is the most commonly used type, while the adoption of N_2 or Ar with H_2 mixture can optimize the growth condition for optimal SWCNT production. The pressure inside the chamber is usually kept between 200 Torr and 500 Torr, as a lower pressure results in the production of amorphous carbon [55]. The chamber temperature is also an important parameter, which affects both the growth yield and CNT quality. A higher temperature than 1000°C is usually adopted to grow crystalline CNTs with minimal defects.

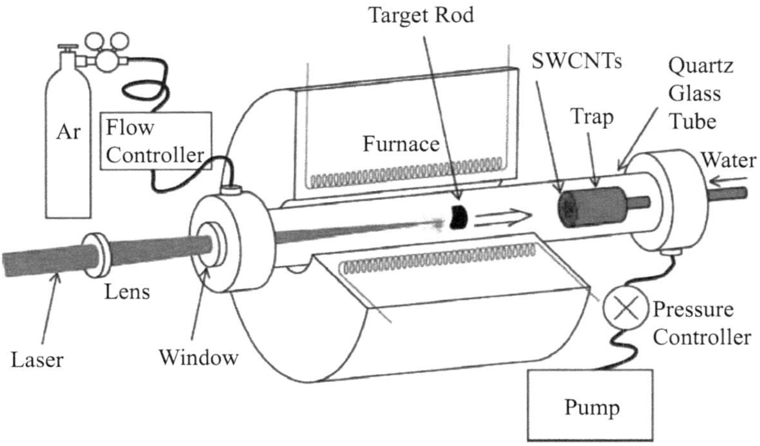

Figure 2.4 Schematic diagram of the laser ablation apparatus for CNT growth. Reprinted from Ref. [27], Copyright 2004, with permission from Elsevier.

Generally, high-quality SWCNTs can be obtained with the laser ablation method [56]. Compared to the arc-discharge method, the CNTs synthesized by the laser ablation method have a higher purity as the metal catalysts could be evaporated from the open end once the CNT growth stops. The yield of SWCNT is 70–90%, which can be further improved by adopting bimetallic catalysts [41]. However, the high production cost because of the high-power laser requirement may be a hurdle for its implementation in large-scale CNT synthesis. This issue could be partly relieved by adopting low-cost energy source such as solar energy and/or alternative low-cost carbon source material such as coal and charcoal instead of pure graphite [55]. Similar to the arc-discharge method, the as-synthesized CNTs on the collector need to be purified to remove the metallic CNTs for high-performance transistor applications.

CNTs can be classified as metallic and semiconducting based on their conductivity, although this is a very rough classification method. For semiconducting CNT, its bandgap is inversely proportional to its diameter. In terms of transistor application, it is desirable to apply the semiconducting SWCNTs with the same bandgap and diameter. A stricter requirement is to apply the CNTs with the same chirality for the sake of transistor uniformity especially when the transistor scales down to the sub-10 nm technology node. A lot of efforts have been devoted to optimize the synthesizing methods to obtain CNTs with uniform diameter or the same chirality [57]. By tuning the chamber temperature in the range of 780–1050°C, the diameter distribution of the CNTs increases with increasing temperatures and generally falls in the range of 0.81~1.51 nm. However, the synthesized CNTs are mostly metallic with a chiral number of (m, m) with m ranges from 6 to 11. Up to now, both the arc-discharge method and the laser ablation method cannot be used directly for CNT-based transistor applications. With the high-quality CNTs synthesized by these methods, most of the studies still try to apply the CNTs for high-performance transistor applications after post-processing like sorting, which is addressed in more detail in Chapter 4.

2.2.3 Chemical Vapor Deposition Method

The carbon sources used in the chemical vapor deposition (CVD) method are different from the above-mentioned methods in that a carbon-containing gas source such as hydrocarbon and CO is adopted for the CNT growth [58–60]. Based on the ways how the gas source is dissociated into carbon and other constituent gas, two types of CVD can be distinguished: thermal CVD and plasma-enhanced CVD (PECVD) [61, 62]. In a typical thermal CVD growth process, a 1–4 inch diameter quartz tube is used as the growth chamber where the growth substrate is mounted inside. The temperature of the chamber is controlled precisely via a tube furnace. The catalyst can be prepared on various kinds of substrate, including the patterned catalyst on silicon wafers for on-chip electron devices integration.

Before the introduction of the carbon source gas, the chamber is cleaned with inert gas and heated to a high temperature. In most cases, the deposited catalyst is a thin metal film that contains Fe, Co, Ni, Pd, or their mixtures [63–67]. During the heating process, the metal film is broken into nanoparticles whose size and distribution are highly dependent on many factors, including the film thickness, chamber temperature and duration, gas pressure, and substrate surface condition [63, 68]. After the formation of metal nanoparticles as the catalysts, the growth of CNT is then initiated by the introduction of the carbon source gas while keeping the chamber at the desired high temperature. The flow rate of the carbon source gas is precisely controlled by a mass flow meter. The growth rate depends on the temperature, flow rate, chamber pressure, and the catalyst type and nanoparticle size. Aided by the presence of catalyst particles, the carbon source gas undergoes thermal dissociation and dissolves in the nanoparticles. When the carbon content is supersaturated in the catalysts particle, it will precipitate on the catalyst particles and forms the microtubule structure [27]. If the areal density of particles on the substrate is high, van der Waal force among CNTs can facilitate vertical self-alignment during growth, which is beneficial for the application, including vertical interconnects for on-chip interconnects and through-wafer vias [69]. When the growth time is reached, the carbon source gas should be stopped immediately to avoid amorphous carbon deposition and change back to the inert gas for cooling down the chamber.

The temperature required to dissociate the carbon source gas is related to the gas species, the chamber pressure, and also the catalyst type. Among the commonly used transition metal catalyst, including Fe, Ni, and Co, Ni needs the lowest energy for dissociating the carbon source, which translates into a low CNT growth temperature [61]. In general, the temperature for thermal CVD is in the range of 500–1000°C, and a higher temperature usually favors the growth of SWCNTs. For many applications, this high-temperature process is detrimental to the existing function devices, including the CMOS transistors or metal interconnects.

PECVD provides another choice for low-temperature synthesis of CNT. The PECVD process is similar to CVD with a key difference being in the mechanism to disassociate hydrocarbon, which is done using an electric field applied in capacitive coupled plasma reactor to produce radicals. The resulting carbon radicals then precipitate on the catalyst sites where CNT growth occurs. In addition to the carbon source gas and inert gas, the chamber can also be flowed with N_2 and H_2 or NH_3 as reducing agent to etch excess carbon from the reactor walls and amorphous carbon from coating the catalyst sites so that continuous CNT growth can be sustained. The applied electric field ensures vertical alignment during CNT growth. It is reported that a temperature down to room temperature can be used to synthesize MWCNTs. Therefore, PECVD CNT growth methods can be compatible to and integrated into conventional IC processing.

There are two growth models to describe the CNT growth process by CVD methods, i.e., base-growth model and tip-growth model. If the carbon species precipitates on the surface of the catalyst particle, then it results in CNT growth on top of the particle, referred to as base growth. If the carbon species precipitates below the catalyst resulting in the particle on the CNT tip, it is called tip growth. The base- and tip-growth processes are shown schematically in Fig. 2.5 [70]. Such growth generally depends on the wettability of catalyst film to the underlayer, as well as that between CNT and catalyst. Meyyappan [71] and Sun [72] reported that catalyst exhibiting strong wettability to underlayer resulted in base growth due to the adherence of the catalyst to the underlayer, and that tip growth for catalyst with poor wettability.

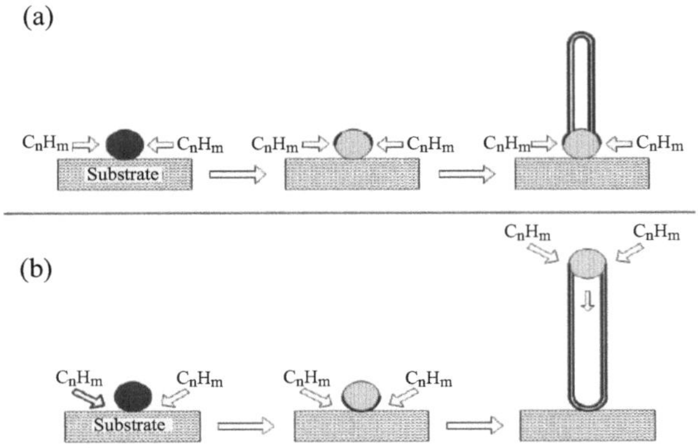

Figure 2.5 (a) Base growth and (b) tip growth of CNT on substrate resulting from catalyst-assisted dissociation of gaseous hydrocarbon source [70].

Now we discuss the recent development of the CVD process for integration with electron devices. Interconnects in the back end of the line (BEOL) process enter the sub-100 nm regime, and the current copper and tungsten metal lines encountered two major problems. One problem is that the metal lines exhibit much higher resistivity than their bulk values due to the grain boundary scattering and surface scattering; the other problem is the reliability issue due to the increased current density. As an alternative on-chip interconnect material, nanocarbon has attracted much attention due to its high mobility, high current-carrying capability, and high stability. To integrate the nanocarbon material into the production line, a low thermal budget process is usually required. What is more, the as-synthesized CNTs and graphene should be readily applied as interconnect structure, and no more post-processing such as purifying and sorting is allowed at this stage.

Ideally, the as-synthesized CNTs should be aligned to a certain direction for interconnect application. In this regard, the PECVD method is perfect for synthesizing vertically aligned CNTs, which is ideal for on-chip vertical interconnect application. For interconnects, it is also desirable to have CNTs in direct contact with the underlayer metal. Therefore, tip growth is preferred so that post-growth contact engineering can be performed to improve the top contact [73]. For example, by performing the standard chemical mechanical polishing (CMP) process, the metal nanoparticles on the tip of the CNTs can be removed to expose the CNT tip for following metallization processes. Ni catalyst is a commonly used CMOS-compatible material for PECVD CNT growth that also results in tip growth on multiple substrate materials, including Ti, Cr, and oxide [71]. The CNT density is a critical parameter determining the interconnect performance and is affected by many factors such as catalyst thickness and substrate underlayer material, growth temperature, and gas flow ratio. Sun [72] conducted extensive experiments on the effect of underlayer and catalyst thicknesses on CNT growth. The results are summarized in Fig. 2.6. One can conclude that for Ni/Ti system, 10 nm Ni is the critical thickness below which sparse growth of CNTs results, attributed to catalyst particles embedded in the underlayer metal grain boundaries [71]. Wilhite [74] developed a CNT growth process for an optimized combination of Ni catalyst and Cr underlayer thicknesses. It was demonstrated from blanket growth on unpatterned substrate that the columnar grain structure of Cr resulted in high CNT areal density ~10^{11} /cm^2, for Ni film thickness down to 4 nm [74]. Aside from yielding a high areal density, Cr has been reported to be an excellent etch stop layer to the reactive ion etching (RIE) process, a key requirement in patterning vias [75]. Ti readily reacts with halogens and oxygen radicals, which are the primary ingredients of the RIE process [76]. Therefore, Ti film can be physically damaged or oxidized during via etch by RIE, resulting in either a discontinuous or oxidized underlayer for subsequent CNT growth in vias. Thus, the Ni/Cr combination is optimal for studying CNT growth inside patterned vias.

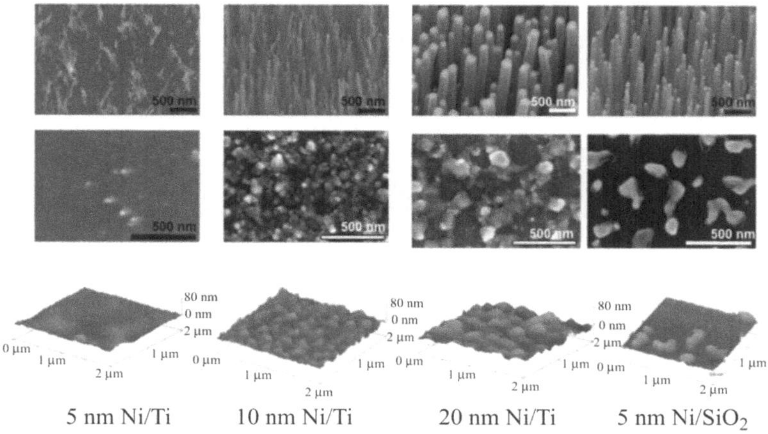

Figure 2.6 Effect of catalyst thickness and underlayer combination on CNT growth. Reprinted from Ref. [72], Copyright 2009, with permission from IOP Publishing Ltd.

Recently, a lot of experimental works have been conducted to find out the optimal growth condition for obtaining a high CNT density [74]. The objective is to achieve the smallest possible average CNT diameter with narrow distribution and the highest possible areal density, key requirements for interconnect application in order to yield low resistance. The first stage of the process is catalyst film dewetting, which determines the diameter distribution and CNT areal density. To achieve the optimum diameter and CNT areal density, various experiments were conducted by varying Ni catalyst thickness (4–18 nm), temperature ramp rate (250–325°C/min), and annealing time (0–4 min), with growth time fixed at 5 min and gas flow rates fixed at 125 sccm and 31 sccm for NH_3 and C_2H_2, respectively [71]. SEM analysis of samples revealed that the average diameter of CNT is primarily determined by catalyst film thickness and annealing time. Longer annealing time resulted in coalescence of some nucleation sites, known as Ostwald ripening mechanism [77], and yielded CNTs with wide diameter distribution and sparse density. Therefore, the growth recipe we use contained no annealing and the highest ramp rate allowed by the deposition system, 325°C/min, to achieve narrow CNT diameter distribution and high areal density. Top-view SEM images of samples with and without annealing are shown in Fig. 2.7.

Figure 2.7 (a) CNT growth after 5 min annealing in ammonia ambient. (b) CNT growth without annealing. Both samples grown using 12 nm Ni film.

Based on the optimized CNT growth recipe, we have recently demonstrated the PECVD process for fabricating on-chip vias with a dimension down to sub-100 nm [20]. Figure 2.8 presents the process flow for fabricating and characterizing on-chip CNT vias, while Figs. 2.9(a) to 2.9(f) show the SEM images at different stages of the process. Starting with a 100 mm SiO_2/Si wafer, a 300 nm chromium (Cr) film is evaporated and then annealed at 400°C to achieve a smoother surface. Next, a 90 nm low-temperature SiO_2 (LTO) film and a 50 nm amorphous Si (a-Si) film are formed by PECVD. E-beam lithography is used to pattern the vias with square cross sections and widths of 150, 120, 90, and 60 nm. During dry etching, the e-beam resist is first used as a mask for a-Si etching, and then the a-Si thin film is used as the hard mask for SiO_2 etching. After dry etching, vias with vertical sidewalls are achieved [see Fig. 2.9(b)]. Upon deposition of a 4 nm Ni film, the excess Ni outside the vias is removed to ensure selective CNT growth inside pre-patterned vias only [see Fig. 2.9(c)].

After CNT growth by PECVD at 700°C, for dielectric filling, a 6 nm Al_2O_3 film is formed by atomic layer deposition (ALD) [see Fig. 2.9(d)], suitable for high-aspect ratio (A/R) structures such as CNTs. Figure 2.9(e) shows the top-view SEM image of a single 60 nm CNT via, revealing multiple CNTs. On average, there are about 7, 15, 26, and 40 CNTs inside the 60 nm, 90 nm, 120 nm, and 150 nm vias, respectively, yielding an average CNT density of about $2 \times 10^{11}/cm^2$.

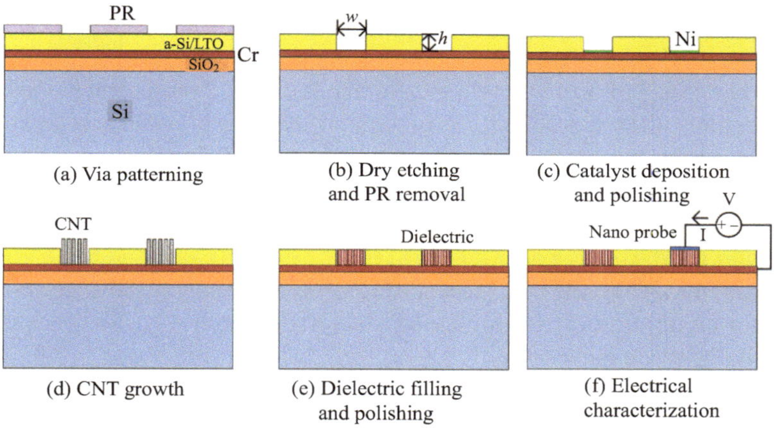

Figure 2.8 Schematic of on-chip CNT via fabrication process: (a) via patterning; (b) dry etching and PR removal; (c) catalyst deposition and polishing; (d) CNT growth; (e) dielectric filling and polishing; (f) electrical characterization. Reprinted with permission from Ref. [20], Copyright 2015, IEEE.

Noticing that the temperature used in the above PECVD CNT growth process is still relatively higher than CMOS-compatible requirement, we have optimized the catalyst design, including both material and thickness, to lower the growth temperature down to 350°C [25, 78]. By modifying the catalyst from a pure Ni single-layer structure to a Ni/Al/Ni multilayer thin film, uniform selective CNT growth on Ti silicide substrate has been achieved. Here the Ti silicide growth substrate is chosen for integrating the CNT via with the transistors as silicide is a commonly used source/drain contact material for Si channel material. The purpose of the Al intermediate layer is to change the surface energy of the Ni nanoparticle by forming Al–Ni compound after annealing. The mobility of the Al–Ni compound is much lower than pure Ni atoms, thus preventing conglomeration of particles and diffusion of Ni nanoparticles [79].

The use of Al to improve the catalyst's catalytic property for CNT growth was also reported in some works with different catalyst and structures [80]. The modified catalyst helps to improve the uniformity of Ni nanoparticle, which eventually leads to uniformly distributed vertically aligned CNTs synthesized at a low temperature. Using the modified catalyst, CNTs are grown at 450°C, and the ramp-up and ramp-down rates are 100°C/min and 22°C/min, respectively. It is noticed that the CMOS transistor performances are almost the same after carbon nanofiber growth under 500°C for 60 min [81]. As the thermal budget of the present study is much lower, we reckon that the CNT growth will not affect the CMOS transistors. High-density CNT forest with very regular structure is observed. The density of the CNTs is found to be around 1.1×10^{11} tubes/cm^2, which is in the same order as recently reported values [20, 82]. The average diameter of the synthesized CNTs is ~25 nm. From the height (1 μm) and synthesis time (6 min), the CNT growth rate is found to be 167 nm/min. These results show the high potential of the PECVD method for integrating vertically aligned CNTs with current CMOS as vertical via interconnects.

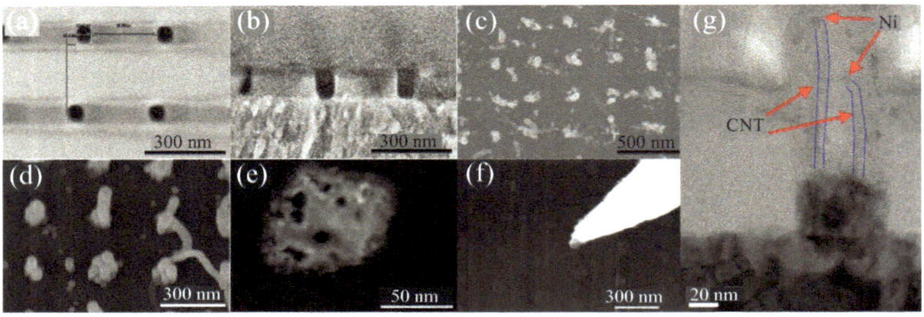

Figure 2.9 SEM images of 60 nm vias: (a) top-view image after via patterning; (b) cross-sectional image after dry etching; (c) top-view image of as-grown CNT vias; (d) 30° angled view image of CNT vias after dielectric filling; (e) top-view of a single CNT after ion milling at glancing angle; (f) nanoprobe landing on single CNT via, making direct contact with CNT tips without prior contact metallization; (g) TEM images of a 60 nm CNT via before ion milling, where Ni catalyst particle at CNT tip is indicated by red arrow and the CNT sidewalls by blue dotted lines. Reprinted with permission from Ref. [20], Copyright 2015, IEEE.

Compared to the vertically aligned CNTs, the development of horizontally aligned CNTs for interconnects has received less attention due to the difficulty of synthesizing horizontally aligned

CNTs with the PECVD method. In interconnect applications, experimental foci have been on the horizontal growth of CNTs, metal-CNT side-bonded and end-bonded contacts [83], electrical characterization [84–86], and circuit integration [87]. In most of the experiments reported, CNTs were grown vertically on another substrate and transferred onto prefabricated test structures by drop-casting or using a polymer sheet. However, there have been successful attempts to grow and characterize CNTs in horizontal configurations [88–93]. A typical example of SWCNTs grown in horizontal configuration on a quartz substrate is shown in Fig. 2.10 [94], with a packing density of 1~2 CNTs/μm.

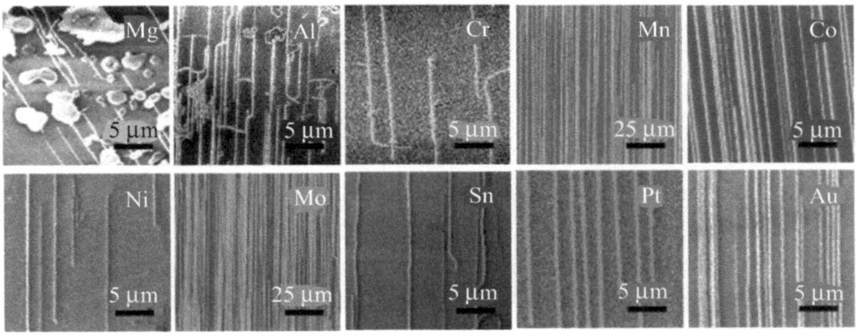

Figure 2.10 SEM images of CNTs grown on quartz substrate using different catalysts. Reprinted with permission from Ref. [94], Copyright 2008, American Chemical Society.

In recent years, significant progress has been made on wafer-scale methods to grow horizontally aligned CNTs with high packing density. Patil [95] reported the first wafer-scale growth of aligned SWCNTs on a 4-inch quartz wafer, with a packing density of 6–7 CNTs/μm, as shown in Fig. 2.11. Hu [96] reported a wafer-scale growth of SWCNTs with a median diameter of 0.9 nm and a packing density of 130 CNTs/μm. In comparison, for 1 nm half-pitch, a packing density of 501 CNTs/μm is required. Therefore, ~4× improvement in wafer-scale packing density of CNTs is needed. It is also essential to control the chirality of CNTs grown on wafer as it determines the conduction property. Recently, Yang [97] reported chirality-controlled synthesis of metallic SWCNTs with chirality of (12,6) and a 92% yield of metallic nanotubes. Further improvement in chirality-controlled synthesis with high yield (>99%) can eventually result in integrating SWCNTs as FET channel and/or on-chip interconnects in integrated circuits.

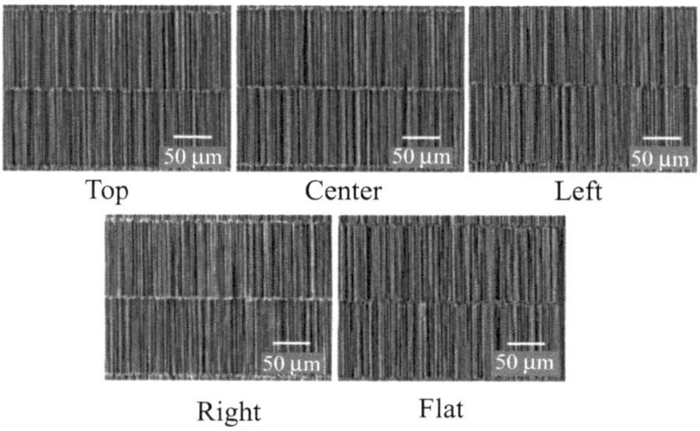

Figure 2.11 SEM images of horizontally aligned SWCNTs grown on quartz substrate. Reprinted with permission from Ref. [95], Copyright 2009, IEEE.

In contrast, the growth of horizontally aligned MWCNTs at lithographically specified locations has been a challenge. Vertical alignment of MWCNTs can be achieved with van der Waals force or with a vertical electric field. However, applying a horizontal electric field between two metal electrodes on a substrate can be challenging. Santini [98] and Yan [99] have demonstrated the growth of horizontal MWCNTs between metal electrodes and between sidewalls of a trench, respectively, as shown in Fig. 2.12. Significant process development is required to achieve high-density array of MWCNTs at pre-defined locations and ultimately on a fully integrated IC.

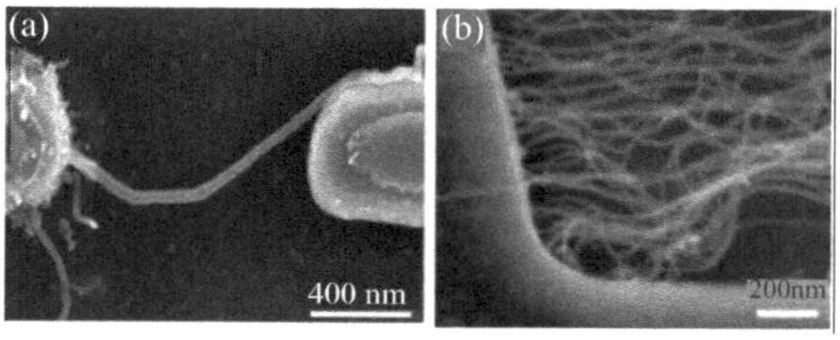

Figure 2.12 (a) Top-view SEM image of a single MWCNT grown between two metal electrodes. (b) Cross-sectional SEM images of a bundle of MWCNTs grown between sidewalls of a trench. Figure (a) reprinted from Ref. [98], Copyright 2011, with permission from Elsevier; and Fig. (b) reprinted from Ref. [99], Copyright 2010, with permission from John Wiley and Sons.

The integration of CVD-grown CNTs for transistor applications is a much more challenging task. The goal is to grow single-chiral semiconducting SWCNTs for transistor applications. At least, it is expected that the grown CNTs are all semiconducting to obtain a transistor with a high on/off ratio. PECVD usually results in MWCNTs that are metallic and thus not suitable for transistor integrations. The thermal CVD method is proven to be an effective method to achieve horizontally aligned SWCNT arrays on crystalline quartz substrate. Xiao [100] obtained aligned SWCNTs on crystalline quartz with different crystal cuts and proposed that the alignment is governed by the angle-dependent van der Waals interactions with the substrate. With patterned stripes of Fe thin film as the catalyst, they were able to grow aligned SWCNT arrays at a high temperature of 925°C. Apart from the commonly used transition metal catalysts, Zhou demonstrated that Cu catalyst can be used to achieve aligned CNTs on Si substrate at a relatively lower temperature of 825–850°C [101]. However, the obtained CNTs are still mostly metallic according to the Raman spectra measurement. The enrichment of semiconducting SWCNTs is achieved by optimizing the carbon source, i.e., the ethanol/methanol ratio in a CVD growth process [102]. During the growth stage, both metallic and semiconducting SWCNTs are formed. The enrichment is mainly due to the preferential etching effects of the metallic SWCNTs with the introduction of etchant species such as H_2O and *H. These etchant species are closely related to the addition of methanol to the carbon source gas. With an optimized gas mixture, a high semiconducting SWCNT yield of nearly 93% is obtained. A thin-film transistor is fabricated based on the horizontally aligned semiconducting SWCNTs on top of a ST-quartz substrate, and an on/off ratio of 35 is obtained, which is one order of magnitude higher than that based on normal SWCNTs without methanol mixture. As the metallic SWCNTs are more chemically reactive, similar preferential etching can be applied to selectively remove the metallic SWCNTs and increase the ratio of the semiconducting SWCNTs, such as plasma etching [103] and irradiation [104–106].

Another popular method to obtain high-purity semiconducting SWCNT is realized by engineering the catalyst particles. Zhang prepared acorn-like Co nanoparticles with thin carbon coating layer as the catalysts [107]. SWCNTs with a narrow diameter distribution were grown by the CVD process using ethanol as the carbon source.

Hydrogen was also introduced due to its preferential etching property during the growth stage. The ratio of the semiconducting SWCNT is as high as 96%, and the diameter of the SWCNT has a very narrow distribution, resulting in a bandgap range of less than 0.08 eV, which is very promising for transistor applications [107]. Recently, Yang prepared heteropolyacid molecular cluster containing W and Co, and then used H_2O vapor to regulate the structures of the nanoparticles to form W_6Co_7 nanoparticles [108]. CVD growth based on these nanoparticles produces mostly semiconducting SWCNTs with a chiral number of (14, 4), well matching the atomic arrangement in the (1, 0, 10) plane of W_6Co_7 crystal, as shown in Fig. 2.13. A high purity of 99.8% semiconducting SWCNT is achieved after treating the as-grown CNTs with water vapor at 550°C by post-etching process taking advantage of the preferential metallic CNT

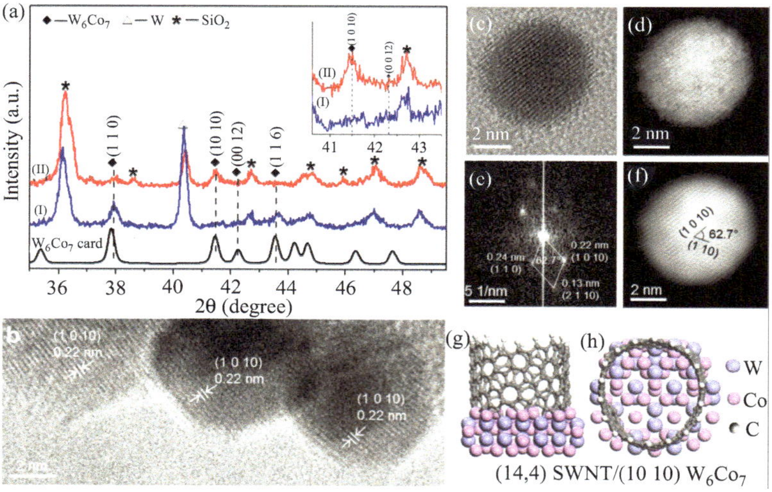

Figure 2.13 Characterization of catalyst structure. (a) XRD patterns obtained on the fresh catalysts reduced by pure H_2 (curve I) at 1030°C and H_2/H_2O at 1050°C (curve II). The W_6Co_7 standard card from the database is also shown. (b) HRTEM image of W–Co catalyst nanoparticles reduced by H_2/H_2O at 1050°C. (c–f) Bright-field (c) and HAADF (d) STEM images, corresponding FFT (e) and inverse FFT (f) of a single W_6Co_7 nanoparticle, spots corresponding to (1 0 10), (1 1 0), and (2 1 10) planes are indicated in (e). (g, h) DFT simulation. Side (g) and top (h) views of interfaces between (14, 4) SWNT and the (1 0 10) plane of W_6Co_7. Reprinted with permission from Ref. [108], Copyright 2017, American Chemical Society.

etching properties. CNT transistors with a high on/off ratio in the range of 10^3–10^6 are demonstrated based on the synthesized high-purity semiconducting SWCNTs. This shows a promising growth method for transistor applications, although the grown CNTs are not aligned on the SiO_2/Si substrate and the growth temperature is still relatively high (>1000°C).

2.3 Graphene Growth and Integration Process

2.3.1 Exfoliation Method

Graphene was discovered by Geim and Novoselov who successfully isolated monolayer graphite using a scotch tape and a highly oriented pyrolytic graphite (HOPG) [109]. This method is known as the mechanical exfoliation method and has now been widely used to obtain various high-quality 2D materials [110]. In their pioneering work, they first prepared ~5 μm thick mesas out of the HOPG and attached the mesas to a photoresist-coated substrate. The mesas left in the photoresist are then pressed with a scotch tape to peel the graphene flakes off the mesa. By repeating the process, monolayer to few-layer graphene are left in the photoresist. The photoresist is then removed by rinsing the substrate in acetone, and the exfoliated graphene flakes are released in the solution at the same time. Finally, the target substrate SiO_2/Si is used to capture the thin flakes from the solution. It is an interesting phenomenon that the exfoliated thin flakes can be directly identified by an optical microscope (Fig. 2.14a) [109]. Usually, an atomic force microscope (AFM) is used to identify the thickness of the graphene flakes especially when the thickness is down to monolayer, which is time consuming and not cost effective for device fabrication. Wang [111] proposed to analyze the optical images of graphene sample by reading the red, green, and blue channel value. They found that the green channel value gives the highest contrast value for graphene on SiO_2 285 nm/Si substrate, and the contrast value is in linear with the layer number (Fig. 2.14b). Li [112] systematically investigated the optical contrast of 2D materials with respect to SiO_2/Si substrates and developed a relatively simple method to identify the thickness of 2D materials in the range of monolayer to 15 layers. This method can be applied to identify various 2D materials, including graphene, MoS_2, WSe_2, etc. [112].

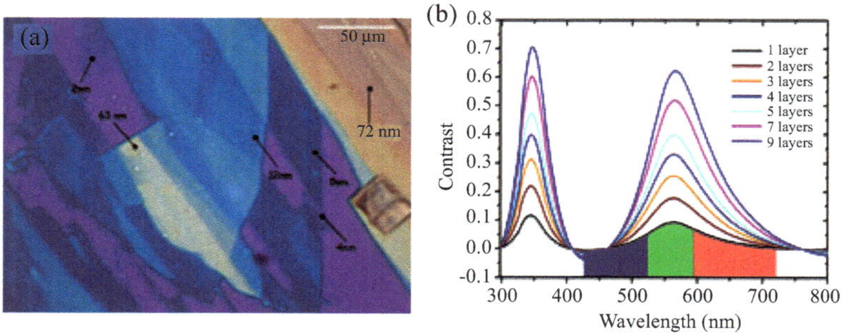

Figure 2.14 (a) Optical photo in the white light of graphitic films of various thicknesses [109] (b) Contrast spectra of graphene samples with different thicknesses on SiO_2 (285 nm)/=Si substrate calculated under normal incidence based on the Fresnel equation. Figure (a) reprinted from Ref. [109], Copyright 2004, AAAS; and Fig. (b) reprinted from Ref. [111], Copyright 2012, with permission from IOP Publishing Ltd.

By the mechanical exfoliation method, a very high-quality graphene can be obtained as evidenced by many reports of high-mobility graphene FETs built on mechanically exfoliated graphene. However, the size of the exfoliated graphene flake is very small, and the yield is very low. Since the advent of the mechanical exfoliation method, various modifications have been made to obtain large size and ideally monolayer graphene. For example, Huang processed the target substrate by oxygen plasma cleaning to remove the ambient adsorbates and then pressed the scotch tape on the substrate at an elevated temperature to increase the uniform contact area [113]. Figure 2.15 shows the modified transfer process, through which monolayer graphene flakes with hundreds of μm side length have been obtained. A gold-mediated exfoliation method was reported to obtain large-size MoS_2 flakes by first depositing gold film on top of a bulk material and then perform the transfer process assisted by thermal release tape [114]. This method takes advantage of the weaker interlayer bonding strength of the 2D materials than that between gold and 2D material and can be readily applied to graphene and other 2D-layered materials. Up to now, the application of the mechanical method has been mainly confined in laboratory pure research work due to its relatively high quality but low yield.

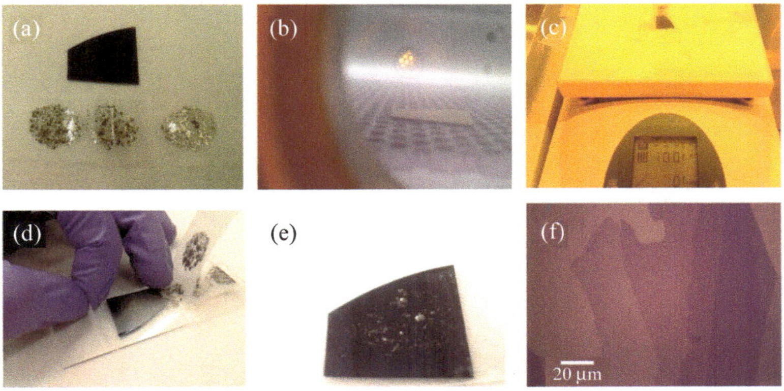

Figure 2.15 Illustration of the modified exfoliation process for layered crystals (shown here for graphene). (a) Optical image of the SiO_2/Si substrate and adhesive tape with graphite flakes, which adhere after contact with an HOPG crystal. (b) Oxygen plasma cleaning of the SiO_2/Si substrate. (c) Contact between the graphite-decorated tape and the substrate surface, followed by heating of the substrate (with tape) on a hot plate in air at ~100°C for 2 min. (d) Removal of the substrate from the hot plate and peeling off of the tape. (e) Optical image of the substrate after graphene exfoliation. (f) Optical micrograph of one of the graphene flakes on the substrate in (e). The flake has a thickness varying in steps between 1 and 4 layers. Reprinted with permission from Ref. [113], Copyright 2015, American Chemical Society.

Exfoliation of graphene can be also processed via graphite intercalation. Actually, intercalation is probably the first method used to obtain graphene flakes [115]. Starting from the graphite intercalation compounds (GIC) with potassium intercalation layer, the K-GIC is then allowed to react with the unsaturated vapor phase hydrocarbons. This reaction results in the expansion of the K-GIC in the c-axis direction, and the molecules of isoprene are introduced into the interlayer spacing and polymerized. Finally, the polymers are removed by heating the sample at 400°C. The sample surface is wavy and rolled up at the edge to decrease the surface energy, which probably consists of suspended graphene layers. Figure 2.16a shows the general process of the intercalation method [116]. Acid treatment and thermal shock can facilitate the exfoliation of thin graphite sheets or graphene fragments. Graphene sheets can also be prepared via an ion exchange process, as shown in Fig. 2.16b [116]. The properties of the liquid-exfoliated graphene sample are usually very poor due to the chemical modification during the process [117].

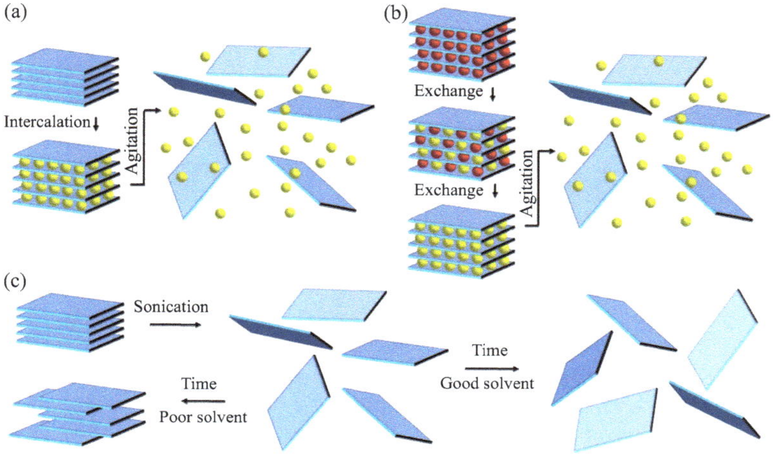

Figure 2.16 Schematic description of the main liquid exfoliation mechanisms. (a) Ion intercalation. Ions (yellow spheres) are intercalated between the layers in a liquid environment, swelling the crystal and weakening the interlayer attraction. Then, agitation (such as shear, ultrasonication, or thermal) can completely separate the layers, resulting in an exfoliated dispersion. (b) Ion exchange. Some layered compounds contain ions between the layers so as to balance surface charge on the layers. These ions (red spheres) can be exchanged in a liquid environment for other, often larger ions (yellow spheres). As above, agitation results in an exfoliated dispersion. (c) Sonication-assisted exfoliation. The layered crystal is sonicated in a solvent, resulting in exfoliation and nanosheet formation. Reprinted from Ref. [116], Copyright 2013, AAAS.

The mass production of graphene can be obtained by sonication-assisted liquid exfoliation method, as shown in Fig. 2.16c [116]. Coleman produced high-quality monolayer graphene flakes by using liquid exfoliation in organic solvents [118]. Graphite powder is dispersed in *N*-methylpyrrolidone (NMP) by bath sonication, which is similar to as used for CNT dispersion [119]. The liquid containing graphite was further centrifuged to remove the aggregates, and monolayer graphene flakes can be obtained. The key to successful liquid exfoliation is the choice of solvent with optimal surface tension to minimize the energetic cost of exfoliation. The monolayer yield for this method is 1 wt.%, which means the graphene concentration in the dispersion is still very low for practical applications. However, unlike the mechanical exfoliation method, the liquid exfoliation method is potentially scalable and can be further improved [120]. The flake size of the exfoliated graphene is relatively smaller than that obtained with mechanical exfoliation, with the early reports of ~μm and recently improved to be about ~5 μm [116]. The quality of the liquid-exfoliated graphene is also much inferior to the mechanical exfoliated samples [121]. Thus, the most possible application of the liquid exfoliation method is probably the composite polymer, electrode in energy storage, and flexible electronics.

2.3.2 Chemical Vapor Deposition Method

The CVD method can be processed at a large scale for mass production of graphene and has attracted intense research effort to obtain large-area, high-crystalline, and low-defect graphene [122, 123]. Just as the case for CNT growth, the graphene growth is also affected by many factors, including the chamber pressure, growth temperature and growth time, substrate materials, carrier gas and carbon source gas, etc. Monolayer large-area graphene has been obtained on copper foils [124]. Li used copper foil as the substrate and obtained single-layer graphene over the copper surface, which is originally very rough compared to the graphene thickness. It is found that the graphene can cross the copper grains and steps to form large-area single-crystalline structure, which is very beneficial for practical applications in electronics. Compared to other commonly used transition metal catalysts for graphene growth, copper has a very low carbon solubility, which results in a self-limiting growth mechanism on copper surface. That is probably the underlying reason why a large percentage (95%) of single-layer coverage is obtained with the copper foil as the growth substrate. For a typical growth process (Fig. 2.17a) [124], the copper foil is placed in a quartz chamber, and then the chamber is heated to 1000°C while flowed with H_2 gas to maintain a pressure of 40 mTorr. The growth of graphene initiates when the carbon source gas CH_4 is introduced and a relatively high pressure of 500 mTorr is maintained. The growth time is also a critical parameter determining the graphene coverage ratio on the copper surface. Figures 2.17b and c show the SEM pictures of the copper foil surface after 1 min and 10 min growth [124]. It is noticed that many single-crystal graphene grains are grown simultaneously, and eventually they merge into one single piece of single-layer graphene. Thus, one could expect that the CVD-grown graphene is a polycrystalline material consisting of many large single-crystalline grains, and the quality is inferior to that obtained by the mechanical exfoliation method.

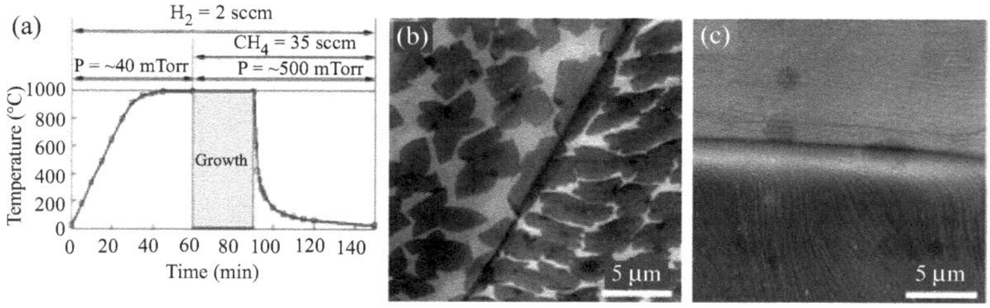

Figure 2.17 (a) Time dependence of experimental parameters: temperature, pressure, and gas composition/flow rate. SEM images of graphene on Cu with different growth times of (b) 1 min, (c) 10 min. Reprinted from Ref. [124], Copyright 2009, AAAS.

After years of development of the CVD graphene growth method, it has become a mature technology for fabricating large-area single-layer graphene. Nowadays, the size of graphene is only limited by the copper foil size that can be placed inside the chamber. Figure 2.18a shows the quartz tube used for CVD graphene growth, and Fig. 2.18b presents the optical picture of the synthesized graphene on copper foil with a size up to 17 inch × 36 inch [125]. A high single-layer graphene yield >98% is obtained under the optimized growth condition, and the graphene can be readily transferred to a PET substrate for flexible electrode application. For industrial production, the area of graphene can be further increased by enlarging the chamber size or by rolling the copper foils in the way that a larger surface area is exposed for CVD graphene growth [126]. Another interesting way to increase the production is a dynamic growth process during which the copper foil is translated for continuous graphene growth [127]. The growth apparatus consists of two concentric quartz tubes, and the copper foil is wrapped around the outer wall of the inner tube, as shown in Fig. 2.19. During the growth process, the inner tube is flowed with a gas mixture of C_2H_4 and H_2, and the gap between the inner tube and the outer tube is exposed to carbon source gas through drilling holes in the inner tube. The metal foil can be translated along the gas flow direction to obtain the dynamic growth of graphene. Even with a high translation speed of 250 mm/min, the metal foil is fully covered with graphene. Thus, the CVD method is quite suitable for mass production of graphene.

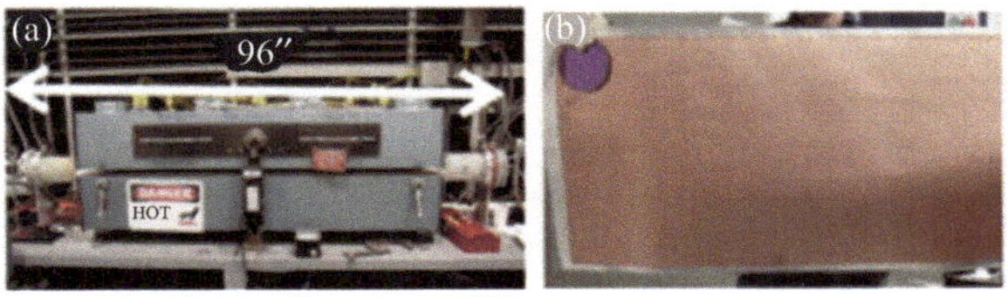

Figure 2.18 ORNL setup for 36 inch × 17 inch monolayer graphene growth using atmospheric pressure CVD. (a) Five-zone furnace with 96-inch-long, 6-inch-diameter quartz tube. (b) Graphene grown on a copper foil; 4 inch wafer is shown for comparison. Reprinted from Ref. [125], Copyright 2013, with permission from Elsevier.

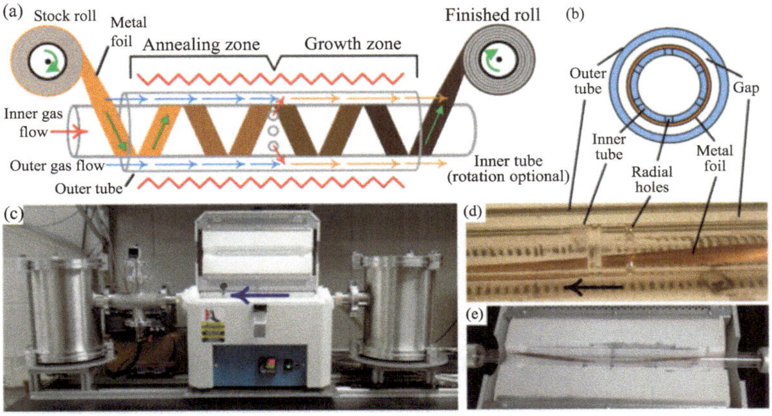

Figure 2.19 Concentric tube CVD (CTCVD) system configured for R2R graphene growth on Cu foil. (a) System schematic showing the helical feed path (left to right), sequential treatment zones, and internal gas injection holes. (b) Cross-sectional view of the concentric tube arrangement. (c) Bench-scale prototype of the CTCVD system (setup where processing is right to left), with rails for alignment with tube furnace. (d) Close-up of (e), showing the gas injection holes used to supply the hydrocarbon gas to the downstream treatment zone. (e) Top view of a Cu foil substrate wrapped through the system. Reprinted from Ref. [127], Copyright 2015, Springer Nature.

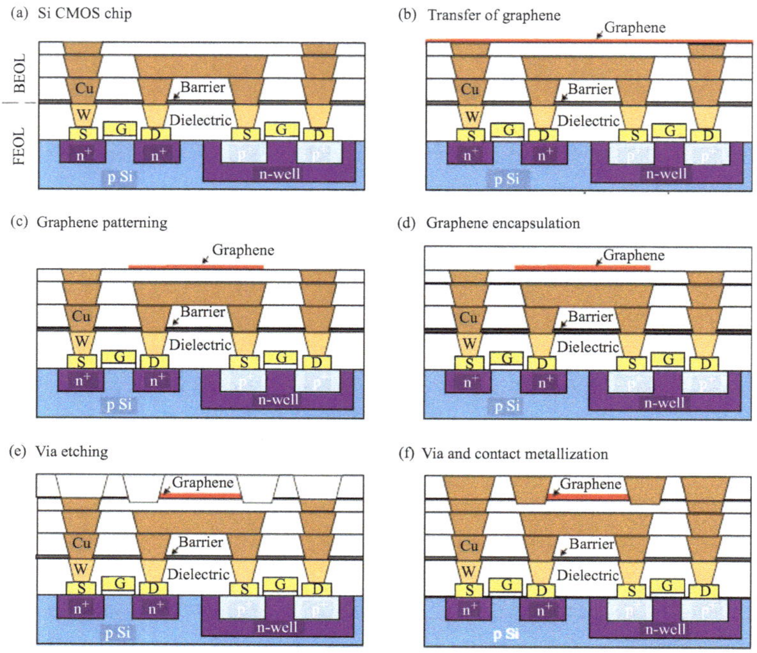

Figure 2.20 Flow chart for graphene integration. (a) Cross-sectional schematic of the CMOS chip before graphene integration, indicating the FEOL and BEOL parts. (b) Transfer of the graphene onto the entire wafer. (c) Patterning of the graphene layer using oxygen plasma. (d) Encapsulation of the graphene using a dielectric layer. (e) Via etching through the top dielectric layer by means of dry etching techniques. (f) Filling of the vias by metal providing edge contacts to the graphene. After this step, further interconnect layers can be fabricated in order to connect the graphene layer to the Si CMOS devices. This integration schematic can be used, for example, for integrating graphene Hall sensors on wafers containing Si CMOS logic. Reprinted from Ref. [24], Copyright 2019, by permission from Springer Nature.

For electron device application, the as-grown graphene cannot be directly used if the graphene is grown on the metal foils. What is more, the growth temperature of ~1000°C in the CVD process is much higher than that can be sustained by CMOS ICs. Thus, one common post-growth process is usually needed to transfer the as-grown graphene to the target substrate. Figure 2.20 shows a process flow for integrating graphene with current CMOS technology after the fabrication of CMOS ICs [24]. The most commonly used transfer processes, including polymer-assisted wet transfer method, dry-transfer method, roll-to-roll transfer, are reviewed by Kang [128]. In a typical polymer (PMMA) assisted transfer process (Fig. 2.21) [129], a thin layer of PMMA is first spin coated on the graphene/growth substrate and cured to harden the PMMA film to increase the adhesion between the PMMA and graphene. Then the growth substrate is removed in the corresponding substrate-etchant solutions, such as $FeCl_3$ solution for copper foil and HNO_3 for Ni foil. The PMMA/graphene is then floated on the solution after removing the substrate, which can be scooped by another substrate for target applications. The PMMA-assisted transfer process usually results in wrinkles and polymer residues on the transferred graphene [130], so rigorous PMMA/graphene washing and PMMA removing are proven to be important to improve the electronic property of graphene. This phenomenon has been revealed by many reports on the mobility improvement after vacuum annealing treatment of the transferred samples [130, 131]. Li [129] proposed a method to avoid the wrinkles by simply coating another layer of PMMA before the PMMA removing. Banszerus developed a dry-transfer method to avoid any polymer contamination during the transfer process and reported a high mobility of ~350,000 $cm^2/(V{\cdot}s)$ at 1.6 K, which rivals the exfoliated graphene samples [132]. However, these transfer methods are not likely to be scaled up for mass production of electron devices. For continuous large-area graphene transfer, a process borrowed from the paper-rolling industry is adopted as the roll-to-roll transfer method, as described in Fig. 2.22 [126]. In this method, the graphene/copper foil is attached to a thermal release tape, while a soft pressure is applied by the two rollers. The copper foil is then removed in copper etchant leaving the graphene/tape for final transfer. The graphene/tape stack is then inserted into the two rollers together with the target substrate. A mild thermal treatment ~90–120°C is applied to release the graphene from the thermal release tape and transfer the graphene to the target substrate. A transfer rate of 150–200 mm/min is obtained and a ~30-inch diagonal length graphene film is transferred to the PET substrate for flexible electrode applications, as shown in Figs. 2.22b and c [126]. With further improvement in the graphene uniformity and transfer process, the application of graphene at the BEOL level, such as interconnects, electrodes, and sensors, can be envisioned in the near future.

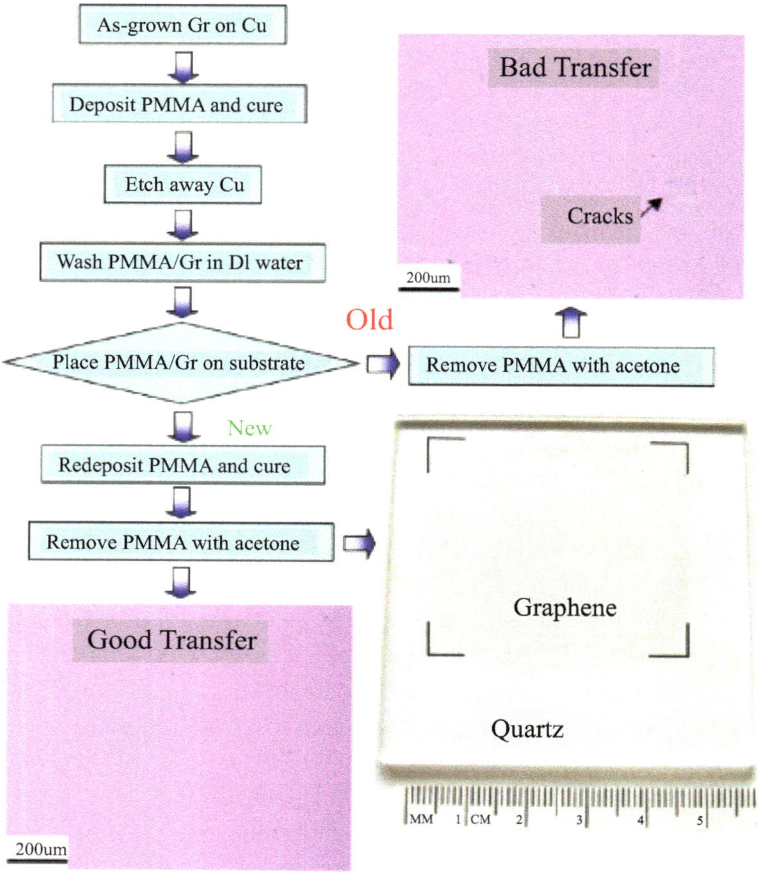

Figure 2.21 Processes for transfer of CVD graphene films using a second layer of PMMA coating and curing. Reprinted with permission from Ref. [129], Copyright 2009, American Chemical Society.

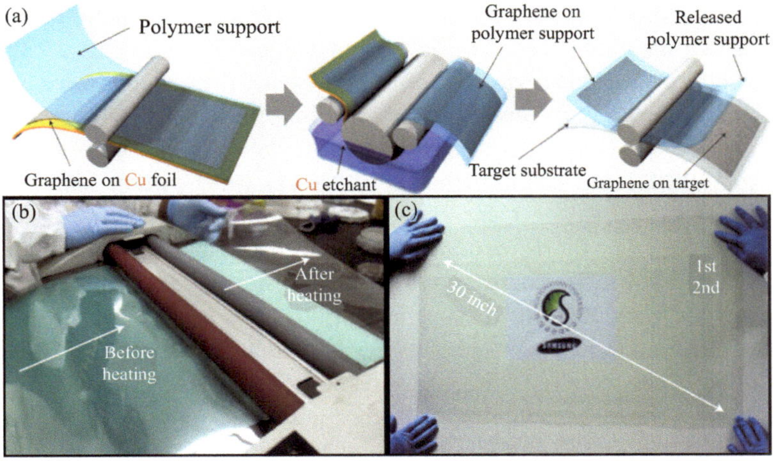

Figure 2.22 (a) Schematic of the roll-based production of graphene films grown on a copper foil. (b) Roll-to-roll transfer of graphene films from a thermal release tape to a PET film at 120°C. (c) A transparent ultralarge-area graphene film transferred on a 35-inch PET sheet. Reprinted from Ref. [126], Copyright 2010, by permission from Springer Nature.

2.3.3 Epitaxial Graphene Growth Method

Graphene can be also prepared via the epitaxial growth on single-crystal substrates. Various substrate materials, including SiC, Ge, and Ru, have been demonstrated for high-crystalline graphene growth, although the growth mechanism is highly dependent on the specific substrate [133, 134]. For SiC substrate, the graphene forms by heating the SiC substrate at a high vacuum chamber, taking advantage of the different sublimation temperature of Si and C. As shown in Fig. 2.23a, Si atoms in the top layers desorb at a high temperature, while the remaining carbon atoms reorganize into the hexagonal structure of graphene. Early work on graphene synthesis based on SiC decomposition reported inhomogeneous graphene flakes formation with a limited flake size [135]. By performing the annealing under Ar environment, Emtsev produced monolayer graphene with a size of >100 μm^2, and a mobility of 2000 cm^2/(V·s) at T = 27 K [133]. Graphene grown on SiC with different terminations, i.e., Si termination or C termination, also exhibits different structures

[136]. Generally, uniform homogeneous few-layer graphene can be grown on the Si face via a layer-by-layer mechanism (Fig. 2.23b), while the thickness can be controlled by heating temperature and time. Figure 2.23c shows the cross-sectional TEM images of graphene grown on Si face under different annealing temperatures from 1350 to 1500°C in a vacuum furnace (10^{-6} Torr). On the other hand, the graphene grown on C face is usually nonhomogeneous multilayer flakes because there are many nucleation sites during the growth on C face (Fig. 2.23b). Compared to the copper substrate used in the CVD method, the SiC substrate is insulating so that the transfer process is not needed to fabricate electron devices like transistors [137]. Up to now, wafer-scale high-frequency transistors and circuits have been demonstrated based on epitaxial graphene layer [18]. However, the high cost of the SiC substrate and the high temperature process also restrict the large-scale application of graphene.

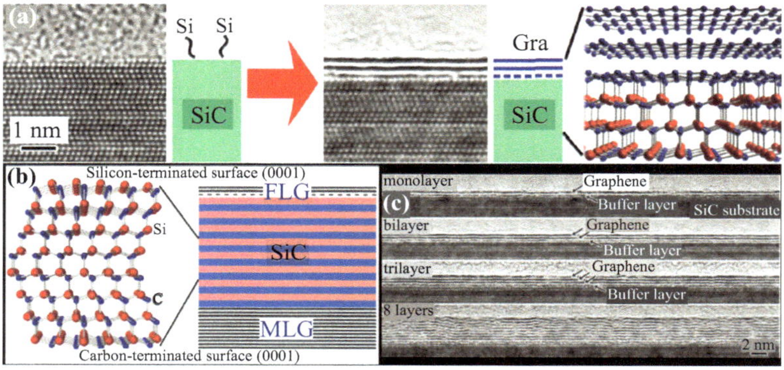

Figure 2.23 (a) Basics of graphene growth by thermal decomposition of SiC, together with the structural model of bilayer graphene on SiC. Shown as the blue broken line is the buffer layer. (b) Crystal structure of 6H-SiC and an illustration of graphene on its surfaces. The large red and small blue balls correspond to the silicon and carbon atoms, respectively. (c) Cross-sectional HRTEM images of graphene on the Si-terminated SiC [138]. Figures (a) and (b) reprinted from Ref. [136], Copyright 2014, with permission from the Royal Society of Chemistry; and Fig. (c) reprinted from Ref. [138], Copyright 2009, with permission from Elsevier.

Wafer-scale monolayer single-crystal graphene can be grown on a germanium wafer, as demonstrated by J-H Lee [134]. Similar to the copper substrate, germanium has an extremely low solubility of carbon, so the self-limiting single-layer growth mechanism on copper surface may also apply to germanium substrates. At the early stage of the graphene growth, many single-crystal graphene flakes are present along the same crystal orientation, which forms the epitaxial relationship with the underlying Ge (110) substrate (Figs 2.24a and b). Due to the self-limiting mechanism and the same arrangements

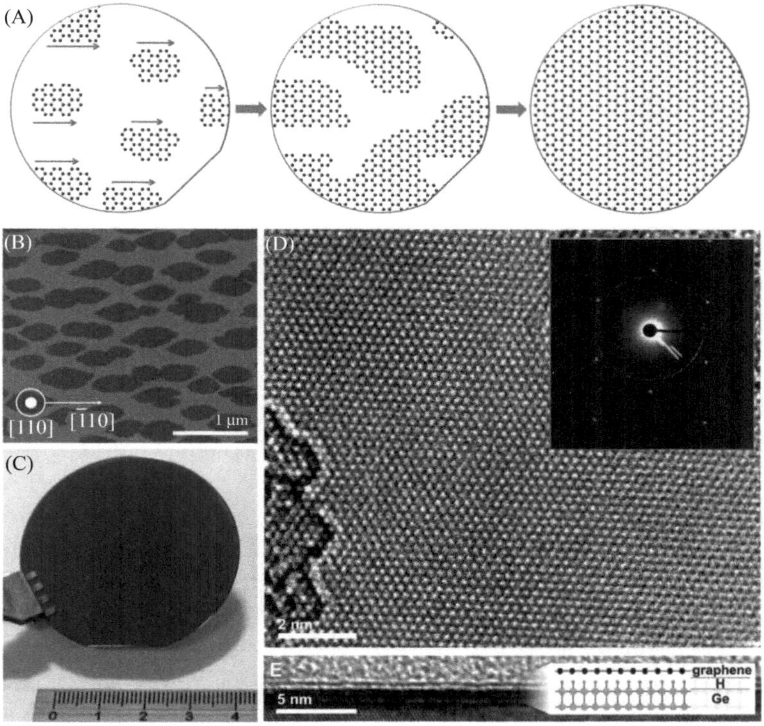

Figure 2.24 Single-crystal monolayer graphene grown on a hydrogen-terminated Ge (110) surface. (a) A schematic illustration of the catalytic growth of single-crystal monolayer graphene from unidirectionally aligned multiple seeds. (b) A typical SEM image of graphene seeds at the early stage of growth. (c) A photograph of graphene grown on a 5.08 cm Ge/Si (110) wafer. (d) An HR-TEM image of the single-crystal monolayer graphene. (Inset) Four overlaid SAED patterns, which were measured across the four different points. The distance between each point is ~2 mm. (e) A cross-sectional TEM image demonstrating that the as-grown graphene is monolayer. (Inset) A schematic illustration of the monolayer graphene grown on the H-terminated Ge surface. Reprinted with permission from Ref. [134], Copyright 2014, AAAS.

of the carbon atoms, the size of the graphene flakes increases and the flakes eventually merge into a single-crystal graphene covering the entire substrate, as shown in Fig. 2.24c. The growth is conducted in a low-pressure CVD chamber flowing with CH_4/H_2 gas mixture. The monolayer and single-crystal characteristics of the grown graphene are confirmed by the TEM analysis (Figs. 2.24d and e). To lower the substrate cost, the germanium can be reused for graphene growth with the same quality after the transfer of grown graphene.

2.4 CNT–Graphene Heterostructure Growth Methods

Although both 1D CNT and 2D graphene are promising in so many applications, it is beneficial to combine them into a single CNT–graphene heterostructure, which not only preserves the excellent properties of the two materials, but also compensates for each other's shortcomings to some extent. Generally, CNT grows along the axial direction, thus forming CNT arrays vertical or parallel to the substrate. Several possible models of joining CNTs and graphene are illustrated in Fig. 2.25. A parallel CNT–graphene heterostructure [139–143] (Fig. 2.25a) can be obtained by drop-casting CNT on the transferred graphene, or graphene can be transferred to cover the CNT network to form a similar structure but with graphene on top [144]. Figure 2.25b shows a typical structure with the CNT axis normal to the graphene plane [145–151], forming a vertical CNT–graphene heterostructure. In certain cases, the graphene can be lifted off during the CNT growth, thus forming a structure as shown in Fig. 2.25c [152–155]. Since both graphene and CNT have the same honeycomb structure, a seamless contact between them appears possible. In principle, the seamless junctions between CNT and graphene can yield a more robust mechanical structure with enhanced interplanar electrical and thermal conduction. Recently, several experimental works claimed to obtain seamless CNT–graphene heterostructures [156, 157] (Figs. 2.25d and e). Multilayered vertical CNT–graphene heterostructure was also envisioned as an all-carbon pillared structure [158] (Fig. 2.25f), in which graphene was used as the platform for CNT growth and the grown CNTs served as pillars to support graphene layers.

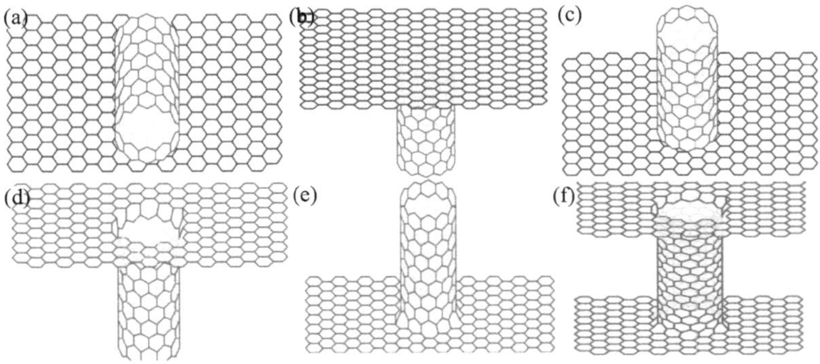

Figure 2.25 Schematic illustrations of the (a) parallel CNT–graphene heterostructure, (b, c) normal CNT–graphene heterostructure, (d, e) seamless CNT–graphene heterostructure, and (f) CNT–graphene–CNT heterostructure.

Parallel CNT-graphene heterostructures are generally formed using CVD [141, 143, 159]. In such a method, a graphene film is synthesized first, followed by catalyst deposition on the graphene and then CNT growth to form the parallel CNT-graphene heterostructure. For example, using $FeCl_3$ solution deposited on graphene as catalyst, CNTs were formed on the dried sample after the introduction of argon/hydrogen/acetylene (30/30/5 sccm) at 750°C [159]. The density of CNTs can be well controlled simply by choosing the corresponding concentration for a targeted density requirement. On the other hand, the use of $FeCl_3$ is avoided in many cases because of the potentially hazardous waste it creates. With CNTs as the template, a CNT spider web was first deposited on a copper substrate [143], then a xylene solution consisting of ferrocene and sulfur was injected into CNT webs, followed by high-temperature annealing. The parallel CNT-graphene heterostructure was then formed with the introduction of a carbon source, and the CNTs served as nucleation centers during the graphene growth, as shown in Fig. 2.26a. This cage-growth method ensures good matching between CNT network-embroidered graphene film and graphene, which contributes to the development of all-carbon devices. An alternative method used graphene as the growth template, on which the CNTs were directly deposited without carbon source gas [141]. It was found that CNTs in the parallel heterostructure lay mainly along the armchair axes of the graphene film (Fig. 2.26b).

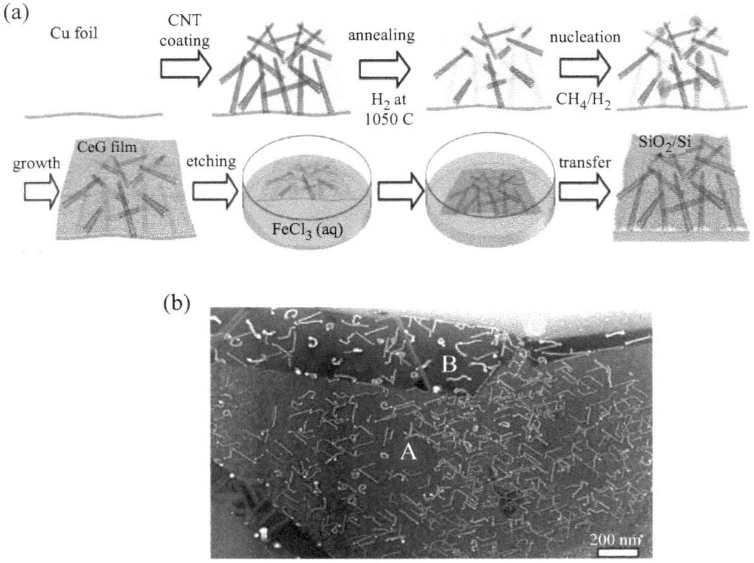

Figure 2.26 Scheme for CNT–graphene parallel heterostructure. (a) Graphene grown by CVD using CNTs as templates. (b) CNTs grown by CVD using graphene film as template. Figure (a) reprinted from Ref. [143], Copyright 2014, with permission from John Wiley and Sons; and Fig. (b) reprinted with permission from Ref. [141], Copyright 2011, American Chemical Society.

Similarly, CVD [160–168] is the most common method for synthesizing vertical CNT–graphene heterostructures, using a two-step process as illustrated in Fig. 2.27 [169]. First, graphene is grown on a metal substrate and subsequently transferred onto a target substrate if needed. Second, the vertical CNT–graphene heterostructure is synthesized after catalyst deposition and introduction of a carbon source gas. There are many factors that affect the growth, and we will focus on the effects of catalysts, temperature, and gases on the characteristics of the resulting heterostructure.

Experiments on the choice of catalyst using different thicknesses of Fe or Ni have been carried out [169]. It was found that if the Fe film thickness could be controlled within a range of 0.2 nm to 1 nm, better quality of CNTs was obtained for thinner films, as confirmed by Raman analyses. When the Fe film thickness was as low as 0.5 nm, SWCNTs were obtained. Otherwise, MWCNTs resulted whether Fe or Ni was used as the catalyst [169]. The diameter of the CNTs is also largely affected by the catalyst film thickness. In addition, the effects of Ni and Fe catalyst on the formation of the CNT heterostructure are quite different, as Ni etches graphene during the growth process, leading to more defects in the final structure. It is worth noting that the etching of the graphene is expected, because in the early stages of CNT growth, graphene is a carbon source in addition to the carbon source gas. And H_2 also plays an important role in the etching of graphene. The growth temperature and gas feedstock can be tuned to minimize graphene etching. By using C_2H_4 as the carbon source, the Ni catalyst film can form higher density nanoparticles at 700°C and the etching becomes less reactive compared to 800°C, resulting in higher density CNTs and less etching of graphene [149].

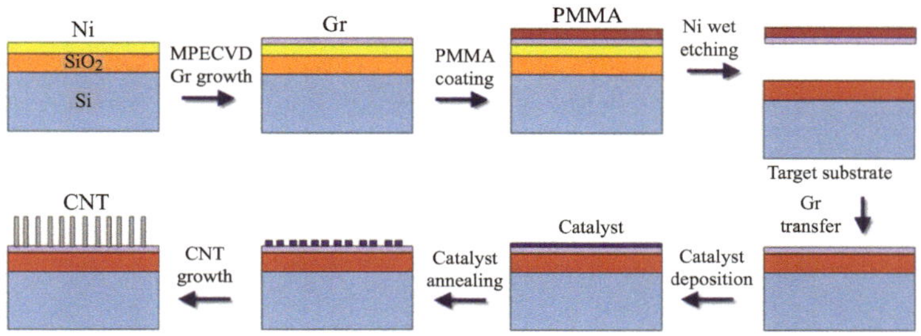

Figure 2.27 A two-step process for CNT–graphene heterostructure preparation. Reprinted from Ref. [169], Copyright 2017, with permission from IOP Publishing Ltd.

Apart from graphene etching, catalytic nanoparticles become embedded in the CNT–graphene junctions in some cases, which limits the properties of the heterostructure. To prevent this effect and form a seamless CNT-graphene heterostructure, Zhu [156] deposited a layer of Al_2O_3 film on the Fe catalyst film as a floating buffer layer (Fig. 2.28a). The floating buffer was designed to transform the bottom growth of the CNTs into tip growth, thus achieving the goal of a seamless CNT-graphene heterostructure with optimal interface properties. Similarly, the heterostructure was successfully [166] synthesized on a porous Ni foam by using the same method but with a better area utilization ratio of the metal substrate, as shown in Fig. 2.28b. Inspired by this method, Jiang [170] realized that CNTs could grow from both sides of graphene, as shown in Fig. 2.28c. Instead of using solid catalyst films, Rodrigo [157] spun a solution of Fe_3O_4/

AlO_x nanoparticles as the catalyst on the graphene/Cu substrate, which is also applicable to curved substrates. Being exposed to hydrogen at 750°C, the catalyst nanoparticles became a mixture of Fe and Al_2O_3. Then a seamless CNT–graphene heterostructure was formed via tip-growth mechanism, with the covalent C–C bonds at the CNT–graphene junction, as shown in Fig. 2.28d. By using Fe catalyst, SWCNTs can be grown at 950°C on an FeMgAl-layered double oxide substrate [171], with the process shown in Fig. 2.28e.

In general, all catalyst film thicknesses in the two-step method are between 1 and 10 nm or thinner to yield good-quality CNTs. With increase in catalyst film thickness, graphene can also be formed. Therefore, CNT and graphene can be grown simultaneously if the thickness of the catalyst can be optimized. Therefore, a one-step CVD method is also feasible for forming vertical CNT–graphene heterostructures, thus simplifying the growth process. Kondo [153] deposited different thickness of Co film on 5 nm TiN to form a mixed catalyst. With the gas ratio of acetylene to argon being 1:9, graphene films were formed first, then the Co catalyst film dewetted to form nanoparticles, followed by CNT growth at 510°C using tip-growth mode and resulting in a vertical CNT–graphene heterostructure on a silica substrate (Fig. 2.29a). Ni/TiN was also reported as a catalyst to synthesize CNTs and graphene [154]. Jousseaume [154] used C_3H_6 as the carbon source gas and prepared a vertical CNT–graphene heterostructure at 400°C using bottom-growth mode. The lower temperature ensures compatibility with chip manufacturing processes for the vertical CNT–graphene heterostructure to serve as part of an on-chip interconnect network. Furthermore, using FeMoMgAl-layered double hydroxides as catalyst, a nitrogen-doped CNT–graphene heterostructure has also been achieved by the one-step method [172]. The schematic of the growth process is shown in Fig. 2.29b. Seamless heterostructures can also be obtained by the one-step CVD method [173]. A typical example is shown in Fig. 2.29c. First, an aluminum wire was exposed to 0.3M oxalic acid solution at 40 V and 3°C so that the external surface could be turned into anodized aluminum oxide. Then without catalyst, CNTs were seamlessly surrounded by a cylindrical graphene layer using CVD. In general, the key in the one-step growth is the proper choice of catalyst (material and thickness) and temperature to form the two constituent nanocarbon materials sequentially or simultaneously.

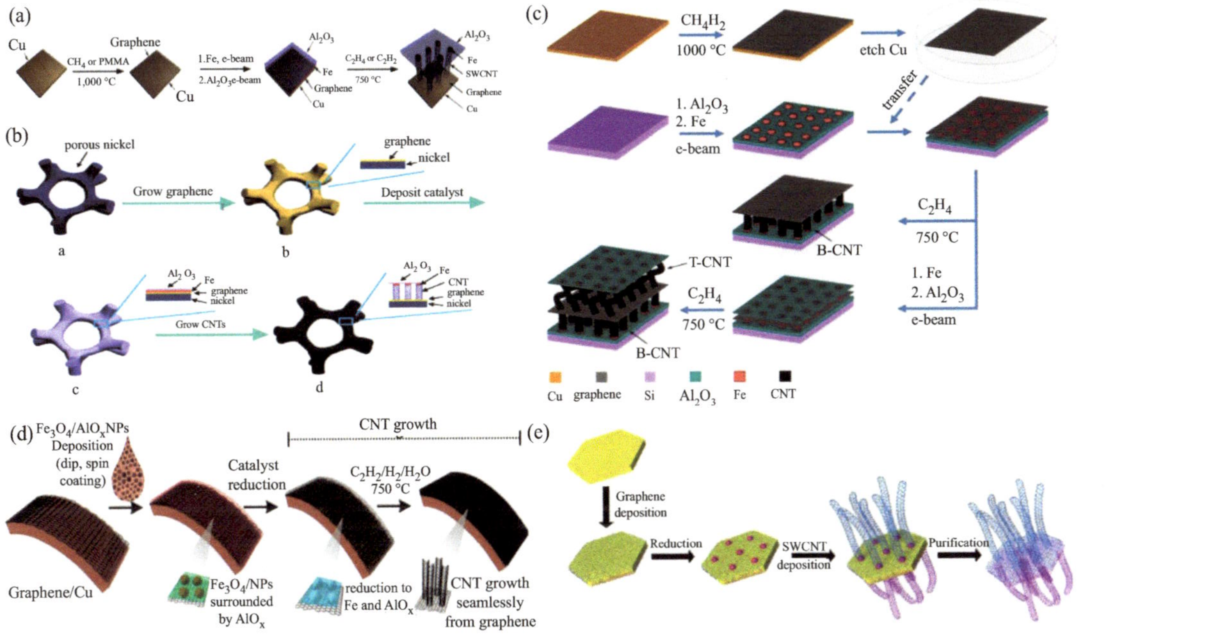

Figure 2.28 Scheme for CNT–graphene vertical heterostructure. (a) Synthesis of CNT carpets directly from graphene by adding a layer of Al_2O_3 film on Cu substrate. (b) Synthesis of seamless CNT–graphene heterostructure by adding a layer of Al_2O_3 film on porous Ni substrate. (c) CNTs grown from both sides of graphene using supporting layer: Al_2O_3 films. (d) Using solution of Fe_3O_4/AlO_x nanoparticles as catalyst. (e) Growing seamless SWCNT–graphene heterostructure on FeMgAl-layered double oxide substrate. Figure (a) reprinted from Ref. [156], Copyright 2012, by permission from Springer Nature and Figs. (b–e) reprinted with permission from Ref. [166], Copyright 2013; Ref. [170], Copyright 2016; Ref. [157], Copyright 2017; Ref. [171], Copyright 2012, respectively, American Chemical Society.

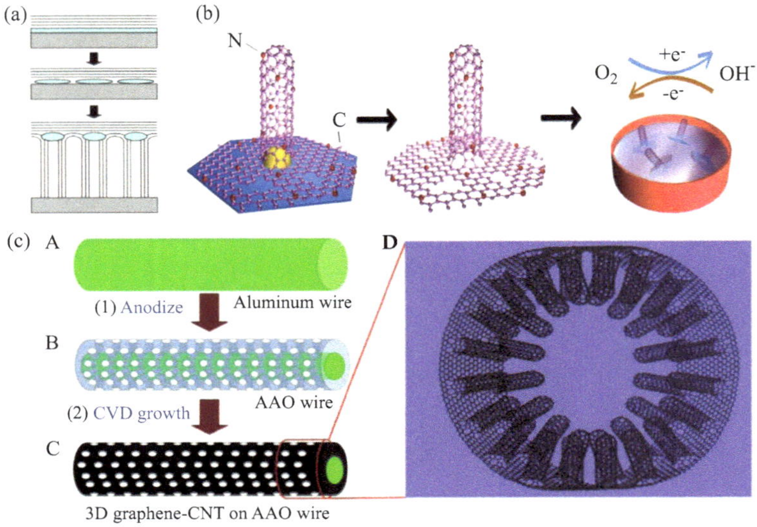

Figure 2.29 (a) Scheme of one-step method process. (b) Process of the nitrogen-doped graphene/carbon nanotube hybrids growth. (c) Schematic of radially aligned CNT growth process. Figure (a) reprinted from Ref. [154], Copyright 2011, with the permission of AIP Publishing; Fig. (b) reprinted from Ref. [172], Copyright 2014, with permission from John Wiley and Sons; Fig. (c) reprinted from Ref. [173], Copyright 2015, The Authors.

It is expected that the seamless heterostructure can provide superior electronic and thermal transport properties through the CNT–graphene junction as well as improved mechanical stability. However, there is still little evidence that the fabricated vertical structure possesses a seamless connection. Therefore, high-resolution transmission electron microscopy is needed to show experimentally how the carbon atoms are connected at the CNT–graphene junction. The study of CNT–graphene heterostructure is still in its infancy, and we hope that with more advanced characterization techniques, the atomic arrangement at the junction can be identified, thus paving the way for future applications in electron devices.

2.5 Summary

In this chapter, the growth methods for CNTs, graphene, and CNT–graphene heterostructures and their integration into basic electronic devices are reviewed. The suitability of each growth method is assessed according to the specific application being considered. The growth mechanisms and parameters for each growth method are discussed in the context of its suitability for specific device integration. For CNTs, growth temperature, alignment, areal density, and diameter distribution are the main considerations, especially for on-chip interconnect applications. For graphene, achieving uniformity and low defect density is generally the primary objective for all applications. For CNT–graphene heterostructure, the single most critical challenge, a bottleneck physically and figuratively, is to obtain a seamless interface between the vertical 1D CNT and the horizontal 2D graphene. In the ensuing chapters, some of these applications are further elaborated and potential solutions to mitigate the challenges are presented.

References

1. Kroto, H. W., J. R. Heath, S. C. O'Brien, R. F. Curl, and R. E. Smalley. 1985. C60: Buckminsterfullerene, *Nature*, **318**, 162–163.
2. De Volder, M. F. L., S. H. Tawfick, R. H. Baughman, and A. J. Hart. 2013. Carbon nanotubes: Present and future commercial applications, *Science*, **339**, 535.
3. Ferrari, A. C., F. Bonaccorso, V. Fal'ko, K. S. Novoselov, S. Roche, P. Bøggild, S. Borini, F. H. L. Koppens, V. Palermo, N. Pugno, J. A. Garrido, R. Sordan, A. Bianco, L. Ballerini, M. Prato, E. Lidorikis, J. Kivioja, C. Marinelli, T. Ryhänen, A. Morpurgo, J. N. Coleman, V. Nicolosi, L. Colombo, A. Fert, M. Garcia-Hernandez, A. Bachtold, G. F. Schneider, F. Guinea, C. Dekker, M. Barbone, Z. Sun, C. Galiotis, A. N. Grigorenko, G. Konstantatos, A. Kis, M. Katsnelson, L. Vandersypen, A. Loiseau, V. Morandi, D. Neumaier, E. Treossi, V. Pellegrini, M. Polini, A. Tredicucci, G. M. Williams, B. Hee Hong, J.-H. Ahn, J. Min Kim, H. Zirath, B. J. van Wees, H. van der Zant, L. Occhipinti, A. Di Matteo, I. A. Kinloch, T. Seyller, E. Quesnel, X. Feng, K. Teo, N. Rupesinghe, P. Hakonen, S. R. T. Neil, Q. Tannock, T. Löfwander, and J. Kinaret. 2015. Science and technology roadmap for graphene, related two-dimensional crystals, and hybrid systems, *Nanoscale*, **7**, 4598–4810.
4. Snow, E. S., F. K. Perkins, and J. A. Robinson. 2006. Chemical vapor detection using single-walled carbon nanotubes, *Chem. Soc. Rev.*, **35**, 790–798.

5. Staii, C., A. T. Johnson, M. Chen, and A. Gelperin. 2005. DNA-decorated carbon nanotubes for chemical sensing, *Nano Lett.*, **5**, 1774–1778.
6. Schedin, F., A. K. Geim, S. V. Morozov, E. W. Hill, P. Blake, M. I. Katsnelson, and K. S. Novoselov. 2007. Detection of individual gas molecules adsorbed on graphene, *Nature Mater.*, **6**, 652.
7. Serp, P. 2003. Carbon nanotubes and nanofibers in catalysis, *App. Catal. A: General.*, **253**, 337–358.
8. Machado, B. F. and P. Serp. 2012. Graphene-based materials for catalysis, *Catal. Sci. Technol.*, **2**, 54–75.
9. Dillon, A. C., K. M. Jones, T. A. Bekkedahl, C. H. Kiang, D. S. Bethune, and M. J. Heben. 1997. Storage of hydrogen in single-walled carbon nanotubes, *Nature*, **386**, 377–379.
10. Pumera, M. 2011. Graphene-based nanomaterials for energy storage, *Energy Environ. Sci.*, **4**, 668–674.
11. An, H. K., W. Kim, Y. Park, Y. C. Choi, S. Lee, D. C. Chung, D. J. Bae, S. C. Lim, and Y. H. Lee. 2001. Supercapacitors using single-walled carbon nanotube electrodes, *Adv. Mater.*, **13**, 497–500.
12. Le, L. T., M. H. Ervin, H. Qiu, B. E. Fuchs, and W. Y. Lee. 2011. Graphene supercapacitor electrodes fabricated by inkjet printing and thermal reduction of graphene oxide, *Electrochem. Commun.*, **13**, 355–358.
13. Lee, S. W., N. Yabuuchi, B. M. Gallant, S. Chen, B.-S. Kim, P. T. Hammond, and Y. Shao-Horn. 2010. High-power lithium batteries from functionalized carbon-nanotube electrodes, *Nat. Nanotechnol.*, **5**, 531
14. Reddy, A. L. M., A. Srivastava, S. R. Gowda, H. Gullapalli, M. Dubey, and P. M. Ajayan. 2010. Synthesis of nitrogen-doped graphene films for lithium battery application, *ACS Nano*, **4**, 6337–6342.
15. Li, H. J., W. G. Lu, J. J. Li, X. D. Bai, and C. Z. Gu. 2005. Multichannel ballistic transport in multiwall carbon nanotubes, *Phys. Rev. Lett.*, **95**, 086601.
16. Bolotin, K. I., K. J. Sikes, J. Hone, H. L. Stormer, and P. Kim. 2008. Temperature-dependent transport in suspended graphene, *Phys. Rev. Lett.*, **101**, 096802.
17. Franklin, A. D., M. Luisier, S.-J. Han, G. Tulevski, C. M. Breslin, L. Gignac, M. S. Lundstrom, and W. Haensch. 2012. Sub-10 nm carbon nanotube transistor, *Nano Lett.*, **12**, 758–762.
18. Lin, Y.-M., A. Valdes-Garcia, S.-J. Han, D. B. Farmer, I. Meric, Y. Sun, Y. Wu, C. Dimitrakopoulos, A. Grill, P. Avouris, and K. A. Jenkins. 2011. Wafer-scale graphene integrated circuit, *Science*, **332**, 1294.

19. Chen, X., D. Akinwande, K. Lee, G. F. Close, S. Yasuda, B. C. Paul, S. Fujita, J. Kong, and H. P. Wong. 2010. Fully integrated graphene and carbon nanotube interconnects for gigahertz high-speed CMOS electronics, *IEEE Trans. Electron Devices*, **57**, 3137–3143.

20. Zhou, C., A. A. Vyas, P. Wilhite, P. Wang, M. Chan, and C. Y. Yang. 2015. Resistance determination for sub-100-nm carbon nanotube vias, *IEEE Electron Device Lett.*, **36**, 71–73.

21. Lee, H., L. Yu, S. Ryu, J. Han, K. Jeon, D. Jang, K. Kim, J. Lee, J. Kim, S. Jeon, G. Lee, J. Oh, Y. Park, W. Bae, H. Lee, J. Yang, J. Yoo, S. Kim, and Y. Choi. Sub-5nm all-around gate FinFET for ultimate scaling, *2006 Symposium on VLSI Technology, 2006. Digest of Technical Papers*, 58–59.

22. Huynh-Bao, T., J. Ryckaert, Z. Tőkei, A. Mercha, D. Verkest, A. V. Thean, and P. Wambacq. 2017. Statistical timing analysis considering device and interconnect variability for BEOL requirements in the 5-nm node and beyond, *IEEE Trans. Very Large Scale Integration (VLSI) Systems.*, **25**, 1669–1680.

23. Hu, C., J. Kelly, H. Huang, K. Motoyama, H. Shobha, Y. Ostrovski, J. H. Chen, R. Patlolla, B. Peethala, P. Adusumilli, T. Spooner, R. Quon, L. M. Gignac, C. Breslin, G. Lian, M. Ali, J. Benedict, X. S. Lin, S. Smith, V. Kamineni, X. Zhang, F. Mont, S. Siddiqui, and F. Baumann. 2018. Future on-chip interconnect metallization and electromigration, *2018 IEEE International Reliability Physics Symposium (IRPS)*. 4F.1-1-4F.1-6.

24. Neumaier, D., S. Pindl, and M. C. Lemme. 2019. Integrating graphene into semiconductor fabrication lines, *Nat. Mater.*, **18**, 525–529.

25. Li, S., S. Raju, C. Zhou, and M. Chan. 2016. Carbon nanotube contact plug on silicide for CMOS compatible interconnect, *IEEE Electron Device Lett.*, **37**, 793–796.

26. Iijima, S. 1991. Helical microtubules of graphitic carbon, *Nature*, **354**, 56–58.

27. Ando, Y., X. Zhao, T. Sugai, and M. Kumar. 2004. Growing carbon nanotubes, *Mater. Today*, **7**, 22–29.

28. Ando, Y. and S. Iijima. 1993. Preparation of carbon nanotubes by arc-discharge evaporation, *Jap. J. Appl. Phys.*, **32**, L107–L109.

29. Zhao, X., M. Ohkohchi, M. Wang, S. Iijima, T. Ichihashi, and Y. Ando. 1997. Preparation of high-grade carbon nanotubes by hydrogen arc discharge, *Carbon*, **35**, 775–781.

30. Ebbesen, T. W. and P. M. Ajayan. 1992. Large-scale synthesis of carbon nanotubes, *Nature*, **358**, 220–222.

31. Gamaly, E. G. and T. W. Ebbesen. 1995. Mechanism of carbon nanotube formation in the arc discharge, *Phys. Rev. B*, **52**, 2083–2089.
32. Iijima, S. and T. Ichihashi. 1993. Single-shell carbon nanotubes of 1-nm diameter, *Nature*, **363**, 603–605.
33. Bethune, D. S., C. H. Kiang, M. S. de Vries, G. Gorman, R. Savoy, J. Vazquez, and R. Beyers. 1993. Cobalt-catalysed growth of carbon nanotubes with single-atomic-layer walls, *Nature*, **363**, 605–607.
34. Ando, Y. and X. Zhao. 2006. Synthesis of carbon nanotubes by arc-discharge method, *New Diamond Frontier Carbon Technol.*, **16**.
35. Shi, Z., Y. Lian, X. Zhou, Z. Gu, Y. Zhang, S. Iijima, L. Zhou, K. T. Yue, and S. Zhang. 1999. Mass-production of single-wall carbon nanotubes by arc discharge method. This work was supported by the National Natural Science Foundation of China, No. 29671030, *Carbon*, **37**, 1449–1453.
36. Hutchison, J. L., N. A. Kiselev, E. P. Krinichnaya, A. V. Krestinin, R. O. Loutfy, A. P. Morawsky, V. E. Muradyan, E. D. Obraztsova, J. Sloan, S. V. Terekhov, and D. N. Zakharov. 2001. Double-walled carbon nanotubes fabricated by a hydrogen arc discharge method, *Carbon*, **39**, 761–770.
37. Journet, C., W. K. Maser, P. Bernier, A. Loiseau, M. L. de la Chapelle, S. Lefrant, P. Deniard, R. Lee, and J. E. Fischer. 1997. Large-scale production of single-walled carbon nanotubes by the electric-arc technique, *Nature*, **388**, 756–758.
38. An, K. H., W. S. Kim, Y. S. Park, Y. C. Choi, S. M. Lee, D. C. Chung, D. J. Bae, S. C. Lim, and Y. H. Lee. 2001. Supercapacitors using single-walled carbon nanotube electrodes, *Adv. Mater.*, **13**, 497–500.
39. Hersam, M. C. 2008. Progress towards monodisperse single-walled carbon nanotubes, *Nat. Nanotechnol.*, **3**, 387.
40. Si, J., L. Xu, M. Zhu, Z. Zhang, and L.-M. Peng. 2019. Advances in high-performance carbon-nanotube thin-film electronics, *Adv. Electron. Mater.*, **5**, 1900122.
41. Thess, A., R. Lee, P. Nikolaev, H. Dai, P. Petit, J. Robert, C. H. Xu, Y. Hee Lee, S.-G. Kim, A. G. Rinzler, D. Colbert, G. E. Scuseria, D. Tomanek, J. E. Fischer, and R. E. Smalley. 1996. Crystalline ropes of metallic carbon nanotubes, *Science*, **273**, 483–487.
42. Guo, T., M. D. Diener, Y. Chai, M. J. Alford, R. E. Haufler, S. M. McClure, T. Ohno, J. H. Weaver, G. E. Scuseria, and R. E. Smalley. 1992. Uranium stabilization of C28: A tetravalent fullerene, *Science*, **257**, 1661.
43. Yudasaka, M., R. Yamada, N. Sensui, T. Wilkins, T. Ichihashi, and S. Iijima. 1999. Mechanism of the effect of NiCo, Ni and Co catalysts on

the yield of single-wall carbon nanotubes formed by pulsed Nd:YAG laser ablation, *J. Phys. Chem. B*, **103**, 6224–6229.

44. Yudasaka, M., N. Sensui, M. Takizawa, S. Bandow, T. Ichihashi, and S. Iijima. 1999. Formation of single-wall carbon nanotubes catalyzed by Ni separating from Y in laser ablation or in arc discharge using a C target containing a NiY catalyst, *Chem. Phys. Lett.*, **312**, 155–160.

45. Zhang, M., M. Yudasaka, and S. Iijima. 2004. Production of large-diameter single-wall carbon nanotubes by adding Fe to a NiCo catalyst in laser ablation, *J. Phys. Chem. B*, **108**, 12757–12762.

46. Yudasaka, M., T. Komatsu, T. Ichihashi, and S. Iijima. 1997. Single-wall carbon nanotube formation by laser ablation using double-targets of carbon and metal, *Chem. Phys. Lett.*, **278**, 102–106.

47. Chrzanowska, J., J. Hoffman, A. Małolepszy, M. Mazurkiewicz, T. A. Kowalewski, Z. Szymanski, and L. Stobinski. 2015. Synthesis of carbon nanotubes by the laser ablation method: Effect of laser wavelength, *Phys. Status Solidi* (b), **252**, 1860–1867.

48. Zhang, M., M. Yudasaka, and S. Iijima. 2001. Single-wall carbon nanotubes: A high yield of tubes through laser ablation of a crude-tube target, *Chem. Phys. Lett.*, **336**, 196–200.

49. Nishide, D., H. Kataura, S. Suzuki, K. Tsukagoshi, Y. Aoyagi, and Y. Achiba. 2003. High-yield production of single-wall carbon nanotubes in nitrogen gas, *Chem. Phys. Lett.*, **372**, 45–50.

50. Yudasaka, M., T. Komatsu, T. Ichihashi, Y. Achiba, and S. Iijima. 1998. Pressure dependence of the structures of carbonaceous deposits formed by laser ablation on targets composed of carbon, nickel, and cobalt, *J. Phys. Chem. B*, **102**, 4892–4896.

51. Puretzky, A. A., H. Schittenhelm, X. Fan, M. J. Lance, L. Allard, and D. B. Geohegan. 2002. Investigations of single-wall carbon nanotube growth by time-restricted laser vaporization, *Phys. Rev. B*, **65**, 245425.

52. Yudasaka, M., F. Kokai, K. Takahashi, R. Yamada, N. Sensui, T. Ichihashi, and S. Iijima. 1999. Formation of single-wall carbon nanotubes: Comparison of CO_2 laser ablation and Nd:YAG laser ablation, *J. Phys. Chem. B*, **103**, 3576–3581.

53. Le Borgne, V., B. Aïssa, M. Mohamedi, Y. Ahm Kim, M. Endo, and M. A. El Khakani. 2011. Pulsed KrF-laser synthesis of single-wall-carbon-nanotubes: Effects of catalyst content and furnace temperature on their nanostructure and photoluminescence properties, *J. Nanopart. Res.*, **13**, 5759–5767.

54. Maser, W. K., A. M. Benito, and M. T. Martínez. 2002. Production of carbon nanotubes: The light approach, *Carbon*, **40**, 1685–1695.
55. Das, R., Z. Shahnavaz, M. E. Ali, M. M. Islam, and S. B. Abd Hamid. 2016. Can we optimize arc discharge and laser ablation for well-controlled carbon nanotube synthesis? *Nanoscale Res. Lett.*, **11**, 510.
56. Scott, C. D., S. Arepalli, P. Nikolaev, and R. E. Smalley. 2001. Growth mechanisms for single-wall carbon nanotubes in a laser-ablation process, *Appl. Phys. A.*, **72**, 573–580.
57. Bandow, S., S. Asaka, Y. Saito, A. M. Rao, L. Grigorian, E. Richter, and P. C. Eklund. 1998. Effect of the growth temperature on the diameter distribution and chirality of single-wall carbon nanotubes, *Phys. Rev. Lett.*, **80**, 3779–3782.
58. Endo, M., K. Takeuchi, S. Igarashi, K. Kobori, M. Shiraishi, and H. W. Kroto. 1993. The production and structure of pyrolytic carbon nanotubes (PCNTs), *J. Phys. Chem. Solids*, **54**, 1841–1848.
59. José-Yacamán, M., M. Miki-Yoshida, L. Rendón, and J. G. Santiesteban. 1993. Catalytic growth of carbon microtubules with fullerene structure, *Appl. Phys. Lett.*, **62**, 202–204.
60. Dai, H., A. G. Rinzler, P. Nikolaev, A. Thess, D. T. Colbert, and R. E. Smalley. 1996. Single-wall nanotubes produced by metal-catalyzed disproportionation of carbon monoxide, *Chem. Phys. Lett.*, **260**, 471–475.
61. Lee, C. J., J. Park, and J. A. Yu. 2002. Catalyst effect on carbon nanotubes synthesized by thermal chemical vapor deposition, *Chem. Phys. Lett.*, **360**, 250–255.
62. Meyyappan, M., L. Delzeit, A. Cassell, and D. Hash. 2003. Carbon nanotube growth by PECVD: A review, *Plasma Sources Sci. Technol.*, **12**, 205–216.
63. Wei, Y. Y., G. Eres, V. I. Merkulov, and D. H. Lowndes. 2001. Effect of catalyst film thickness on carbon nanotube growth by selective area chemical vapor deposition, *Appl. Phys. Lett.*, **78**, 1394–1396.
64. Wei, S., W. P. Kang, J. L. Davidson, and J. H. Huang. 2006. Aligned carbon nanotubes fabricated by thermal CVD at atmospheric pressure using Co as catalyst with NH_3 as reactive gas, *Diam. Relat. Mater.*, **15**, 1828–1833.
65. Dikonimos Makris, T., L. Giorgi, R. Giorgi, N. Lisi, and E. Salernitano. 2005. CNT growth on alumina supported nickel catalyst by thermal CVD, *Diam. Relat. Mater.*, **14**, 815–819.

66. Wong, Y. M., S. Wei, W. P. Kang, J. L. Davidson, W. Hofmeister, J. H. Huang, and Y. Cui. 2004. Carbon nanotubes field emission devices grown by thermal CVD with palladium as catalysts, *Diam. Relat. Mater.*, **13**, 2105–2112.

67. Li, Y., J. Liu, Y. Wang, and Z. L. Wang. 2001. Preparation of monodispersed Fe–Mo nanoparticles as the catalyst for CVD synthesis of carbon nanotubes, *Chem. Mater.*, **13**, 1008–1014.

68. Moshkalyov, S. A., A. L. D. Moreau, H. R. Guttiérrez, M. A. Cotta, and J. W. Swart. 2004. Carbon nanotubes growth by chemical vapor deposition using thin film nickel catalyst, *Mater. Sci. Eng. B*, **112**, 147–153.

69. Murakami, Y., S. Chiashi, Y. Miyauchi, M. Hu, M. Ogura, T. Okubo, and S. Maruyama. 2004. Growth of vertically aligned single-walled carbon nanotube films on quartz substrates and their optical anisotropy, *Chem. Phys. Lett.*, **385**, 298–303.

70. Hermann, S. 2011. Growth of carbon nanotubes on different support/catalyst systems for advanced interconnects in integrated circuits, Doctoral Thesis.

71. Meyyappan, M. 2009. A review of plasma enhanced chemical vapour deposition of carbon nanotubes, *J. Phys. D: Appl. Phys.*, **42**, 213001.

72. Sun, X., K. Li, R. Wu, P. Wilhite, T. Saito, J. Gao, and C. Yang. 2009. The effect of catalysts and underlayer metals on the properties of PECVD-grown carbon nanostructures, *Nanotechnology*, **21**, 045201.

73. Lim, S. C., J. H. Jang, D. J. Bae, G. H. Han, S. Lee, I.-S. Yeo, and Y. H. Lee. 2009. Contact resistance between metal and carbon nanotube interconnects: Effect of work function and wettability, *Appl. Phys. Lett.*, **95**, 264103.

74. Wilhite, P., A. Vyas, J. Tan, C. Yang, P. Wang, J. Park, H. Ai, and M. Narasimhan. 2013. Electrical and structural analysis of CNT-metal contacts in via interconnects, *ICQNM 2013, The Seventh International Conference on Quantum, Nano and Micro Technologies*, Barcelona, Spain.

75. Solak, H. H., C. David, J. Gobrecht, V. Golovkina, F. Cerrina, S. O. Kim, and P. F. Nealey. 2003. Sub-50 nm period patterns with EUV interference lithography, *Microelectron. Eng.*, **67–68**, 56–62.

76. d'Agostino, R., F. Fracassi, C. Pacifico, and P. Capezzuto. 1992. Plasma etching of Ti in fluorine-containing feeds, *J. Appl. Phys.*, **71**, 462–471.

77. Alloyeau, D., G. Prévot, Y. Le Bouar, T. Oikawa, C. Langlois, A. Loiseau, and C. Ricolleau. 2010. Ostwald ripening in nanoalloys: When thermodynamics drives a size-dependent particle composition, *Phys. Rev. Lett.*, **105**, 255901.

78. Li, S., C. Zhou, S. Raju, and M. Chan. 2017. Catalyst design for high-density and low-temperature CNT synthesis on conductive Ti silicide substrate, *Diam. Relat. Mater.*, **75**, 39–43.

79. Cao, A., R. Lu, and G. Veser. 2010. Stabilizing metal nanoparticles for heterogeneous catalysis, *Phys. Chem. Chem. Phys.*, **12**, 13499–13510.

80. Vollebregt, S., F. D. Tichelaar, H. Schellevis, C. I. M. Beenakker, and R. Ishihara. 2014. Carbon nanotube vertical interconnects fabricated at temperatures as low as 350°C, *Carbon*, **71**, 249–256.

81. Ghavanini, F. A., H. Le Poche, J. Berg, A. M. Saleem, M. S. Kabir, P. Lundgren, and P. Enoksson. 2008. Compatibility assessment of CVD growth of carbon nanofibers on bulk CMOS devices, *Nano Lett.*, **8**, 2437–2441.

82. Yang, J., S. Esconjauregui, A. W. Robertson, Y. Guo, T. Hallam, H. Sugime, G. Zhong, G. S. Duesberg, and J. Robertson. 2015. Growth of high-density carbon nanotube forests on conductive TiSiN supports, *Appl. Phys. Lett.*, **106**, 083108.

83. Cao, Q., S.-J. Han, J. Tersoff, A. D. Franklin, Y. Zhu, Z. Zhang, G. S. Tulevski, J. Tang, and W. Haensch. 2015. End-bonded contacts for carbon nanotube transistors with low, size-independent resistance, *Science*, **350**, 68.

84. Fedorov, A. G., S. Kim, M. Henry, D. Kulkarni, and V. V. Tsukruk. 2014. Focused-electron-beam-induced processing (FEBIP) for emerging applications in carbon nanoelectronics, *Appl. Phys. A*, **117**, 1659–1674.

85. Mu, W., S. Sun, D. Jiang, Y. Fu, M. Edwards, Y. Zhang, K. Jeppson, and J. Liu. 2015. Tape-assisted transfer of carbon nanotube bundles for through-silicon-via applications, *J. Electron. Mater.*, **44**, 2898–2907.

86. Srivastava, N., H. Li, F. Kreupl, and K. Banerjee. 2009. On the applicability of single-walled carbon nanotubes as VLSI interconnects, *IEEE Trans. Nanotechnol.*, **8**, 542–559.

87. Shulaker, M. M., G. Hills, N. Patil, H. Wei, H.-Y. Chen, H. S. P. Wong, and S. Mitra. 2013. Carbon nanotube computer, *Nature*, **501**, 526.

88. Huang, S., X. Cai, and J. Liu. 2003. Growth of millimeter-long and horizontally aligned single-walled carbon nanotubes on flat substrates, *J. Am. Chem. Soc.*, **125**, 5636–5637.

89. Huang, L., X. Cui, B. White, and S. P. O'Brien. 2004. Long and oriented single-walled carbon nanotubes grown by ethanol chemical vapor deposition, *J. Phys. Chem. B*, **108**, 16451–16456.

90. Phokharatkul, D., Y. Ohno, H. Nakano, S. Kishimoto, and T. Mizutani. 2008. High-density horizontally aligned growth of carbon nanotubes

with Co nanoparticles deposited by arc-discharge plasma method, *Appl. Phys. Lett.*, **93**, 053112.

91. Watanabe, T., E. L. H. Sadki, T. Yamaguchi, and Y. Takano. 2014. Electrical transport properties of small diameter single-walled carbon nanotubes aligned on ST-cut quartz substrates, *Nanoscale Res. Lett.*, **9**, 374.

92. Kang, L., Y. Hu, L. Liu, J. Wu, S. Zhang, Q. Zhao, F. Ding, Q. Li, and J. Zhang. 2015. Growth of close-packed semiconducting single-walled carbon nanotube arrays using oxygen-deficient TiO_2 nanoparticles as catalysts, *Nano Lett.*, **15**, 403–409.

93. Inoue, T., D. Hasegawa, S. Chiashi, and S. Maruyama. 2015. Chirality analysis of horizontally aligned single-walled carbon nanotubes: Decoupling populations and lengths, *J. Mater. Chem. A*, **3**, 15119–15123.

94. Yuan, D., L. Ding, H. Chu, Y. Feng, T. P. McNicholas, and J. Liu. 2008. Horizontally aligned single-walled carbon nanotube on quartz from a large variety of metal catalysts, *Nano Lett.*, **8**, 2576–2579.

95. Patil, N., A. Lin, E. R. Myers, K. Ryu, A. Badmaev, C. Zhou, H. P. Wong, and S. Mitra. 2009. Wafer-scale growth and transfer of aligned single-walled carbon nanotubes, *IEEE Trans. Nanotechnol.* **8**, 498–504.

96. Hu, Y., L. Kang, Q. Zhao, H. Zhong, S. Zhang, L. Yang, Z. Wang, J. Lin, Q. Li, Z. Zhang, L. Peng, Z. Liu, and J. Zhang. 2015. Growth of high-density horizontally aligned SWNT arrays using Trojan catalysts, *Nat. Commun.*, **6**, 6099.

97. Yang, F., X. Wang, D. Zhang, J. Yang, D. Luo, Z. Xu, J. Wei, J.-Q. Wang, Z. Xu, F. Peng, X. Li, R. Li, Y. Li, M. Li, X. Bai, F. Ding, and Y. Li. 2014. Chirality-specific growth of single-walled carbon nanotubes on solid alloy catalysts, *Nature*, **510**, 522.

98. Santini, C. A., A. Volodin, C. Van Haesendonck, S. De Gendt, G. Groeseneken, and P. M. Vereecken. 2011. Carbon nanotube–carbon nanotube contacts as an alternative towards low resistance horizontal interconnects, *Carbon*, **49**, 4004–4012.

99. Yan, F., C. Zhang, D. Cott, G. Zhong, and J. Robertson. 2010. High-density growth of horizontally aligned carbon nanotubes for interconnects, *Phys. Status Solidi (b)*, **247**, 2669–2672.

100. Xiao, J., S. Dunham, P. Liu, Y. Zhang, C. Kocabas, L. Moh, Y. Huang, K.-C. Hwang, C. Lu, W. Huang, and J. A. Rogers. 2009. Alignment controlled growth of single-walled carbon nanotubes on quartz substrates, *Nano Lett.*, **9**, 4311–4319.

101. Zhou, W., Z. Han, J. Wang, Y. Zhang, Z. Jin, X. Sun, Y. Zhang, C. Yan, and Y. Li. 2006. Copper catalyzing growth of single-walled carbon nanotubes on substrates, *Nano Lett.*, **6**, 2987–2990.

102. Ibrahim, I., J. Kalbacova, V. Engemaier, J. Pang, R. D. Rodriguez, D. Grimm, T. Gemming, D. R. T. Zahn, O. G. Schmidt, J. Eckert, and M. H. Rümmeli. 2015. Confirming the dual role of etchants during the enrichment of semiconducting single wall carbon nanotubes by chemical vapor deposition, *Chem. Mater.*, **27**, 5964–5973.

103. Zhang, G., P. Qi, X. Wang, Y. Lu, X. Li, R. Tu, S. Bangsaruntip, D. Mann, L. Zhang, and H. Dai. 2006. Selective etching of metallic carbon nanotubes by gas-phase reaction, *Science*, **314**, 974.

104. Zhang, Y., Y. Zhang, X. Xian, J. Zhang, and Z. Liu. 2008. Sorting out semiconducting single-walled carbon nanotube arrays by preferential destruction of metallic tubes using xenon-lamp irradiation, *J. Phys. Chem. C.*, **112**, 3849–3856.

105. Huang, H., R. Maruyama, K. Noda, H. Kajiura, and K. Kadono. 2006. Preferential destruction of metallic single-walled carbon nanotubes by laser irradiation, *J. Phys. Chem. B*, **110**, 7316–7320.

106. Song, J. W., H. W. Seo, J. K. Park, J. E. Kim, D. G. Choi, and C. S. Han. 2008. Selective removal of metallic SWNTs using microwave radiation, *Curr. Appl. Phys.*, **8**, 725–728.

107. Zhang, F., P.-X. Hou, C. Liu, B.-W. Wang, H. Jiang, M.-L. Chen, D.-M. Sun, J.-C. Li, H.-T. Cong, E. I. Kauppinen, and H.-M. Cheng. 2016. Growth of semiconducting single-wall carbon nanotubes with a narrow band-gap distribution, *Nat. Commun.*, **7**, 11160.

108. Yang, F., X. Wang, J. Si, X. Zhao, K. Qi, C. Jin, Z. Zhang, M. Li, D. Zhang, J. Yang, Z. Zhang, Z. Xu, L.-M. Peng, X. Bai, and Y. Li. 2017. Water-assisted preparation of high-purity semiconducting (14,4) carbon nanotubes, *ACS Nano*, **11**, 186–193.

109. Novoselov, K. S., A. K. Geim, S. V. Morozov, D. Jiang, Y. Zhang, S. V. Dubonos, I. V. Grigorieva, and A. A. Firsov. 2004. Electric field effect in atomically thin carbon films, *Science*, **306**, 666.

110. Li, H., G. Lu, Y. Wang, Z. Yin, C. Cong, Q. He, L. Wang, F. Ding, T. Yu, and H. Zhang. 2013. Mechanical exfoliation and characterization of single- and few-layer nanosheets of WSe2, TaS2, and TaSe2, *Small*, **9**, 1974–1981.

111. Wang, Y. Y., R. X. Gao, Z. H. Ni, H. He, S. P. Guo, H. P. Yang, C. X. Cong, and T. Yu. 2012. Thickness identification of two-dimensional materials by optical imaging, *Nanotechnology*, **23**, 495713.

112. Li, H., J. Wu, X. Huang, G. Lu, J. Yang, X. Lu, Q. Xiong, and H. Zhang. 2013. Rapid and reliable thickness identification of two-dimensional nanosheets using optical microscopy, *ACS Nano*, **7**, 10344–10353.

113. Huang, Y., E. Sutter, N. N. Shi, J. Zheng, T. Yang, D. Englund, H.-J. Gao, and P. Sutter. 2015. Reliable exfoliation of large-area high-quality flakes of graphene and other two-dimensional materials, *ACS Nano*, **9**, 10612–10620.

114. Desai, S. B., S. R. Madhvapathy, M. Amani, D. Kiriya, M. Hettick, M. Tosun, Y. Zhou, M. Dubey, J. W. Ager Iii, D. Chrzan, and A. Javey. 2016. Gold-mediated exfoliation of ultralarge optoelectronically-perfect monolayers, *Adv. Mater.*, **28**, 4053–4058.

115. Shioyama, H. 2001. Cleavage of graphite to graphene, *J. Mater. Sci. Lett.*, **20**, 499–500.

116. Nicolosi, V., M. Chhowalla, M. G. Kanatzidis, M. S. Strano, and J. N. Coleman. 2013. Liquid exfoliation of layered materials, *Science*, **340**, 1226419.

117. Li, X., X. Wang, L. Zhang, S. Lee, and H. Dai. 2008. Chemically derived, ultrasmooth graphene nanoribbon semiconductors, *Science*, **319**, 1229.

118. Hernandez, Y., V. Nicolosi, M. Lotya, F. M. Blighe, Z. Sun, S. De, I. T. McGovern, B. Holland, M. Byrne, Y. K. Gun'Ko, J. J. Boland, P. Niraj, G. Duesberg, S. Krishnamurthy, R. Goodhue, J. Hutchison, V. Scardaci, A. C. Ferrari, and J. N. Coleman. 2008. High-yield production of graphene by liquid-phase exfoliation of graphite, *Nat. Nanotechnol.*, **3**, 563.

119. Furtado, C. A., U. J. Kim, H. R. Gutierrez, L. Pan, E. C. Dickey, and P. C. Eklund. 2004. Debundling and dissolution of single-walled carbon nanotubes in amide solvents, *J. Am. Chem. Soc.*, **126**, 6095–6105.

120. Khan, U., A. O'Neill, M. Lotya, S. De, and J. N. Coleman. 2010. High-concentration solvent exfoliation of graphene, *Small*, **6**, 864–871.

121. Torrisi, F., T. Hasan, W. Wu, Z. Sun, A. Lombardo, T. S. Kulmala, G.-W. Hsieh, S. Jung, F. Bonaccorso, P. J. Paul, D. Chu, and A. C. Ferrari. 2012. Inkjet-printed graphene electronics, *ACS Nano*, **6**, 2992–3006.

122. Mattevi, C., H. Kim, and M. Chhowalla. 2011. A review of chemical vapour deposition of graphene on copper, *J. Mater. Chem.*, **21**, 3324–3334.

123. Zhang, Y., L. Zhang, and C. Zhou. 2013. Review of chemical vapor deposition of graphene and related applications, *Acc. Chem. Res.*, **46**, 2329–2339.

124. Li, X., W. Cai, J. An, S. Kim, J. Nah, D. Yang, R. Piner, A. Velamakanni, I. Jung, E. Tutuc, S. K. Banerjee, L. Colombo, and R. S. Ruoff. 2009. Large-area synthesis of high-quality and uniform graphene films on copper foils, *Science*, **324**, 1312.

125. Vlassiouk, I., P. Fulvio, H. Meyer, N. Lavrik, S. Dai, P. Datskos, and S. Smirnov. 2013. Large scale atmospheric pressure chemical vapor deposition of graphene, *Carbon*, **54**, 58–67.

126. Bae, S., H. Kim, Y. Lee, X. Xu, J.-S. Park, Y. Zheng, J. Balakrishnan, T. Lei, H. Ri Kim, Y. I. Song, Y.-J. Kim, K. S. Kim, B. Özyilmaz, J.-H. Ahn, B. H. Hong, and S. Iijima. 2010. Roll-to-roll production of 30-inch graphene films for transparent electrodes, *Nat. Nanotechnol.*, **5**, 574.

127. Polsen, E. S., D. Q. McNerny, B. Viswanath, S. W. Pattinson, and A. John Hart. 2015. High-speed roll-to-roll manufacturing of graphene using a concentric tube CVD reactor, *Sci. Rep.*, **5**, 10257.

128. Kang, J., D. Shin, S. Bae, and B. H. Hong. 2012. Graphene transfer: Key for applications, *Nanoscale*, **4**, 5527–5537.

129. Li, X., Y. Zhu, W. Cai, M. Borysiak, B. Han, D. Chen, R. D. Piner, L. Colombo, and R. S. Ruoff. 2009. Transfer of large-area graphene films for high-performance transparent conductive electrodes, *Nano Lett.*, **9**, 4359–4363.

130. Park, H., C. Lim, C.-J. Lee, J. Kang, J. Kim, M. Choi, and H. Park. 2018. Optimized poly(methyl methacrylate)-mediated graphene-transfer process for fabrication of high-quality graphene layer, *Nanotechnology*, **29**, 415303.

131. Pirkle, A., J. Chan, A. Venugopal, D. Hinojos, C. W. Magnuson, S. McDonnell, L. Colombo, E. M. Vogel, R. S. Ruoff, and R. M. Wallace. 2011. The effect of chemical residues on the physical and electrical properties of chemical vapor deposited graphene transferred to SiO_2, *Appl. Phys. Lett.*, **99**, 122108.

132. Banszerus, L., M. Schmitz, S. Engels, J. Dauber, M. Oellers, F. Haupt, K. Watanabe, T. Taniguchi, B. Beschoten, and C. Stampfer. 2015. Ultrahigh-mobility graphene devices from chemical vapor deposition on reusable copper, *Sci. Adv.*, **1**, e1500222.

133. Emtsev, K. V., A. Bostwick, K. Horn, J. Jobst, G. L. Kellogg, L. Ley, J. L. McChesney, T. Ohta, S. A. Reshanov, J. Röhrl, E. Rotenberg, A. K. Schmid, D. Waldmann, H. B. Weber, and T. Seyller. 2009. Towards wafer-size graphene layers by atmospheric pressure graphitization of silicon carbide, *Nat. Mater.*, **8**, 203.

134. Lee, J.-H., E. K. Lee, W.-J. Joo, Y. Jang, B.-S. Kim, J. Y. Lim, S.-H. Choi, S. J. Ahn, J. R. Ahn, M.-H. Park, C.-W. Yang, B. L. Choi, S.-W. Hwang, and

D. Whang. 2014. Wafer-scale growth of single-crystal monolayer graphene on reusable hydrogen-terminated germanium, *Science*, **344**, 286.

135. Hibino, H., H. Kageshima, F. Maeda, M. Nagase, Y. Kobayashi, and H. Yamaguchi. 2007. Microscopic thickness determination of thin graphite films formed on SiC from quantized oscillation in reflectivity of low-energy electrons, *Phys. Rev. B*, **77**, 075413.

136. Norimatsu, W. and M. Kusunoki. 2014. Epitaxial graphene on SiC{0001}: Advances and perspectives, *Phys. Chem. Chem. Phys.*, **16**, 3501–3511.

137. Berger, C., Z. Song, X. Li, X. Wu, N. Brown, C. Naud, D. Mayou, T. Li, J. Hass, A. N. Marchenkov, E. H. Conrad, P. N. First, and W. A. de Heer. 2006. Electronic confinement and coherence in patterned epitaxial graphene, *Science*, **312**, 1191.

138. Norimatsu, W. and M. Kusunoki. 2009. Transitional structures of the interface between graphene and 6H–SiC (0001), *Chem. Phys. Lett.*, **468**, 52–56.

139. Wu, S., E. Shi, Y. Yang, W. Xu, X. Li, and A. Cao. 2015. Direct fabrication of carbon nanotube-graphene hybrid films by a blown bubble method, *Nano Res.*, **8**, 1746–1754.

140. Shi, E., H. Li, L. Yang, J. Hou, Y. Li, L. Li, A. Cao, and Y. Fang. 2015. Carbon nanotube network embroidered graphene films for monolithic all-carbon electronics, *Adv. Mater.*, **27**, 682–688.

141. Hunley, D. P., S. L. Johnson, J. K. Stieha, A. Sundararajan, A. T. Meacham, I. N. Ivanov, and D. R. Strachan. 2011. Crystallographically aligned carbon nanotubes grown on few-layer graphene films, *ACS Nano*, **5**, 6403–6409.

142. Yan, Z., Z. Peng, G. Casillas, J. Lin, C. Xiang, H. Zhou, Y. Yang, G. Ruan, A.-R. O. Raji, E. L. G. Samuel, R. H. Hauge, M. J. Yacaman, and J. M. Tour. 2014. Rebar graphene, *ACS Nano*, **8**, 5061–5068.

143. Shi, E., H. Li, L. Yang, J. Hou, Y. Li, L. Li, A. Cao, and Y. Fang. 2014. Carbon nanotube network embroidered graphene films for monolithic all-carbon electronics, *Adv. Mater.*, **27**, 682–688.

144. Maarouf, A. A., A. Kasry, B. Chandra, and G. J. Martyna. 2016. A graphene–carbon nanotube hybrid material for photovoltaic applications, *Carbon*, **102**, 74–80.

145. Choi, H., H. Kim, S. Hwang, M. Kang, D.-W. Jung, and M. Jeon. 2011. Electrochemical electrodes of graphene-based carbon nanotubes grown by chemical vapor deposition, *Scripta Materialia*, **64**, 601–604.

146. Kim, Y. S., K. Kumar, F. T. Fisher, and E. H. Yang. 2012. Out-of-plane growth of CNTs on graphene for supercapacitor applications, *Nanotechnology*, **23**, 015301.

147. Li, X., G. Zhu, and Z. Xu. 2012. Nitrogen-doped carbon nanotube arrays grown on graphene substrate, *Thin Solid Films*, **520**, 1959–1964.

148. Nguyen, D. D., N. H. Tai, S. Y. Chen, and Y. L. Chueh. 2012. Controlled growth of carbon nanotube-graphene hybrid materials for flexible and transparent conductors and electron field emitters, *Nanoscale*, **4**, 632–638.

149. Kumar, K., Y.-S. Kim, X. Li, J. Ding, F. T. Fisher, and E.-H. Yang. 2013. Chemical vapor deposition of carbon nanotubes on monolayer graphene substrates: Reduced etching via suppressed catalytic hydrogenation using C_2H_4, *Chem. Mater.*, **25**, 3874–3879.

150. Rao, R., G. Chen, L. M. Arava, K. Kalaga, M. Ishigami, T. F. Heinz, P. M. Ajayan, and A. R. Harutyunyan. 2013. Graphene as an atomically thin interface for growth of vertically aligned carbon nanotubes, *Sci. Rep.*, **3**, 1891.

151. Ryu, J. H., G. J. Lee, W. S. Kim, H. E. Lim, M. Mativenga, K. C. Park, and H. K. Park. 2014. All-carbon electrode consisting of carbon nanotubes on graphite foil for flexible electrochemical applications, *Materials*, **7**, 1975–1983.

152. Ramos, R., A. Fournier, M. Fayolle, J. Dijon, C. P. Murray, and J. McKenna. 2016. Nanocarbon interconnects combining vertical CNT interconnects and horizontal graphene lines, *2016 IEEE International Interconnect Technology Conference / Advanced Metallization Conference (IITC/AMC)*, 48–50.

153. Kondo, D., S. Sato, and Y. Awano. 2008. Self-organization of novel carbon composite structure: Graphene multi-layers combined perpendicularly with aligned carbon nanotubes, *Appl. Phys. Express*, **1**, 074003.

154. Jousseaume, V., J. Cuzzocrea, N. Bernier, and V. T. Renard. 2011. Few graphene layers/carbon nanotube composites grown at complementary-metal-oxide-semiconductor compatible temperature, *Appl. Phys. Lett.*, **98**, 123103.

155. Choi, J. W., S. K. Youn, and H. G. Park. 2013. Carbon micronymphaea: Graphene on vertically aligned carbon nanotubes, *J. Nanomater.*, **2013**, 1–7.

156. Zhu, Y., L. Li, C. Zhang, G. Casillas, Z. Sun, Z. Yan, G. Ruan, Z. Peng, A. R. Raji, C. Kittrell, R. H. Hauge, and J. M. Tour. 2012. A seamless

three-dimensional carbon nanotube graphene hybrid material, *Nat. Commun.*, **3**, 1225.

157. Salvatierra, R. V., D. Zakhidov, J. Sha, N. D. Kim, S. K. Lee, A. O. Raji, N. Zhao, and J. M. Tour. 2017. Graphene carbon nanotube carpets grown using binary catalysts for high-performance lithium-ion capacitors, *ACS Nano*, **11**, 2724–2733.

158. Dimitrakakis, G. K., E. Tylianakis, and G. E. Froudakis. 2008. Pillared graphene: A new 3-D network nanostructure for enhanced hydrogen storage, *Nano Lett.*, **8**, 3166–3170.

159. Van Chuc, N., C. T. Thanh, N. Van Tu, V. T. Q. Phuong, P. V. Thang, and N. T. Thanh Tam. 2015. A simple approach to the fabrication of graphene-carbon nanotube hybrid films on copper substrate by chemical vapor deposition, *J. Mater. Sci. Technol.*, **31**, 479–483.

160. Lin, X., P. Liu, Y. Wei, Q. Li, J. Wang, Y. Wu, C. Feng, L. Zhang, S. Fan, and K. Jiang. 2013. Development of an ultra-thin film comprised of a graphene membrane and carbon nanotube vein support, *Nat. Commun.*, **4**, 2920.

161. Engels, S., P. Weber, B. Terres, J. Dauber, C. Meyer, C. Volk, S. Trellenkamp, U. Wichmann, and C. Stampfer. 2013. Fabrication of coupled graphene-nanotube quantum devices, *Nanotechnology*, **24**, 035204.

162. Matsumoto, T. and S. Saito. 2002. Geometric and electronic structure of new carbon-network materials: Nanotube array on graphite sheet, *J. Phys. Soc. Jap.*, **71**, 2765–2770.

163. Dang, V. T., D. D. Nguyen, T. T. Cao, P. H. Le, D. L. Tran, N. M. Phan, and V. C. Nguyen. 2016. Recent trends in preparation and application of carbon nanotube–graphene hybrid thin films, *Adv. Nat. Sci. Nanosci. Nanotechnol.*, **7**, 033002.

164. Ping, L., P. X. Hou, C. Liu, J. Li, Y. Zhao, F. Zhang, C. Ma, K. Tai, H. Cong, and H. M. Cheng. 2017. Surface-restrained growth of vertically aligned carbon nanotube arrays with excellent thermal transport performance, *Nanoscale*, **9**, 8213–8219.

165. Rout, C. S., A. Kumar, T. S. Fisher, U. K. Gautam, Y. Bando, and D. Golberg. 2012. Synthesis of chemically bonded CNT–graphene heterostructure arrays, *RSC Adv.*, **2**, 8250.

166. Yan, Z., L. Ma, Y. Zhu, I. Lahiri, M. G. Hahm, Z. Liu, S. Yang, C. Xiang, W. Lu, Z. Peng, Z. Sun, C. Kittrell, J. Lou, W. Choi, P. M. Ajayan, and J. M. Tour. 2013. Three-dimensional metal–graphene–nanotube multifunctional hybrid materials, *ACS Nano*, **7**, 58–64.

167. Lee, D. H., J. E. Kim, T. H. Han, J. W. Hwang, S. Jeon, S. Y. Choi, S. H. Hong, W. J. Lee, R. S. Ruoff, and S. O. Kim. 2010. Versatile carbon hybrid

films composed of vertical carbon nanotubes grown on mechanically compliant graphene films, *Adv. Mater.*, **22**, 1247–1252.

168. Jyothirmayee Aravind, S. S., V. Eswaraiah, and S. Ramaprabhu. 2011. Facile synthesis of one-dimensional graphene wrapped carbon nanotube composites by chemical vapour deposition, *J. Mater. Chem.*, **21**, 15179.

169. Zhou, C., R. Senegor, Z. Baron, Y. Chen, S. Raju, A. A. Vyas, M. Chan, Y. Chai, and C. Y. Yang. 2017. Synthesis and interface characterization of CNTs on graphene, *Nanotechnology*, **28**, 054007.

170. Jiang, J., Y. Li, C. Gao, N. D. Kim, X. Fan, G. Wang, Z. Peng, R. H. Hauge, and J. M. Tour. 2016. Growing carbon nanotubes from both sides of graphene, *ACS Appl. Mater. Interface.*, **8**, 7356–7362.

171. Zhao, M.-Q., X.-F. Liu, Q. Zhang, G.-L. Tian, J.-Q. Huang, W. Zhu, and F. Wei. 2012. Graphene/single-walled carbon nanotube hybrids: One-step catalytic growth and applications for high-rate Li–S batteries, *ACS Nano*, **6**, 10759–10769.

172. Tian, G.-L., M.-Q. Zhao, D. Yu, X.-Y. Kong, J.-Q. Huang, Q. Zhang, and F. Wei. 2014. Nitrogen-doped graphene/carbon nanotube hybrids: In situ formation on bifunctional catalysts and their superior electrocatalytic activity for oxygen evolution/reduction reaction, *Small*, **10**, 2251–2259.

173. Xue, Y., Y. Ding, J. Niu, Z. Xia, A. Roy, H. Chen, J. Qu, Z. L. Wang, and L. Dai. 2015. Rationally designed graphene-nanotube 3D architectures with a seamless nodal junction for efficient energy conversion and storage, *Sci. Adv.*, **1**, 2375–2548.

Chapter 3

Electronic Transport in Nanocarbon Interconnects

Changjian Zhou,[a] Yu Zheng,[b] Zubair Ahmed,[c] Patrick Wilhite,[b] and Cary Y. Yang[b]

[a]*School of Microelectronics, South China University of Technology, Guangzhou 510641, China*

[b]*Center for Nanostructures, Santa Clara University, Santa Clara, California 95053, USA*

[c]*Department of Electronic and Computer Engineering, Hong Kong University of Science and Technology, Hong Kong, China*

cyang@scu.edu

3.1 Introduction

Carbon-based nanostructures (or nanocarbons) such as carbon nanotubes (CNTs) [1–7], carbon nanofibers (CNFs) [8–10], and graphene [11–14] are candidate materials for next-generation semiconductor devices and integrated circuits (ICs) due to their tolerance to electromigration under high currents and excellent electrical, thermal, and mechanical properties [15–25]. Nanocarbons show increasing promise to replace copper (Cu) for

Nanocarbon Electronics
Edited by Changjian Zhou, Min Zhang, and Cary Y. Yang

ISBN 978-981-4877-11-4 (Hardcover), 978-1-003-04308-9 (eBook)
www.jennystanford.com

on-chip interconnects and through-silicon vias in three-dimensional chips [26–27], largely because of their excellent controllability for directional growth and high current capacity. In this section, we present a brief overview of the electronic structures of graphene and CNT that underlie their electronic transport properties in various interconnect configurations discussed in subsequent sections.

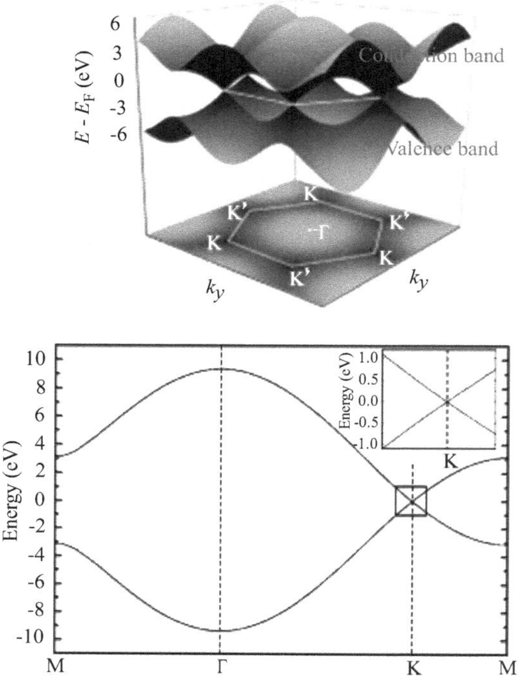

Figure 3.1 (a) Complete energy-band structure of monolayer graphene obtained using tight-binding approach [29]. The hexagonal first Brillouin zone is also shown. (b) *E*(*k*) behavior along the MΓKM direction with inset zoomed in around K. Reprinted with permission from Ref. [30], Copyright 2006, by the American Physical Society.

Since CNT can be structurally viewed as a rolled-up sheet of graphene, a brief discussion of graphene is helpful to understand carrier transport of CNT. The structure of graphene consists of sp^2 hybridized C–C atoms in a planar honeycomb lattice. The energy-band structure of a graphene monolayer is given in Fig. 3.1a, showing the valence and conduction bands. For a material to be classified as conductor, electrons must populate the conduction band. In graphene, the valence and conduction bands meet at K and K′ points, which allow electrons to easily populate the conduction band at finite temperatures and participate in conduction. The point in reciprocal lattice where valence and conduction bands meet is called the Dirac point. In graphene, it happens to coincide with a K or K′ point. As there is no bandgap between conduction and valence bands, monolayer graphene is considered a zero-bandgap material [28].

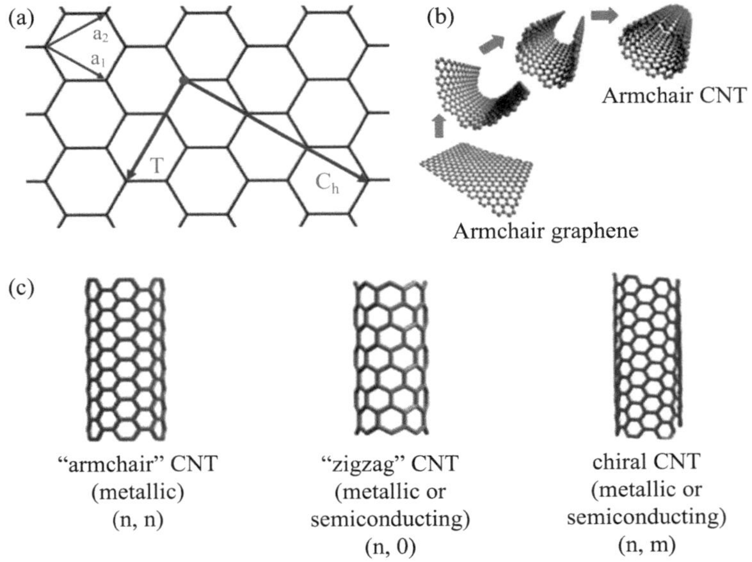

Figure 3.2 (a) Schematic of honeycomb graphene sheet showing chiral and translational vectors, $\mathbf{C_h}$ = (3,0) and **T** = (1,-2). (b) Illustration of a single graphene sheet rolled into SWCNT in armchair configuration. (c) Schematic of SWCNTs with different chiralities, with armchair and zigzag being the only achiral nanotubes. Adapted from Ref. [33], Copyright 1994, the Physical Society of Japan.

The CNT structure is a sheet of graphene rolled up into a cylindrical shell. CNT can be broadly classified into two types: single-walled CNT (SWCNT) and multi-walled CNT (MWCNT).

MWCNT consists of multiple concentric cylindrical CNT shells with a hollow interior. A third type, known as CNF, which can appear under certain growth conditions discussed in Chapter 2, consists of stacked cup-shape shells and a solid interior. The orientation and atomic arrangements of a single CNT shell can be specified using two vectors: chiral vector $\mathbf{C_h}$ and translational vector $\mathbf{T}$. The chiral vector $\mathbf{C_h}$ is the vector along which the graphene sheet is rolled into a cylinder to form a CNT cylinder, superimposing both ends of $\mathbf{C_h}$ in the process. The translational vector $\mathbf{T}$ is defined as the unit lattice vector in CNT, along the nanotube axis and normal to the chiral vector. The chiral vector (n, m) determines the configuration of SWCNT such as armchair and zigzag, as well as chiral nanotubes, as shown in Fig. 3.2. The armchair nanotube was deduced to be metallic and the zigzag configuration semiconducting [31]. However, contacts with conventional metals can affect the transport properties of actual CNT structures [32].

The SWCNT electronic structure, similar to its atomic structure, can be derived from that of graphene. CNT can be considered a 1D material in only the axial direction. The electronic structure of SWCNT can be derived from graphene using a zone-folding (ZF) approximation [34], in which the 1D energy sub-bands of CNT are constructed by "cutting" into the 2D graphene bands in its Brillouin zone (BZ) with the set of cutting lines. An example of ZF for an armchair and a zigzag CNT is shown in Figs. 3.3a and c. A number of cutting lines for an armchair (5,5) CNT are projected onto the 2D BZ of graphene as shown in Fig. 3.3c, and the corresponding energy band obtained from each cutting line is shown in Fig. 3.3a. The quantized purple line crosses the Fermi level at K and K′ points of the graphene BZ, thus resulting in zero-bandgap and metallic behavior for armchair CNT. For a zigzag (9,0) CNT, the projection of cutting lines onto graphene BZ is shown in Fig. 3.3d, and the resulting CNT band structure having a zero bandgap is shown in Fig. 3.3b. It should be noted that the (9,0) CNT satisfies the condition $n - m = 3i$ (for integer i), which renders it nearly metallic with a very small bandgap due to the CNT curvature. If a CNT has a significant nonzero bandgap, with $n - m \neq 3i$, it is semiconducting. Thus, the chirality of a CNT strongly affects its electronic transport.

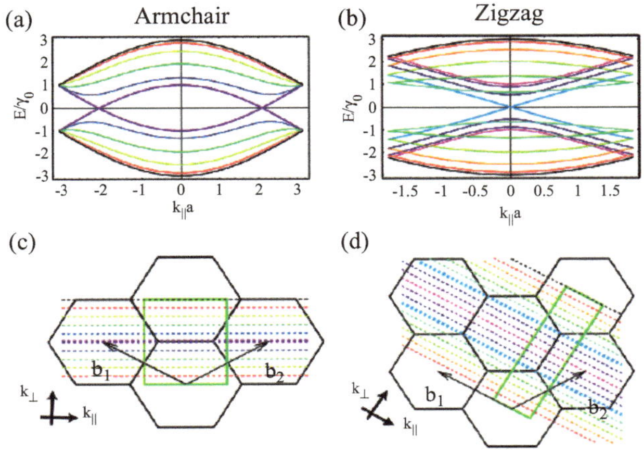

Figure 3.3 (a) Band structure of (5, 5) CNT constructed using ZF. The purple bands crossing the Fermi energy stem from the purple cutting line in (c), which crosses K and K′ points [35]. (b) Band structure of (9,0) CNT, showing double degeneracy at the valence and conduction band crossing due to the two blue thick cutting lines in (d) [35]. (c) The 10 cutting lines of (5,5) CNT projected onto graphene BZ, with the same color coding as corresponding bands in (a). (d) Similar to (c) for (9,0) zigzag CNT [35].

The ZF method provides a good approximation of obtaining the electronic band structure of CNT from graphene for CNTs having large diameter, when curvature effects can be neglected. For CNTs with diameter <1.5 nm [29], the curvature effects cannot be neglected. Whereas the bandgap of a metallic CNT is zero, the bandgap of a semiconducting CNT is inversely proportional to its diameter [36]. A semiconducting CNT with 1 nm diameter has a bandgap approximately 0.7 eV [34]. For CNTs with diameter >10 nm, as in the case of the outer shells of MWCNT, the bandgap becomes vanishingly small. Therefore, for all practical purposes, all SWCNTs with diameters >10 nm can be considered metallic, even if chirality predicts a semiconducting behavior. Further, a MWCNT consisting of shells with one or more chiralities can be considered metallic as well, since the conduction is dominated by the shells with the lesser resistance, which are usually the outer shells with the larger diameters.

CNT generally contains defects [37–39], and the types of defects in CNT affect its electrical and mechanical properties. Point defects in CNT can manifest themselves as pentagonal and heptagonal distortions of the hexagonal network. Structural defects comprise caps, bends, branches, and helical structures. In general, MWCNTs are relatively more prone to defects as multiple shells are present in each nanotube. In particular, instead of concentric cylindrical shells, cup-shaped structures in CNF are frequently observed and can be considered a defective form of MWCNT [29]. In all cases, defects invariably lead to reduction in carrier mean free path and degradation of electrical and thermal conduction.

Resistance of an ideal metallic CNT can be derived from the ZF method of cutting lines on the graphene BZ. The result for SWCNT can then be extended to MWCNT. Each cutting line that passes through K or K′ points of the graphene BZ represents a channel for electron to travel along the CNT axis. The number of channels M_E depends on the chirality vector $\mathbf{C_h}$. Further, the probability that a continuous channel exists for electron to travel unimpeded between two electrodes is denoted by ψ. Therefore, the CNT resistance can be expressed as Eq. 3.1 in terms of a combined quantum resistance (i.e., having ballistic transport of carriers) of all parallel channels between the electrodes [32].

$$R_{\mathrm{CNT}} = \frac{R_0}{M_{\mathrm{E}}\psi} = \frac{h}{2e^2}\frac{1}{M_{\mathrm{E}}\psi} \tag{3.1}$$

Here, h is Planck's constant, e is the electron charge, and R_o is the quantum resistance of a 1D conductor with ballistic transport, where ideal electrode contacts are assumed. If a channel is noncontinuous between electrodes, then an electron travelling along the CNT axis experiences scattering, giving rise to a mean free path, λ_e, and the probability ψ of such an event in terms of λ_e and CNT length, L_{CNT}, can be expressed as [32]

$$\psi = \frac{1}{\left(1+\frac{L_{\mathrm{CNT}}}{\lambda_{\mathrm{e}}}\right)}. \tag{3.2}$$

For a "short" CNT, $\lambda_e >> L_{\mathrm{CNT}}$, yielding $\psi \approx 1$ and near-ballistic transport. Thus, a metallic CNT with $M_E = 2$, near-ballistic transport, and ideal contacts would yield a resistance of $h/4e^2$ = 6.45 kΩ [3].

MWCNT with N_{shells} shells can be treated as a parallel combination of resistors. Using Fermi statistics and the energy dispersion relation of graphene, Naeemi [40] obtained an approximate expression (Eq. 3.3) for the average number of channels per MWCNT shell in terms of shell diameter D_{CNT}.

$$M_E(D_{CNT}) = 0.062\ D_{CNT} + 0.425, \quad 3\ \text{nm} < D_{CNT} < 85\ \text{nm} \qquad (3.3)$$

Such expression predicts that for MWCNT, the outer shells yield lower resistance than inner shells, since more conduction channels exist in the outer shells. Further, each conducting shell has a finite contact resistance R_C at each electrode [41–44]. Therefore, resistance of an MWCNT with finite contact resistance at both contacts can be expressed as

$$R_{CNT} = \sum_{\text{all shells}} \left[\frac{h}{2e^2 M_E(D_{CNT})} \left(1 + \frac{L_{CNT}}{\lambda e} \right) \right] + \frac{2R_c}{N_{shells}}. \qquad (3.4)$$

λ_e is the average distance an electron travels within the material between two successive scattering events, which can be due to electron–phonon interactions, contacts, impurities, and imperfections in the crystal structure [29].

For CNT interconnects with submicron lengths, defects are the primary factor in determining the electron mean free path [34]. CNT is prone to defects such as missing C atom in graphene lattice, extra C atom in lattice, local bonding between shells, and abrupt shell terminations along the length of CNT [45–50]. These defects result in the modification of chiral property of CNT and deviation of its electrical properties from that of an ideal CNT structure. For a CNT with negligible defects, λ_e can be as high as 24 μm [29]. For a highly defective MWCNT, λ_e can be as low as 5 nm [51]. A short electron mean free path generally results in high resistance. The minimum contact resistance of a CNT is its quantum resistance, but as the CNT structure deviates from ideal, contact resistance increases. The imperfect metal–CNT interface and impurities can create contact resistance that dominates the total resistance. For MWCNT, contact resistance was reported to be in the kΩ range [51–58]. Arc-discharged CNTs are generally of the highest quality, having the longest electron mean free path. However, as the CNT growth temperature is lowered to below 1000°C using other techniques such as chemical vapor deposition (CVD), more defects are introduced, which degrade electrical conduction. Chai [59] conducted a comprehensive series of experiments to analyze electron mean free path and contact resistance of horizontally grown MWCNTs at 850°C on a SiO_2 substrate using plasma-enhanced CVD (PECVD), where CNTs with three different diameter distributions were produced by varying catalyst thickness. The reported results are shown in Fig. 3.4 [59].

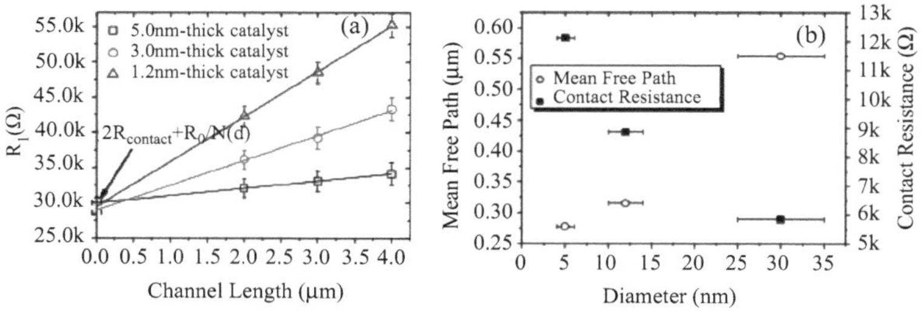

Figure 3.4 (a) Resistance of MWCNT as a function of length. (b) Mean free path and contact resistance (excluding quantum resistance) of MWCNT as a function of diameter. Reprinted from Ref. [59], Copyright 2010, with permission from IOP Publishing Ltd.

The results of length dependence of MWCNT resistance given in Fig. 3.4a are consistent with ohmic conduction. Figure 3.4b shows an electron mean free path of 0.55 μm for MWCNTs with diameters around 30 nm, compared with 25 μm for a 100 nm diameter MWCNT produced by the arc-discharge method [60]. The contact resistance given in Fig. 3.4b has an expected diameter dependence. These results strongly suggest that the MWCNTs used in these experiments exhibited ohmic transport as opposed to near-ballistic transport, which can be attributed to structural degradation due to lower growth temperature. For MWCNTs with diameters around 12 nm, the contact resistance from Fig. 3.4b is approximately 9 kΩ, which is a significant fraction of the total CNT resistance. Therefore, it is critical to understand CNT contact resistance and factors contributing to it in order to enhance CNT interconnect performance. An SWCNT with ballistic conduction and no scattering or tunneling at the SWCNT–metal interface exhibits a contact resistance equal to the quantum resistance R_0 = 6.45 kΩ [61]. However, in a real system, even if defects can be neglected, contact resistance depends on the work-function difference between CNT and contact metal, contact configuration (end contact versus side contact) [62], wettability of contact metal [63] , and interface nanostructure [63].

3.2 Side Contact versus End Contact

The key performance-limiting factor for metallic SWCNTs, MWCNT (generally metallic), and CNFs is the high contact resistance originating from the nanocarbon's interface with metal electrodes [64]. While not unique to nanocarbons, the effect of contact resistance is enhanced as devices shrink [65] and is particularly problematic in the nanoscale [66]. In addition, it is essential to control the chirality of CNTs grown on wafer as it determines the conduction property. Yang [67] reported chirality-controlled synthesis of metallic SWCNTs with the chirality of (12,6) and a 92% yield of metallic nanotubes. Further improvement in chirality-controlled synthesis with high yield (>99%) can eventually result in integrating SWCNTs as FET channel and/or on-chip interconnects in integrated circuits.

To realize nanocarbons in general and CNT in particular as on-chip interconnect materials, the contact resistance stemming from the metal–CNT interface must be well understood and minimized. Understanding the complex mechanisms at the interface can lead to effective contact resistance reduction. The two metal–CNT contact geometries, sidewall or side contact and end contact, have different ranges of contact resistance due to different mechanisms. A schematic representation of the side-contact and end-contact CNT configurations is shown in Fig. 3.5.

Side contacts (Fig. 3.5b) typically result in contact resistances that are above 1 kΩ, whereas end contacts (Fig. 3.5a), such as that for as-grown vertically aligned CNTs on a metal underlayer, can have significantly lower contact resistance, often below 1 kΩ. This difference in contact resistance is due to the different carrier transport mechanisms between the end-contact CNT and the side-contact CNT. For the end-contact CNT, the lower contact resistance is largely due to strong bonding between edge carbon atoms with atoms on the metal surface (Fig. 3.5c), as unsaturated carbon bonds at the base of the CNT can easily latch onto the metal surface atoms, resulting in a relatively low contact resistance. For the side-contact CNT, carrier transport across a side-contacted interfacial gap via tunneling (Fig. 3.5d) is generally associated with high contact resistance. The structures of these two metal–nanocarbon contacts, and their respective carrier transport mechanisms, are elaborately reviewed by Wilhite et al. [68].

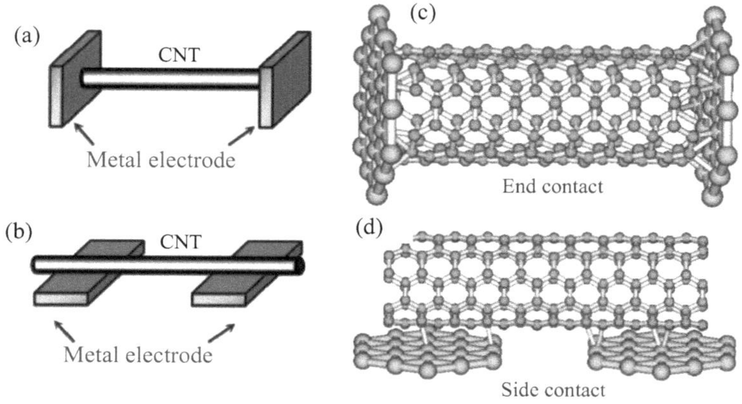

Figure 3.5 Schematic representation of two different metal–CNT contact configurations, (a) end contact and (b) side contact [62]. End-contact configuration exists for vertically grown CNTs and generally has a lower contact resistance, often below 1 kΩ. Side-contact CNT may be made by drop casting and generally has a higher contact resistance, often above 1 kΩ. For end-contact CNT, the bonding between edge carbon atoms with atoms on the metal surface is shown in (c) [32]. For side-contact CNT, the carrier transport across a side-contacted interfacial gap via tunneling is shown in (d) [32]. Figures (a) and (b) reprinted from Ref. [62], Copyright 2010, with permission of the American Chemical Society and Figs. (c) and (d), reprinted from Ref. [32], Copyright 2003, with permission of the American Physical Society.

3.3 Horizontal CNT Interconnects with Side Contacts

A typical side-contacted structure is a CNT horizontally bridging two metal electrodes, shown schematically in Fig. 3.5b. A widely used fabrication method consists of suspending a purified mixture of CNTs in an alcohol solution, which is then sonicated to de-bundle the CNTs, and then drop casted on pre-patterned electrodes. Typical as-fabricated resistance values are high, ranging from 10^4 to 10^9 Ω. Our experiments show that the initial current–voltage (I–V) behavior tends to be highly nonlinear and asymmetric, and these effects are accentuated for devices with higher resistance. These initial resistances drop to the kΩ range after stressing with a current of 10^5–10^6 A/cm^2. As no appreciable degradations are observed in the CNT nanostructure using high-resolution electron microscopy, this drastic reduction in the total resistance is likely caused by a decrease in the contact resistance induced by localized Joule heating at the interface [15, 69]. A single-junction semi-empirical tunneling model supports this explanation, where an interfacial gap between CNT (or CNF) and metal is reduced by current stressing [70]. When the gap separation shrinks, the I–V behavior becomes linear as the device resistance decreases and approaches ohmic, as illustrated in Fig. 3.6.

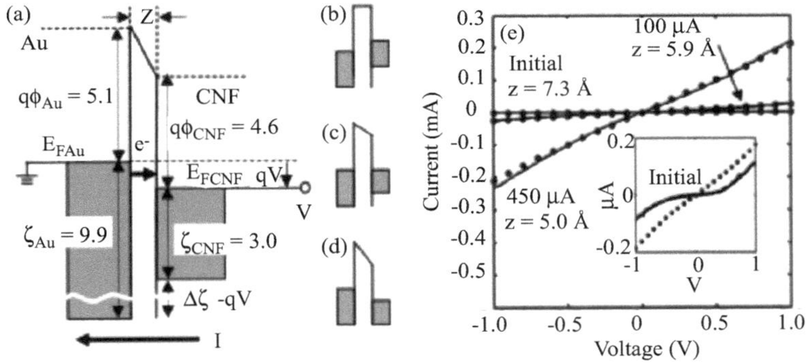

Figure 3.6 (a) An energy-band model for Au-CNF (assuming CNT band parameters) contact with a tunneling gap of spacing z and voltage V between them, $q\varphi$ is work function, ζ is Fermi energy measured from the bottom of the Au valence band, and EF is Fermi-level position. $\Delta\zeta = \zeta_{FAu} - \zeta_{FCNF}$ is the difference between Au and CNF Fermi energies. The energy-band diagram is shown for (b) $V = 0$, (c) $V = 0$, and (d) $V > 0$. (e) Measured I–V before and after multiple current-stressing cycles, with current-stressing conditions shown. Solid circles are modeling results at 300 K, fitted to measured I–V data, resulting in z = 7.3, 5.9, and 5.0 Å. The inset shows improvement in I–V linearity upon current stressing, with solid line indicating initial behavior. Reprinted from Ref. [68], Copyright 2014, with permission from IOP Publishing Ltd.

The existence of an interfacial gap is likely due to contact asperities (unevenness of surface) and/or defects such as trapped adsorbates or residual impurities from device fabrication. The transport across the gap is similar to that through a Schottky barrier at a metal-semiconducting SWCNT contact [71]. Whereas for metal–semiconductor contact, tunneling is the direct result of inherent material properties, namely, work function difference, tunneling across the metal–CNT interface is most likely due to a combination of contact asperities and defects.

A second important consideration for a side-contacted CNT is its ambient reactivity. Considering a monocrystalline sp^2 graphitic carbon structure with no dangling bonds, CNTs are highly susceptible to their environment due to the non-localized carbon π-electrons, resulting in van der Waals bonding out of the graphene plane [72–75]. This weak bonding from electrostatic interactions between the CNT sidewall and the adsorbates may change its density of states (DOS) [74, 76–78] and has measurable impact on the conductivity of CNTs [75, 79]. This local interaction resulting from π–d orbital coupling has been correlated to an increase in the DOS of graphene, which caused a drop in contact resistivity [80]. Further, adsorbates from the atmosphere such as H_2O, O_2, CO_2, and hydrocarbons, or from suspension solutions such as those used in drop casting, can be trapped on the surface leading to the formation of a detrimental interfacial layer between the CNT sidewall and the metal electrodes, resulting in a high contact resistance. It is likely that the Joule heating by current stressing or contact improvement by rapid thermal annealing (RTA) in inert environments can lead to desorption of trapped chemical species [81, 82]. Maeda [81] reported that the current stressing of side-contacted nanocarbons in vacuum may substantially reduce the average resistance.

A third consideration is the metal–CNT contact area. A larger contact area leads to a lower contact resistance. Two-point probe (2PP) measurements on very long CNTs were used to determine an effective contact length (proportional to contact area) at which the resistance reached a minimum [83, 84], indicating a minimum contact length above which the total resistance is contact-area independent. A semi-empirical model based on contact geometry was then used to extrapolate the resistivity of the CNTs and contact resistivity using Ti/Au electrodes [83, 84]. Saito [20] reported depositing tungsten (W) on a side-contact CNF to fully cover the region where the CNF is in contact with the Au electrode, as shown in Fig. 3.7. Contacts fabricated with this technique have a typical resistance in the kΩ range, which is one to three orders of magnitude smaller than those without the tungsten deposition. This reduction in contact resistance is due largely to an increase in the metal–nanocarbon contact area from the W deposits. The metal deposition around the nanocarbon side-contact area can be implemented using electron-beam-induced deposition (EBID) or ion-beam-induced deposition (IBID). Generally, photolithography and electron beam lithography may be used to form metal contacts [8].

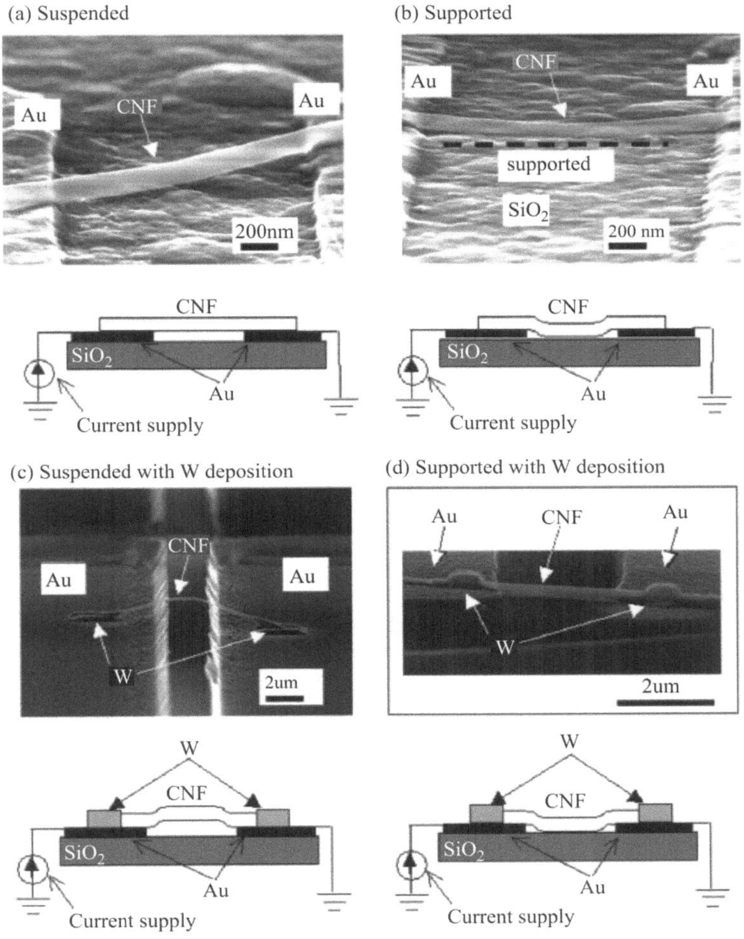

Figure 3.7 Current-stressing experiments: (a) CNF suspended on gold electrodes; (b) CNF supported by SiO_2 substrate; (c) CNF suspended over SiO_2 substrate with W deposited on contacts; and (d) CNF supported by SiO_2 substrate with W deposited on contacts. Upper figures show SEM image of a CNF device at 75° tilted-angle view, and lower figures illustrate schematic of electrical measurement setup for current-stressing experiments. Reprinted from Ref. [20], Copyright 2008, with the permission of AIP Publishing.

Alternatively, contacts can be formed by IBID using a focused ion beam (FIB) [20], or by EBID with an electron beam from an SEM [81]. IBID-W is generally more effective in reducing the contact resistance, but EBID-W produces similar contact resistance reductions, which can be further improved by current stressing.

Recent high-frequency two-port measurements on CNT test devices revealed a significant frequency-dependent capacitive component in the large contact impedance for drop-casted contacts. Upon contact metallization with separately IBID-Pt and IBID-W and further current stressing, the contact impedance decreased drastically and became frequency independent as well as virtually resistive, with a negligible reactive component [85, 86].

Figure 3.8 shows the total resistance of a horizontal CNF interconnect test device versus applied stress current. The benefits of depositing W on the contacts are evident, especially for the case of IBID-W, which show low resistance that remains invariant under current stressing, suggesting that the contact resistance is minimized [20]. While these results are particular to W, the technique is applicable to other metals, and the final resistance value may be lower for metals with matching work functions to that of CNT and higher wettability [63].

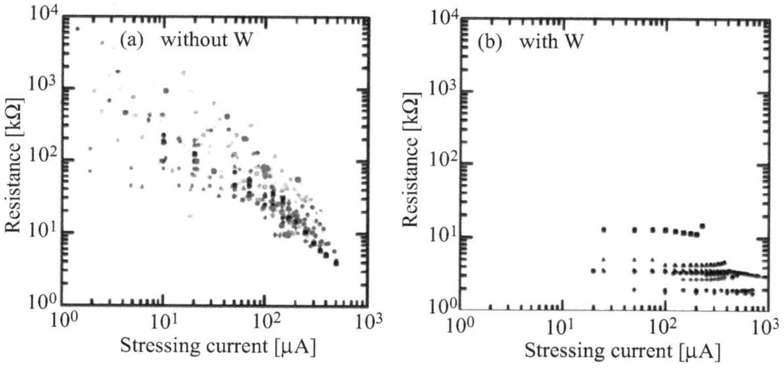

Figure 3.8 Resistance of two-terminal CNF test devices versus stress current (a) without W and (b) with IBID-W deposited on CNF/Au electrode contacts. Reprinted from Ref. [20], Copyright 2008, with the permission of AIP Publishing.

3.4 Vertical CNT Interconnects with End Contacts

Despite progress made in contact engineering to improve side-contacted metal–CNT interfaces, further improvements are critically needed for CNTs to be incorporated in functional devices. As reported by numerous groups, including ours, the better way to make metal–CNT contacts is to bind the metal surface atoms to the unsaturated C-bonds at the edge of the CNT wall, namely, making end contacts [87]. For vertically aligned CNTs, it is natural to make end contacts at the as-grown interface with the metal underlayer, yielding lower contact resistance than that for side-contacted CNTs.

Figure 3.9 shows a series of cross-sectional TEM images of metal–CNT (CNF in this case) interfaces grown on a Ti underlayer using PECVD. The interface regions display little evidence of impurities and clearly show the graphitic planes nearly normal to the metal surface. From these images, it appears that the graphitic planes make direct contact to the Ti surface atoms, providing the necessary bonding to facilitate carrier transport and resulting in low contact resistance. While the formation of carbides at this interface has been reported [88–90], it is unclear if this occurs for all PECVD-grown CNTs. Generally, when the wettability between the CNT and the metal underlayer is high, such as that for Ti or Cr, which correlates with strong bonding between carbon and metal surface atoms, it is likely that metal carbide may form at the interface. A comparison of side-contacted and end-contacted CNTs with interfacial tungsten carbide formed by electron beam irradiation was made, which showed significantly more reduction in resistance for the end-contacted device [90]. Similar results have been observed [87] or predicted [62], if contact areas are taken into account. Wang [91] reported a contact geometry in which only the 1D edge of a graphene layer was metallized. This has noticeable impact on graphene performance, drastically reducing contact resistance and improving carrier mobility [91].

In general, end-contacted geometries are advantageous over side contacts, because of the stronger CNT–metal bonding across the interface and the absence of an interfacial layer or gap through which carrier tunneling occurs. A typical end-contacted nanocarbon structure is the CNT interconnect via.

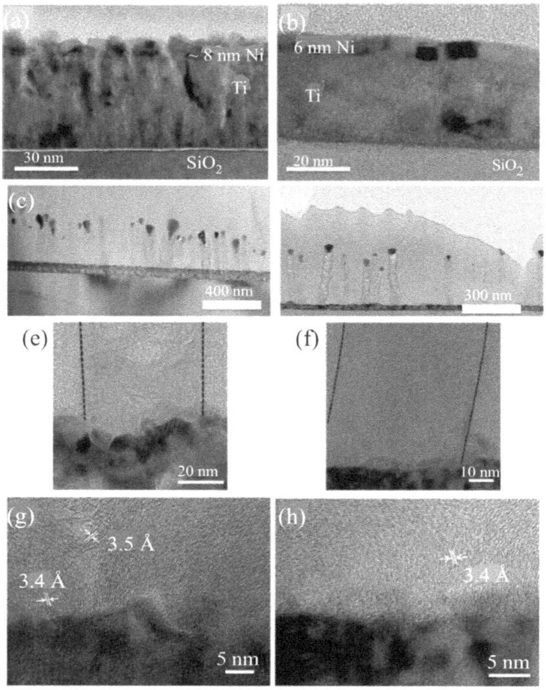

Figure 3.9 Comparison between two similar catalyst/underlayer metal structures deposited by magnetron sputtering with (a) $<10^{-6}$ torr base pressure, 100 W DC and (b) $<10^{-8}$ torr base pressure, 100 W RF. The resulting CNFs grown at the same time for each sample have significant different structures with (c) cup-shape CNF morphology and (d) bamboo-shape CNF morphology. Interfacial images show clean contacts (e) and (f). (g) and (h) Interfaces showing graphitic interplanar spacing, and graphene planes nearly normal to the underlayer metal surface. Reprinted from Ref. [68], Copyright 2014, with permission from IOP Publishing Ltd.

3.4.1 CNT Interconnect Vias

The DC performance of CNT vias can be defined by three fundamental electrical parameters: via resistance, CNT resistivity, and contact resistance. Via reliability is quantified using the maximum current-carrying capacity of the CNT-filled via [92].

A simple yet effective approach to investigate CNT as vias is a 2PP setup with a nanoprober on patterned vias, the schematic and SEM image of which are shown in Fig. 3.10. The 2PP structure consists of contributions from the nanoprobes, metal underlayer, and contacts between probes and metal pads, as well as CNT contacts and CNTs. Each of these contributions will be addressed in both the test structure design and the method to analyze the measured results.

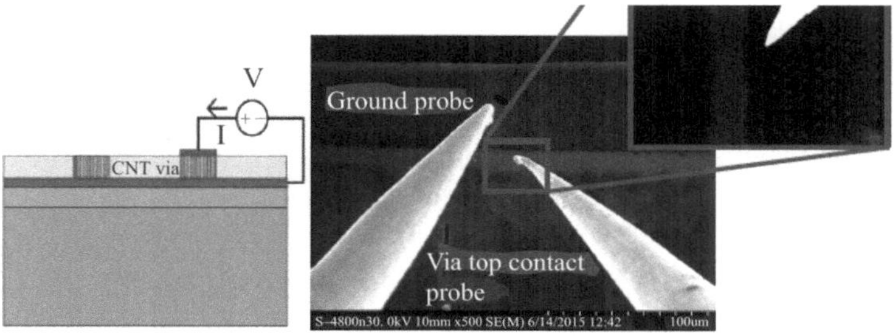

Figure 3.10 Schematic of 2PP measurement setup for direct probing of a single CNT via [93] (a) and SEM image of the 2PP setup [94] (b). Figure (a) reprinted with permission from Ref. [93], Copyright 2015, IEEE.

In an IC, a via connects two horizontal interconnect lines on adjacent levels separated by a dielectric layer. The 2PP CNT via test structure is designed to resemble vias in an IC, where each via connects the underlayer metal to the top contact. In our investigation, the CNT via test structure with Cr underlayer yields CNT areal density ~$5 \times 10^{11}/\text{cm}^2$ and narrow diameter distribution with median diameter of 13 nm. Therefore, a 300 nm Cr film is used as the underlayer to yield high CNT areal density, with little contribution to the total resistance in the 2PP measurement [94].

The choice of top-contact metal and contact configuration is critical to achieve low contact resistance. Pt is selected for via top-contact metallization, as it does not oxidize in air at room temperature [95] during sample transfer from metal deposition to nanoprobing. As will be shown later, the Pt-metallized via top contacts form end-bonded contacts to exposed CNTs inside the via and, together with the Cr underlayer and the CNTs, constitute a realistic model for the building block of an on-chip 3D interconnect network. Direct probing of individual CNT vias with different heights and widths allows us to analyze the measured resistance data to accurately extract various resistance contributions. The 2PP method used in our work, with a simple test structure design and fabrication process, is well suited for the study of performance and reliability of CNT interconnect vias.

The choice of dielectric material in which CNT vias are patterned is critical to ensure structural and electrical integrity of CNT vias. Since our CNT growth process is carried out at 700°C in ammonia plasma ambient, the dielectric material should be able to withstand ion bombardment in plasma and prevent diffusion of the underlayer Cr into the dielectric during CNT growth. Structural damage to the dielectric material, and/or diffusion of Cr into the dielectric, would compromise the electrical measurements of CNT vias, because it would provide an alternative conduction path in addition to the CNT. It has been reported that SiO_2 can withstand ion bombardment in plasma and temperature up to 800°C [96–102]. Therefore, SiO_2 is employed as the dielectric surrounding the vias.

To ensure that the CNTs inside the via make contact with the metal underlayer and participate in electrical conduction, it is critical to suppress CNT growth from the dielectric sidewalls of the via because such growth would limit the growth of CNTs from the base of the via by reducing the diffusion of carbon precursors deep into the via. A key design consideration is patterning the vias with vertical sidewalls and using directional catalyst deposition to minimize the deposition of the carbon precursors on the via sidewalls.

Based on all the above considerations, a test structure consisting of vias with square cross sections, and with widths of 150 nm, 120 nm, 90 nm, 60 nm, and 40 nm, is designed and fabricated [93]. Standard silicon IC fabrication techniques were employed in the CNT via test structure fabrication, as described in detail by Zhou [93] and Vyas [94]. A lightly doped Si wafer with a 200 nm thermal oxide was used as the substrate. A 300 nm Cr film was evaporated on the oxide, which serves as the underlayer for CNT growth. Then a 100 nm SiO_2 film (functioning as a regular mask to pattern the vias) was deposited on the Cr film using PECVD process with SiH_4 and O_2 precursors. In order to achieve a vertical via profile, a layer of 50 nm a-Si film (functioning as a hard mask to prevent overexposure of the top region of vias to reactants) was deposited on top of the 100 nm SiO_2 layer. Next, a 300 nm photoresist film was spin-coated on the entire wafer and subjected to baking to evaporate the excess resist. After baking, the resist film was irradiated by the e-beam using a mask, where the area of resist under each patterned via was exposed to e-beam, weakening the resist polymer and making it soluble in organic solvents. The wafer was then soaked in amyl acetate solution to etch the resist from the e-beam exposed regions. This process transferred the via pattern from mask to the resist film.

The e-beam-exposed resist film was first used as a mask for the a-Si film (hard mask) etching, using ICP with SF_6 and O_2. Subsequently, the underlying SiO_2 film (regular mask) etching was performed using reactive ion etching (RIE) with CHF_3 and O_2 as reactants.

By introducing the a-Si layer to prevent overexposure of top region of via to etchants, vertical walls were obtained for vias with width 150 nm to 60 nm. Further, a 50 nm SiO_2 layer and a 40 nm a-Si film were employed to fabricate the 40 nm vias. All samples with patterned vias were subjected to HF rinse to remove any native oxidation on exposed underlayer Cr. A 4 nm Ni film was deposited on the vias using e-beam evaporation. A recipe for CNT growth by PECVD at 700°C, with acetylene C_2H_2 as the carbon precursor, was used for both the 60 to 150 nm (Sample A) via and the 40 nm via (Sample B) samples. A summary of the two via samples is presented in Table 3.1.

Table 3.1 Summary of via samples prepared for CNT growth

Specification	Sample A	Sample B
Via height or depth	130 nm	80 nm
Via width(s)	60 nm, 90 nm, 120 nm, 150 nm	40 nm
Catalyst, film thickness (nm)	Ni, 4 nm	Ni, 4 nm

An average CNT height of approximately 1 μm, which is much larger than the via height, is observed with SEM. CNT growth occurs in vias and on top of oxide. Top-view SEM images reveal the presence of voids between CNT vias. To ensure robustness and vertical alignment of CNTs, the voids between the CNTs and between the CNT vias were filled with aluminum oxide (Al_2O_3), a dielectric material. This dielectric filling creates support and adhesion to CNTs inside vias for them to withstand polishing, facilitates dissipation of heat generated from Joule heating of CNTs during current stressing, and provides a barrier to metal diffusion during a later fabrication step of top-contact metal deposition.

A polishing process was applied to planarize the sample surface and to remove the excess Al_2O_3 as well as to expose the CNTs inside vias. These CNTs, having exposed graphitic shells from the removal of catalyst particles at their tips, make contacts with nanoprobes directly or through deposited metal in an end-bonded configuration. Sample A was polished using ion milling at a glancing angle to prepare CNT vias with a fixed height. Sample B was subjected to mechanical polishing (using SiO_2 nanoparticles as slurry) at an angle to fashion a wedge structure with an approximately 1° angle. It creates vias of multiple heights on the same sample, which can then be measured to yield the height dependence of CNT via resistance. Both samples were then ready for nanoprobing directly or for top-contact metallization.

The final step in CNT via fabrication process is metal deposition on vias to realize end-bonded contacts for CNTs inside vias. Selective deposition of Pt contacts on individual CNT vias was made using EBID carried out in an SEM chamber. EBID-Pt depositions (90 nm × 90 nm) were made on 60 nm vias on Sample A. For the 40 nm vias on Sample B at five different heights, areas of 70 nm × 70 nm are deposited. SEM images confirm that Pt contacts indeed exist on individual vias and each deposit is confined at the target without overlapping adjacent vias.

3.4.2 CNT Via Nanostructure Characterization

After top-contact metallization, the nanostructure of CNT vias is characterized to confirm their structural integrity, as well as to examine the metal–CNT interfaces. The methods used are cross-sectional TEM, energy-dispersive X-ray spectroscopy (EDS), and Raman spectroscopy.

Cross-sectional TEM imaging is performed on vias and the CNTs within to analyze their nanostructure. TEM image for a 60 nm via after Al_2O_3 dielectric fill is shown in Fig. 3.11a. Cr protrusion occurs inside the via but is limited to less than half the via height. The Al_2O_3 layer is shown to be uniformly and conformally deposited around CNTs and fills the inter-CNT voids. Diffusion of top-contact metal Pt inside the via is not evident from the TEM image shown in Fig. 3.11b, which also reveal a CNT embedded in Al_2O_3 filler making contact with the top metal Pt cap. Thus, CNTs inside via do form an end-bonded via top contact with Pt, as well as base contact with Cr. Such contacts with metals were also demonstrated by Matsuda [62, 103] and Lim [63] to yield lower contact resistance compared to side-bonded CNT contacts.

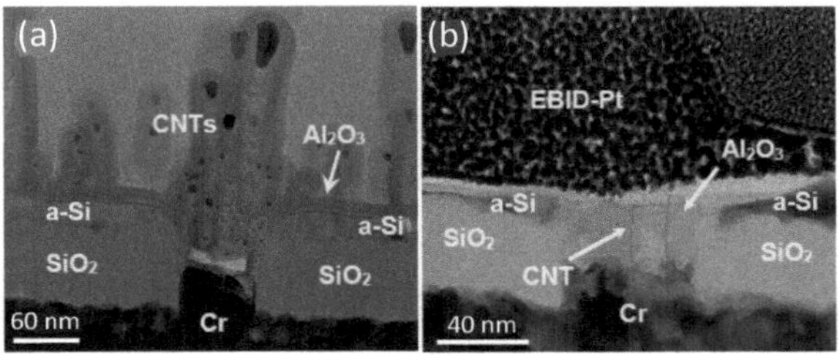

Figure 3.11 (a) Cross-sectional TEM image of 60 nm vias before polishing. (b) Cross-sectional TEM image of 40 nm via with EBID-Pt "end-bonded" top contact [94].

To determine the locations of various elements inside the via, EDS is performed on the cross section of a 60 nm via [104]. The resulting elemental maps confirm that catalyst element Ni is detected only inside the via in the form of catalyst particles on CNT tips, and not in the Si or oxide regions, thus ruling out the formation of any electrically conductive NiSi. The elemental map of Cr shows that Cr is confined below CNTs inside the via. Thus, no Cr diffusion into either SiO_2 or Al_2O_3 occurs. Carbon signal is traced all the way down to the via bottom through the protruded Cr, which confirms that the CNT base is not on the surface of the protruded Cr underlayer. This last finding is critical to an accurate determination of CNT height inside via, which can now be equated to the via height.

In Fig. 3.11a, the internal structure of CNTs reveals a "bamboo-shaped" configuration, which is typical of CNFs. Such apparent structural discontinuities increase the number of scattering events for electron transport [105]. The resulting electron mean free path is expected to be significantly smaller than the CNT length, and ohmic conduction most likely occurs, instead of ballistic conduction that occurs in pure CNTs.

Raman analysis is performed for our as-grown CNTs on unpatterned substrates, prepared at three different growth temperatures, to investigate the electron mean free path. Dressellhaus [49, 105–112] and Vollebregt [113] have extensively characterized CNTs using

Raman spectroscopy to reveal long-range order and determine electron mean free path. The Raman spectroscopy yields a G-band that corresponds to the active mode of graphitic material, and a D-band indicates presence of defects. The physical origin of the D-band is attributed to the graphitic shell terminations on the CNT outermost wall that result in the bamboo-like structure shown in Fig. 3.11a [113–126], missing C atoms, and the presence of heptagons in hexagonal honeycomb lattice [46–49]. These defects contribute to electron scattering and result in reduction of electron mean free path. The intensity ratio of D-band to G-band can be used to estimate the defect density and electron mean free path inside the CNT [98, 101, 113, 127]. Based on the intensities I_D and I_G of the D and G peaks, electron mean free path can be obtained with [101]

$$\frac{I_D}{I_G} = \frac{C(\lambda)}{L_a}, \tag{3.5}$$

where $C(\lambda) = 4.4$ nm is a fitting parameter for graphite and L_a is the electron mean free path.

Our investigation reveals that L_a decreases (5.2, 4.8, and 4.4 nm) with lower growth temperature (700, 600, and 500°C). For CNTs grown in vias at 700°C, the electron mean free path of 5.2 nm is significantly smaller than via heights of 130 nm (Sample A) and 80 nm (Sample B). We conclude that ohmic transport is the predominant conduction mechanism. Therefore, we may model a CNT via as a resistor consisting of various components, including contact resistance, and validate the analyses of measured resistance versus via width and height data [128].

3.4.3 Electrical Characterization of Carbon Nanotube Vias

The design considerations for test structure discussed above suggest that CNT vias can be accurately characterized with a 2PP method by minimizing the stray resistance contribution, primarily from the underlayer and the probes. Toward this end, a 300 nm thick Cr layer serves as an underlayer in CNT via test structure to minimize its resistance. Stray resistance contribution from the nanoprobing system is minimized by employing a low-resistance nanoprober.

The nanoprober is installed inside the SEM chamber with sample resting at the center of the nanoprober stage. The system is allowed to evacuate for 4 h to allow outgassing of air molecules adsorbed on nanoprober and sample. Achieving high vacuum (approximately 10^{-7} Torr) is key to achieve accurate measurements as stray oxygen can have an adverse effect on both performance and reliability determinations [81]. The resistances of the nanoprobing system, probe–metal contact, and underlayer resistance are measured in situ under high vacuum. The intrinsic resistance of the nanoprobing system, which appears as a parasitic series resistance in 2PP measurements, is obtained by electrically shorting two probes, which forms a short circuit across the nanoprobing system. The SEM image of two probes in contact is shown in Fig. 3.12a with the corresponding I–V and resistance behaviors shown in Fig. 3.12b. The contribution from the nanoprobing system is 5 Ω.

The combined probe–metal contacts and underlayer resistance are extracted by first applying 2.5 V initially between the two probes with a current compliance of 50 nA, while both probes are made to land firmly on an exposed area of the Cr underlayer, serving as the ground. Once both probes make contact with Cr, the voltage drops, and the current reaches its maximum value set by the compliance, as shown in Fig. 3.13a, with the resulting I–V and resistance behaviors shown in Fig. 3.13b.

Figure 3.13b reveals that the total contribution from the nanoprobing system, probe–metal contacts, and underlayer is 9 Ω, out of which 5 Ω is from the nanoprobing system. Resistances of CNT vias for widths 30–200 nm were reported to be in the 1–5 kΩ range [55–57, 63]. Assuming similar resistances for our samples A and B, the error introduced by neglecting all series parasitic resistances in 2PP measurements is at most 1%. Thus, the use of 2PP to measure CNT via resistance as discussed above is justified.

CNT via resistance measurements are carried out on samples A and B, and their specifications are summarized in Table 3.2. I–V behaviors of at least 10 vias in Sample A are obtained for each linewidth. The total via resistance is then determined by taking the reciprocal slope of the I–V plot at zero voltage. The data for the four via widths are then analyzed to project sub-60 nm via resistance.

Contact improvement for each via is achieved by current stressing prior to recording its resistance value. Sample B also undergoes I–V measurements to determine resistance for 40 nm CNT vias at five different heights along the wedge. CNT vias with and without EBID-Pt contacts are measured to extract their contact resistance and resistivity. Finally, current-carrying capacities for 60 nm and 40 nm vias are obtained using current-stressing experiments.

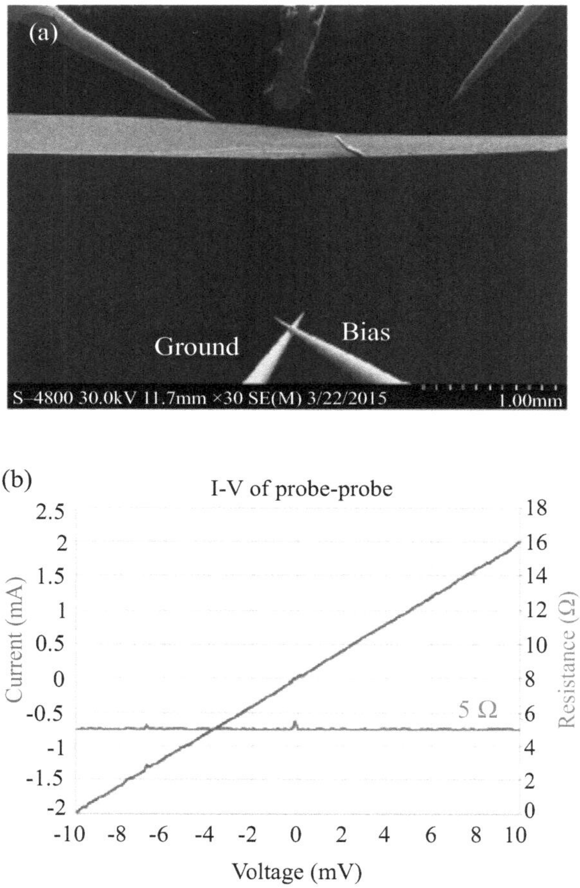

Figure 3.12 (a) SEM image of shorting two nanoprobes. (b) Resulting I–V and resistance behaviors [94].

Table 3.2 Summary of CNT via sample parameters and resistance measurements

Parameter/Task	Sample A	Sample B	Description
Via height (nm)	130 (fixed)	80 (wedge)	Obtained from via etching during fabrication process
Via width(s) (nm)	60, 90, 120, 150	40	Targeted linewidths for end-of-roadmap nodes
Top-contact metallization	Vias with and without EBID-Pt	Vias with and without EBID-Pt	Effect of metallized top contact
Electrical measurements and analyses	I–V	I–V	Determine I–V linearity and via resistance
	Resistance versus via width	NA	Resistance projection for sub-60 nm linewidths
	NA	Resistance versus via height	Extract contact resistance and resistivity
	Current stressing	Current stressing	Determine current-carrying capacity
	NA	Voltage sweep with incremental current compliance	Verify current-carrying capacity

The cross-sectional schematic of a 2PP I–V measurement setup is shown in Fig. 3.14a. One nanoprobe lands on the Cr underlayer serving as ground, while the other probe makes a mechanically robust contact with the top of the CNT via, as illustrated in the SEM image shown in Fig. 3.14b for a 60 nm CNT via. Initial contact between nanoprobe and unmetallized via top contact yields high resistance, which is improved by current stressing; and the final value is recorded when no further reduction in resistance occurs. A typical CNT via I–V plot is shown in Fig. 3.14c. Linearity of the I–V behavior confirms the ohmic transport mechanism as suggested by the TEM and Raman analyses discussed previously.

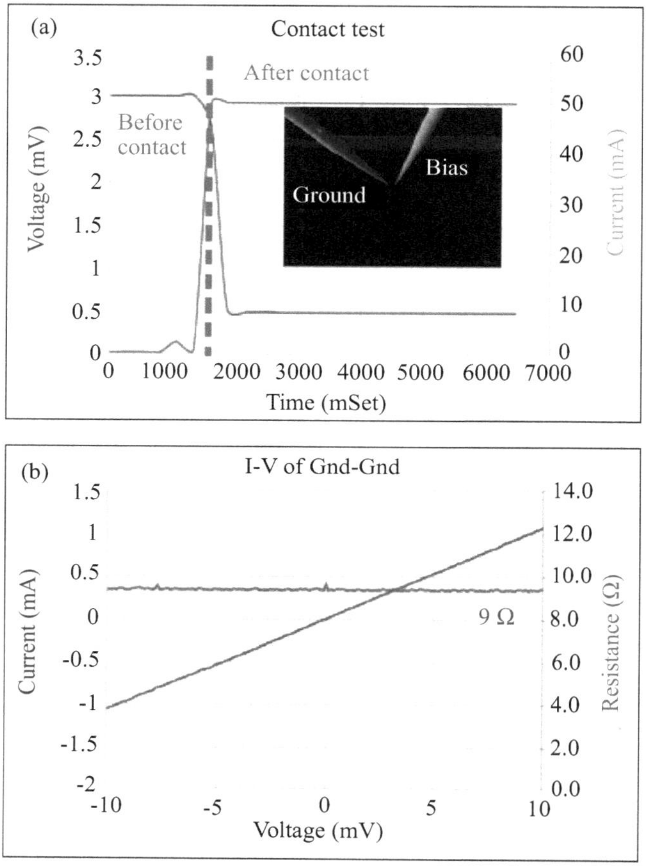

Figure 3.13 (a) Voltage between nanoprobes before and after making contact with Cr, with inset showing SEM image of nanoprobes on Cr underlayer. (b) I–V and resistance behaviors when nanoprobes are in contact with Cr underlayer [94].

I–V measurements on the CNT vias without metallized top Pt contacts in Sample A are conducted. The purpose is to extrapolate the via resistance in the sub-60 nm regime from statistical analysis of the resistance versus via width data for the 60 nm to 150 nm vias. Such projection is useful for assessing the potential of CNT as an interconnect material for the current interconnect pitch [92]. With the 2PP setup in Fig. 3.14a, the measured resistance of a single CNT via is given by

$$R_{\text{via}} = R_{\text{Cr}} + R_{\text{probe}} + R_{\text{probe-Cr}} + R_{\text{probe-CNT}} + R_{\text{Cr-CNT}} + R_{\text{CNT}} \quad (3.6)$$

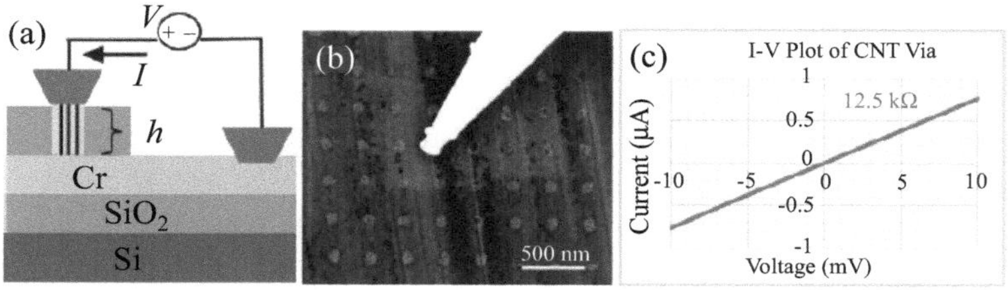

Figure 3.14 (a) Schematic of I–V measurement setup. (b) SEM image showing nanoprobe in contact with 60 nm CNT via without metallized top Pt contact. (c) Typical I–V plot of a 60 nm via without metallized top Pt contact [94].

R_{Cr}, R_{probe}, and $R_{\text{probe-Cr}}$ denote the resistances of the Cr underlayer, nanoprobes, and probe–Cr contact resistance, respectively. Their sum R_{M} is estimated to be approximately 9 Ω. $R_{\text{Cr-CNT}}$ and $R_{\text{probe-CNT}}$ are lumped together to represent the total contact resistance R_{C}, and R_{CNT} is the intrinsic CNT resistance. The total CNT via resistance becomes

$$R_{\text{via}} = R_{\text{M}} + R_{\text{C}} + R_{\text{CNT}}. \quad (3.7)$$

R_{C} is estimated by dividing R_{Ci}, contact resistance of a single CNT, by an average number of CNTs inside a via that make contact with the nanoprobe, which is the product of average CNT areal density N_{CNT} and via cross-sectional area w^2. The CNT resistance can then be expressed in terms of CNT resistivity and via dimensions, assuming ohmic conduction along the CNT height.

$$R_{\text{via}} = R_{\text{M}} + \frac{R_{\text{Ci}}}{N_{\text{CNT}} \times w^2} + \frac{\rho_{\text{CNT}} \times h}{N_{\text{CNT}} \times w^2 \times A_{\text{CNT}}} \quad (3.8)$$

Here, ρ_{CNT} is the CNT resistivity, h is the CNT height, and A_{CNT} is the average total CNT cross-sectional area inside the via.

The CNT via resistance versus via width plot is shown in Fig. 3.15, with the mean and standard deviation for each width given in the inset. Note that the standard deviation significantly increases for sub-100 nm linewidths, suggesting large variations at the probe-

CNT interface among the vias. Such variations among the vias with the same via width are mostly due to the differences in interface morphology between the nanoprobe and the CNT tips, resulting in different numbers of graphene shells making contact with the nanoprobe. For the via top contact, it is essential that the metal (nanoprobe or contact metal layer) creates a current path through as many CNT walls as possible. A nanoprobe with tungsten tip is used. The W tip can be softened by current annealing, effectively increasing the contact area and resulting in carrier flow through the maximum number of CNT walls. Based on such considerations and for the average areal density obtained from our CNT growth, if all process and probing steps are approaching optimum, the via resistance can be as low as 150 Ω for a 60 nm via, as shown in Fig. 3.15. However, variations in contacts (both the base and the top contacts) among the vias, compounded by the fact that a different percentage of CNTs reach the top of each via, create a substantial spread in the measured CNT via resistance data. Nevertheless, by measuring multiple devices and performing a standard statistical analysis, we can systematically study the downward scaling trend of CNT via resistance.

Since R_M (approximately 9 Ω) is at least an order of magnitude lower than R_{via} in all cases, it is neglected in rewriting Eq. 3.8 as

$$R_{via} = \frac{R_{Ci}}{N_{CNT} \times w^2} + \frac{\rho_{CNT} \times h}{N_{CNT} \times w^2 \times A_C} \quad (3.9)$$

Now, taking logarithm of both sides of Eq. 3.9, we obtain

$$\log_{10}(R_{via}) = \log_{10}\left(\frac{R_{ci}}{N_{CNT}} + \frac{\rho_{CNT} \times h}{N_{CNT} \times A_C}\right) - 2 \times \log_{10} w . \quad (3.10)$$

Based on Eq. 3.10 and statistical linear regression analysis of the measured data plotted in a log–log scale in Fig. 3.15, the slope of the fitted line is −1.9, compared to the ideal case of −2, where N_{CNT}, A_C, and ρ_{CNT} are width-independent, thus supporting this analysis for examining the downward scaling trend of via performance. This result also suggests that the dependence of N_{CNT}, A_C, and ρ_{CNT} on w may not be significant, which further attests to uniformity in CNT growth. By extrapolating the fitted line, the projected resistance for a 30 nm CNT via with a height of 130 nm is 6.3 kΩ and can be as low as 295 Ω in the best (or minimum resistance) case.

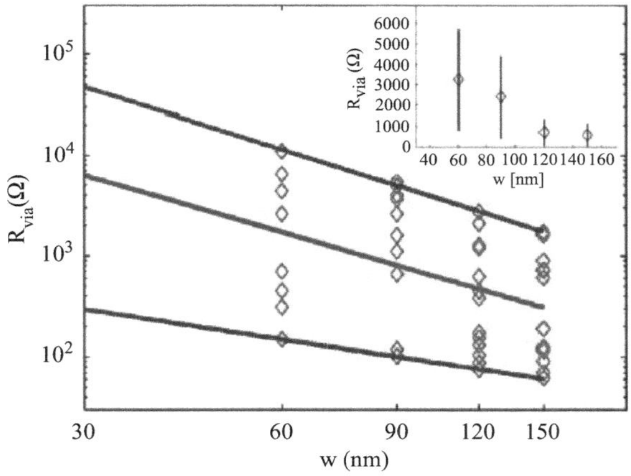

Figure 3.15 Measured CNT via resistance versus via width (data points) and statistical linear regression fit of the measured data (solid line). Slope of the fitted solid line is −1.9, compared to the ideal case of −2. The dashed lines show the fits based on the lowest and highest resistances, respectively, for each width. The lowest resistance obtained for a 60 nm via is 150 Ω, and the lowest extrapolated resistance for a 30 nm via is 295 Ω. Based on the log–log plot, the statistical average R_{via} for 60 nm vias is 1.7 kΩ with standard deviation between 420 Ω and 7.1 kΩ. As a comparison, the inset shows the arithmetical mean (data point) and standard deviation (vertical bar) for each via width. Reprinted with permission from Ref. [93], Copyright 2015, IEEE.

To examine the contributions of the contact resistance R_C and the intrinsic CNT resistance R_{CNT}, we have computed the CNT areal density by directly counting the numbers of CNTs inside a large number of vias of different widths, yielding an average of approximately $2 \times 10^{11}/cm^2$. Using R_{ci} = 300 Ω [129], R_C is estimated to be 42 Ω for a 60 nm via. By including R_M = 9 Ω and the best measured R_{via} value of 150 Ω for a 60 nm via, R_{CNT} = 150 − 42 − 9 = 99 Ω for the 60 nm via. Although there are considerable differences among literature reports in fabrication processes, measurement techniques, and data analysis methods, our best result of 150 Ω for a 60 nm via represents the lowest reported measured resistance for sub-100 nm CNT vias [29, 63, 100, 127, 130–132] to date, albeit at a relatively high CNT growth temperature.

The effect of top-contact metallization on R_{via} is examined by conducting I–V measurements on two sets of twenty-each 60 nm CNT vias, with and without metallized top Pt contacts, as it is the smallest width on sample A comparable to the linewidth in the current technology node size [92]. The measured via resistances on a cumulative frequency distribution (CFD) plot are shown in Fig. 3.16.

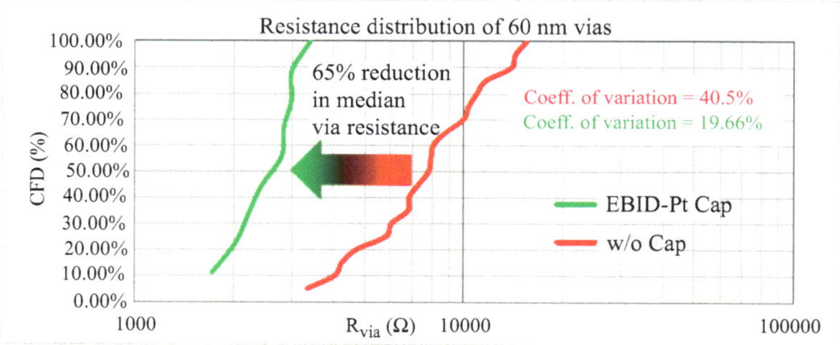

Figure 3.16 Cumulative frequency distribution plot for 60 nm CNT via resistance with and without EBID-Pt top-contact metallization [94].

The median R_{via} value and coefficient of variation (CV) for each set reveal a 65% decrease and a 20% decrease, respectively, confirming significant improvement of via resistance by top-contact metallization. Top-contact metallization serves to reduce the contribution of nanoprobe–via contact to the overall via resistance as suggested by the CFD plot in Fig. 3.16. Since the analysis of data shown in Fig. 3.15 is consistent with the assumption that the dependence of N_{CNT}, A_C, and ρ_{CNT} on w is not significant, the improvement in via resistance shown in Fig. 3.16 lends further support for the conclusion that variations in nanoprobe–via contacts among vias are principally responsible for the large scatter in measured via resistance data. The other component of R_C, from the via base contact between Cr and CNTs inside via formed during CNT growth, is not expected to have significant variations among vias due to uniformity of CNT growth. Another source for the data spread is from height variations among vias, which can be attributed to nonuniformity in the polishing process and/or imperfect CNT alignments inside via. Nevertheless, the data shown in Figs. 3.15 and 3.16 provide strong evidence to support the finding that improved via top contact enhances the overall CNT via performance.

For Sample B, I–V measurements are carried out on 40 nm CNT vias along the wedge, as summarized in Fig. 3.17, with identical techniques used for Sample A. A typical CNT via I–V plot is shown in Fig. 3.17d. Linearity of the I–V behavior confirms ohmic transport suggested by TEM and Raman analyses presented previously. Resistances of at least 10 vias for each of the five different via heights are obtained using the technique illustrated in the schematic in Fig. 3.17f.

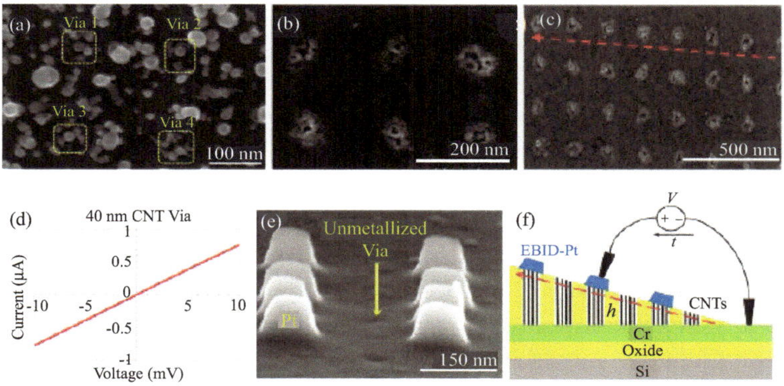

Figure 3.17 (a) Top-view SEM images of 40 nm vias after CNT growth. (b) CNT vias of 40 nm after Al_2O_3 filling and polishing. (c) Top-view of wedge formed after polishing and before EBID-Pt deposition. The dashed arrow indicates the direction of the wedge from low to high. (d) Typical measured I–V behavior that yields via resistance R_{via} before EBID-Pt top-contact metallization. (e) Glancing view of CNT vias after EBID-Pt top-contact metallization. Alternate vias are selectively metallized in order to avoid overlaps. (f) Cross-sectional schematic of wedge with nanoprobing setup showing varying via height h used for contact resistance extraction. The dashed arrow indicates the direction of the wedge as in (c). Reprinted from Ref. [104], Copyright 2016, with permission from IOP Publishing Ltd.

The range of via heights is selected such that the shortest height is at least half the maximum via height in order to avoid accidental shorting of the top and bottom electrodes. Since the via height range is 48–80 nm, and the electron mean free path in as-grown CNTs estimated from analysis of Raman spectra is about 5 nm, the assumption of ohmic transport for the electrons is still valid. Thus, each via resistance can be expressed as

$$R_{via} = R_M + R_C + \frac{\rho_{CNT} \times h}{A_{CNT}}. \quad (3.11)$$

As in the analysis of Sample A data, R_M represents the total resistance contributions from the Cr underlayer, the nanoprobes, and the probe–Cr contacts, and its value is estimated by landing both nanoprobes on the exposed Cr area, as described previously. As before, R_M is negligible compared to the lowest measured R_{via}. R_C is the total contact resistance of the via and varies among the vias according to the number and diameter of the CNTs inside each via, as well as the nanoprobe–via contact. A_{CNT} is the total CNT cross-sectional area inside via and also varies among the vias.

The results of via resistance versus via height are shown in Fig. 3.18. The data spread for each via height is likely due to the variations of R_C and A_{CNT} as discussed above. The via height h is fixed among the vias of the same height using the wedge structure shown in Fig. 3.17f, assuming uniformity in the wedge-formation process. The CNT resistivity ρ_{CNT} depends on the number of defects in the graphitic structure and the number of walls. Relative to the variations of A_{CNT} and R_C among the vias, and considering the small range of via heights used, the variation of ρ_{CNT} is expected to be small, and an average value for the sample is assumed in the analysis. Thus, using Eq. 3.11, the intercept of the statistical linear fit to the R_{via} versus h data amounts to an average R_C that takes into account all variations among the vias. The fitted straight line in Fig. 3.18a yields a slope of 142.5 Ω/nm and 61 Ω/nm for CNT vias without and with metallized top contacts. The coefficient of determination R^2 exceeds 0.9 in either case, supporting the assumption that electron transport through the entire CNT via structure is indeed ohmic, with an average CNT linear resistivity ρ_{CNT}/A_{CNT} = 142.5 Ω/nm and 61 Ω/nm for unmetallized and metallized CNT vias, respectively. The mean R_{via} and standard deviation for each height are given in Fig. 3.18b.

Therefore, despite significant variations among vias, such statistical analysis can be employed to extract an average via contact resistance and an average CNT resistivity, with R_C = 4.1 kΩ and ρ_{CNT} = 4.3 mΩ·cm for unmetallized CNT vias, and R_C = 2.8 kΩ and ρ_{CNT} = 1.8 mΩ·cm for the metallized ones, showing a significant improvement in contact resistance and resistivity by metallizing the via's top contact. For this calculation, an average total area of CNT (A_{CNT}) of approximately 300 nm^2 is obtained from SEM image analysis of eight unmetallized 40 nm vias in the vicinity of the ones which undergo electrical measurements. A sample set of vias used in such image analysis is shown in Fig. 3.19. The resistivity value extracted for either data set is about an order of magnitude lower than the ones obtained for the lower CNT growth temperature [133], but still more than two orders of magnitude higher than those of Cu and W with similar linewidths [133].

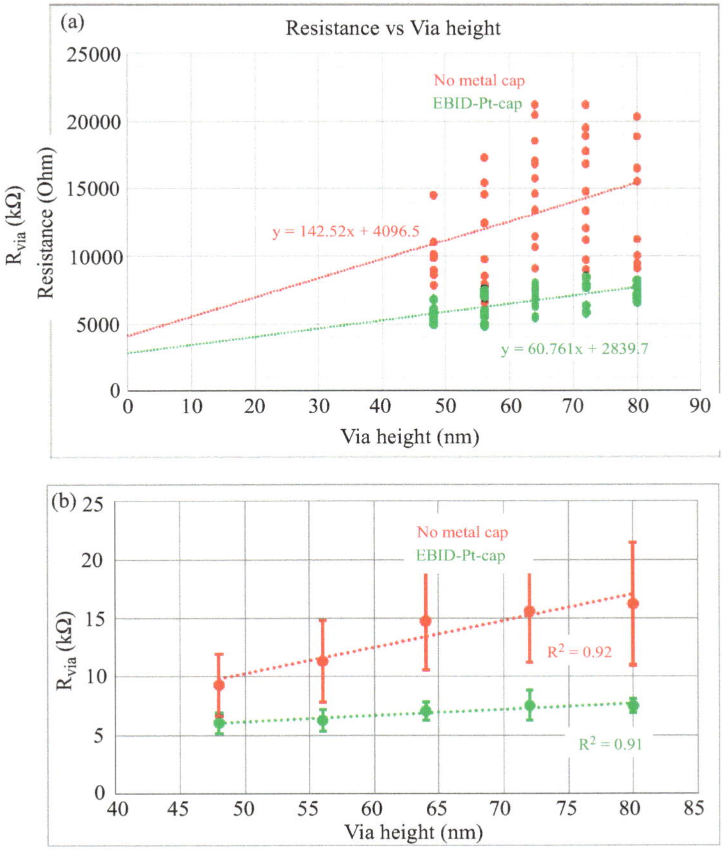

Figure 3.18 (a) Measured via resistance R_{via} versus via height for 40 nm CNT vias with and without metallized top contacts. Statistical linear fit yields an average contact resistance R_C from the R_{via} intercept. (b) Mean and standard deviation of measured resistances for each via height with $R^2 > 0.9$ for both data sets [94].

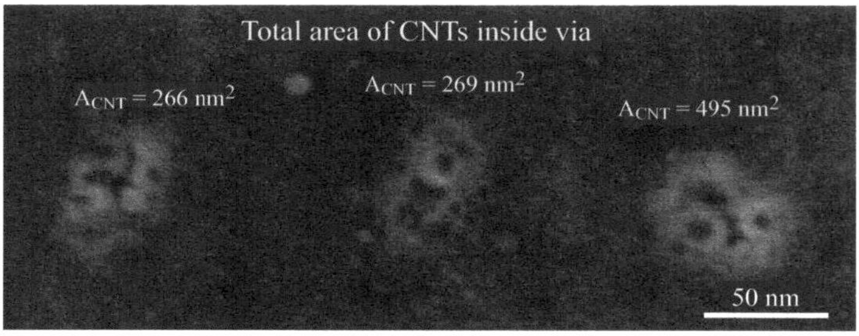

Figure 3.19 SEM images of vias used to estimate average total area of CNTs inside 40 nm via [94].

To examine the variations among the vias further, a statistical analysis is performed on the variations in the diameter and number of CNTs inside each via, as well as the resulting variations in the extracted contact resistance. The diameters and number of CNTs making contacts with the top and bottom electrodes define the total conducting cross-sectional area in a via, and variations in that area among the vias manifest themselves in variations in the measured R_{via} and extracted R_C. For the set of vias used in this experiment, the mean (μ), standard deviation (σ), and coefficient of variation (CV = σ/μ) for each of these parameters are summarized in Table 3.3.

Table 3.3 Statistics of 40 nm CNT via parameters, which contribute to variations in measured via resistance and extracted contact resistance

Parameter in a set of 40 nm CNT vias	μ	σ	CV (%)
CNT diameter across the entire set (nm)	9.03	3.62	40.1
Number of CNTs in via, N_{CNT}	16	2	12.5
Total area of CNTs in via, A_{CNT} (nm^2)	310	101	32.6
Extracted R_C (kΩ) for metallized CNT vias from Fig. 3.18a	2.84	0.8	28.6
Extracted R_C (kΩ) for unmetallized CNT vias from Fig. 3.18a	4.1	4.05	98.7

Note: For the diameter and the number of CNTs, the data are obtained from SEM images of unfilled and unpolished vias such as those in Fig. 3.17a. The total CNT area in via is obtained from SEM images of polished samples shown in Fig. 3.17b. For each parameter, the mean μ, standard deviation σ, and coefficient of variation CV are obtained, showing similarities in variations CV for A_{CNT} and R_C in metallized vias.

For CNT vias exhibiting ohmic conduction described by Eq. 3.11, both contact resistance R_C and CNT resistance R_{CNT} are inversely proportional to A_{CNT}. Thus, one would expect the extracted R_C to have variations similar to those for A_{CNT}. However, for the CNT via with unmetallized top contacts, R_C has an additional variation from $R_{probe\text{-}CNT}$, which as we have demonstrated for 60 nm vias (see Fig. 3.16), is principally responsible for the resistance variations among vias of the same geometry. If we were to use the average CNT diameter and average number of CNTs inside via given in Table 3.3 to deduce the average A_{CNT}, it would be approximately 1000 nm^2. This value is clearly an overestimation, since the SEM image of the unpolished vias in Fig. 3.17a reveals CNTs with lengths lower than the via heights, which do not contribute to conduction. On the other hand, the presence of oxide surrounding the vias as well as the exposed dielectric filler material in the polished sample results in insufficient contrast for accurate individual CNT diameter and number determination, but probably adequate for total area estimation. The resulting A_{CNT} statistics given in Table 3.3 correlate well with those for R_C for metallized CNT vias, thus supporting the analysis used to extract the latter. Despite the fact that R_C varies inversely as A_{CNT}, in the range of A_{CNT} obtained for 40 nm vias, the magnitudes of R_{CNT} and A_{CNT} variations are approximately linearly related. Thus, correlation of CV data between these two parameters given in Table 3.3 can be justified.

3.4.4 Current-Carrying Capacity Experiment

Electronic transport in CNT can be further understood by examining its current-carrying capacity J_{MAX}. At the same time, J_{MAX} serves to quantify the reliability of CNT vias. As long as J_{MAX} remains higher than the required current density for interconnects, they can operate without degradation in performance. We have recently demonstrated [42, 104] the advantage of growing CNTs at temperatures higher than those considered to be acceptable for IC manufacturing [92], where the resulting high CNT quality leads to lower resistivity and higher current-carrying capacity [134]. To obtain J_{MAX} for CNT vias, two complementary methods—current-stress test and voltage-sweep test—are employed.

For the first method, a constant current is applied to the via in incremental steps and the via resistance is recorded after each stress cycle until device breakdown. The resistance versus stress current behaviors for 40 nm and 60 nm CNT vias with and without metallized top contacts are shown in Fig. 3.20. It is observed that in either case, the vias with Pt top contacts can carry much higher current than those without, and with much lower resistance prior to breakdown. The Pt contact allows all the exposed CNT tips and the graphene shells inside each CNT to interface with the deposited metal. For vias without the metallized top contact, the via resistance is very sensitive to variations in the probe–via contact, which generally does not cover all the CNTs inside the via, thereby resulting in higher contact resistance and increased Joule heating from current stressing. Furthermore, heat dissipation from the CNTs under current stressing is less efficient without the top-contact metal, so the via breaks down at a lower current density. An additional benefit of metallized top contact is a further improvement in the top and base contacts by Joule heat from current stressing, which facilitates fusion of nanocrystalline grains in the metal, as observed by Wilhite [68, 135] for EBID-W contacts and were reported by Wang [136] for EBID-Pt nanocrystals. Thus, current stressing provides local annealing of individual vias instead of having to thermally anneal the entire sample.

Figure 3.20a shows the via resistance behavior versus stress current for two 40 nm vias with EBID-Pt metallization, each with breakdown at 1 mA. Using A_{CNT} of approximately 300 nm^2 obtained for 40 nm vias, J_{MAX} is approximately 330 MA/cm^2. For the 60 nm via with EBID-Pt metallization in Fig. 3.20b, with A_{CNT} of approximately 700 nm^2, J_{MAX} is approximately 170 MA/cm^2. These via current-carrying capacities are more than two orders of magnitude higher than those for Cu and W with similar dimensions and are consistent with reported values [133]. The improvements in both via resistance and current-carrying capacity are evident for each linewidth. The breakdown in CNTs tends to occur at defect sites and can be exacerbated by ambient oxygen adsorption. Therefore, the improvement in long-range order in CNTs by reducing defects and using non-oxide-based dielectric fillers can result in further enhancement in J_{MAX} as well as a reduction in via resistance. Nonetheless, the J_{MAX} values obtained in this experiment and reported by others [5, 96, 134, 137] are more than sufficient for functionalized on-chip CNT interconnects to last through to the end of semiconductor technology nodes, IF the line resistance can be reduced to approach its Cu and W counterparts.

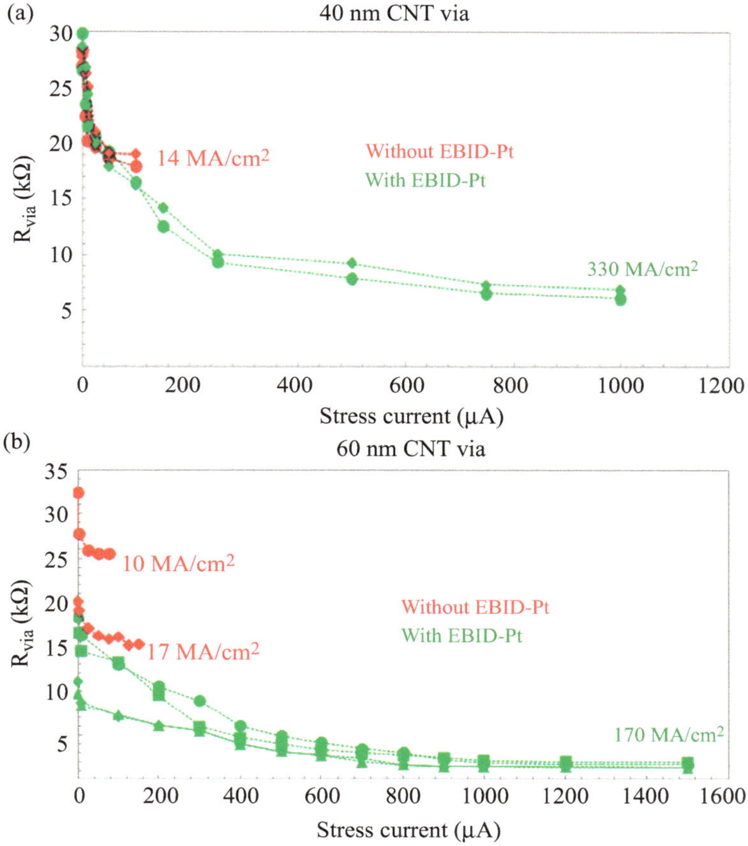

Figure 3.20 (a) Via resistance versus stress current behaviors for two 40 nm vias with and without EBID-Pt top contacts. Metallization and current stressing improve and stabilize top via contact by making conformal interfaces between metal contact and the exposed CNT tips, resulting in lower via resistance, and improve the current-carrying capacity compared to unmetallized vias. (b) Resistance versus stress current behavior for three 60 nm vias with and without EBID-Pt top contacts, showing the same trend as the 40 nm vias [94].

To verify independently the current-carrying capacity of the CNT vias obtained with the current-stressing experiment described above, a voltage-sweeping experiment is performed. The voltage across each via is swept between a predefined or compliance range of currents. If a continuous linear I–V is obtained, it implies that the CNTs inside the via remain intact. An array of Pt-metallized vias after the voltage sweep is shown in Fig. 3.21a. At the outset, a current compliance range of ±1 mA is used with a voltage sweep from −4 V to 4 V. As shown in Fig. 3.21b, continuous I–V behavior is obtained for this compliance range. The via meeting this compliance of ±1 mA remains intact, while those with higher compliance ranges (±1.25 mA; ±1.5 mA; ±2 mA) break down, as shown in Fig. 3.21a. Thus, the current-carrying capacity shown in Fig. 3.20a is verified.

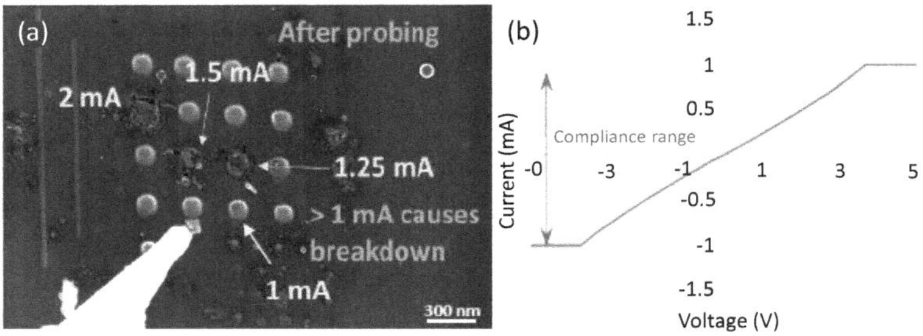

Figure 3.21 (a) SEM image of 40 nm CNT vias after voltage-sweep test. Different current ranges are used to reveal via current-carrying capacity. Only the device with current compliance of ±1 mA remains intact and those with higher compliances (±1.25 mA; ±1.5 mA; ±2 mA) break down, confirming the measured current-carrying capacity of 330 MA/cm^2. (b) The I–V behavior with current capped at ±1 mA [94].

3.4.5 Comparison of CNT Vias with Cu and W Vias

A comparison of R_{via} and J_{MAX} of CNT, Cu, and W vias is given in Table 3.4, showing how CNT areal density correlates with via current-carrying capacity. However, in order to achieve higher areal density, smaller CNT diameters are needed to pack more of them into each via, which could impact the via resistance. Also defects in CNTs can reduce the via current-carrying capacity. Even for 700°C CNT growth, the defect ratio (I_D/I_G) from Raman analysis is 0.85. Therefore, taller vias are susceptible to reduced current-carrying capacity as more defects are present along the CNT length, consistent with the data obtained in this work and by others.

Table 3.4 Comparison of reported CNT vias results and with those for Cu and W vias of similar dimensions

References	CNT growth temperature (°C)	Via dimensions (width × height)	Via resistance (Ω)	CNT resistivity (mΩ·cm)	Contact resistance (Ω)	CNT areal density (#/cm^2)	Current-carrying capacity (MA/cm^2)
[42, 93, 94, 104,	700	40 nm × 80 nm	7500	1.8	2800	~10^{12}	330
138]	700	60 nm × 130 nm	2795	NA	NA	2×10^{11}	170
[5, 97, 132, 139]	700	30 nm × 150 nm	7800	NA	NA	~10^{11}	400
[130, 137, 140]	400	70 nm × 100 nm	11000	NA	NA	5×10^{11}	100
[131]	400	160 nm × 120 nm	63	NA	NA	3×10^{11}	5
[141]	425	100 nm × 190 nm	1400	NA	NA	~10^{11}	NA
[142]	700	100 nm × 4000 nm	9000	0.44	NA	8×10^{8}	7
[63]	600	100 nm × 200 nm	NA	NA	3100 (Pt)	4×10^{11}	NA
[6]	750	100 nm × 2000 nm	300	NA	NA	1.6×10^{9}	10000
[29, 143–145]	600	1000 nm × 5500 nm	NA	NA	6	2×10^{11}	NA
[146]	540	150 nm × 400 nm	2200	5.1	500	~10^{11}	NA
[55–57]	470	300 nm × 450 nm	9500	3.7	1160	2×10^{11}	NA
[98, 99, 127, 147]	350	2000 nm × 3000 nm	25	10	8	×	NA
Copper [133]	NA	30 nm × 130 nm	25	0.007	15	NA	2.5
Tungsten [133]	NA	30 nm × 130 nm	60	0.038	5	NA	1

As in horizontal CNT interconnects discussed earlier, dissipation of Joule heat from current stressing is less efficient for a taller via, contributing to variations in J_{MAX} among vias with different heights. Katagiri [140] reported a resistance of 11 kΩ and a current-carrying capacity of 100 MA/cm^2 for a 70 nm wide via with CNT growth at 450°C, while an extensive study by Vollebregt [98, 127] conclusively demonstrated that the number of defects in CNTs increases as the growth temperature was reduced. Graham [96, 97] reported that 700°C CNT growth and subsequent annealing resulted in a current-carrying capacity of 400 MA/cm^2 for a 30 nm via with a CNT areal density of 1.5×10^{11} cm^{-2}. This higher J_{MAX} is probably due to a multi-walled structure with a hollow interior, as evident in the cross-sectional TEM image of a CNT in Ref. [97], whereas the CNTs in our vias mostly have a "bamboo" structure, where interlayer defects can lower the J_{MAX}. In all cases, while the current-carrying capacities well exceed those projected for end-of-roadmap technology nodes, via resistance to approach those of Cu and W remains a daunting challenge, though our results suggest that further innovations in contact engineering and improving the long-range order could begin to overcome such challenge. One such example of contact engineering, using an all-carbon nanostructure, is presented next.

3.5 Electronic Transport in CNT–Graphene Heterostructure

As both graphene and CNT are nanocarbon materials, one a 2D sheet and the other a 1D nanowire, their structures can complement each other in some potential applications. For example, although both have excellent electrical conductivity in their respective in-plane directions, the conductivity significantly drops for out-of-plane electronic transport [148]. This limits their use in applications that require optimal transport in all directions, such as interconnects in integrated circuits [148]. One way to mitigate this would be to stitch the CNTs and the graphene sheet together in a 3D all-carbon heterostructure [149], to exploit the transport properties of both materials. Since both graphene and CNT have the same honeycomb structure, a seamless contact between them might be realizable [150]. The resulting heterostructure can then exhibit excellent conduction in both horizontal and vertical directions, as well as minimal contact resistance. Such nanostructure can then be the building block of an all-carbon interconnect network in either a silicon-based or carbon-based IC.

Ideally, the CNT–graphene contact is a logical means to achieve the best isotropic electrical conductivity. However, forming a seamless contact in practice, experimentally or even theoretically, is a daunting challenge. Here, we briefly review both approaches in attempting to form such a contact between a vertical CNT and a horizontal graphene layer.

3.5.1 Vertical CNT–Graphene Heterostructure

In general, CVD is the most common method for synthesizing vertical CNT–graphene heterostructures [151], using a two-step process [150]. First, graphene is grown on a metal substrate and subsequently transferred onto a target substrate if needed. Second, the vertical CNT–graphene heterostructure is synthesized after catalyst deposition and introduction of a carbon source gas. There are many factors such as catalysts, temperature, and gases that affect the growth and the characteristics of the resulting heterostructure, as discussed in detail by Du [148].

It is expected that a seamless heterostructure can provide superior electronic and thermal transport properties through the CNT–graphene junction as well as improved mechanical stability. However, there is still little evidence that the fabricated vertical structure possesses a seamless connection. Several proposed connection topologies between CNT and graphene are presented here along with first-principle calculations. However, high-resolution transmission electron microscopy is needed to show experimentally how the carbon atoms are connected at the CNT–graphene junction. We hope that with more advanced characterization techniques, the atomic arrangement at the junction can be identified and provide an experimentally confirmed structure for theoretical calculations.

3.5.2 Geometry Definition and Contact Optimization

Conventional TCAD models fail in studying the behavior of the contact as the atomistic details of the interface are not taken into account. Hence, first-principle simulations alongside molecular dynamics concepts are needed to gain a better understanding of the heterostructure contact. To form a seamless CNT–graphene heterostructure, a least-square method is utilized to achieve C–C bond lengths or bond angles as close as possible to those of the ideal case [152, 153]. Moreover, Euler's theorem is utilized to select the polygons for the contact stitching process [152–157]. Figure 3.22a shows the possible connections that the eight open bonds of a (4,4) CNT or (8,0) CNT can form with the underlying graphene sheet [152]. Many theoretical calculations have adopted the same rules to form seamless CNT–graphene heterostructures [158, 159]. After identifying the bond contact spots on the graphene surface and the CNT, molecular dynamic simulations are performed to minimize the total binding energy of the heterostructure. For structure relaxation, a full quantum mechanical optimization, including force-field relaxation of the nuclei as well as the electrons, is necessary. However, such optimization requires prohibitive amount of time and computational resources [160]. Therefore, another approach is employed to achieve force-field convergence using classical molecular dynamics approach, which neglects electron–electron interactions [150, 161, 162]. The advantage of such an approach is drastically reduced computational time, albeit with less accuracy of the final optimized structure.

3.5.3 Electronic Transport Properties

With the optimized heterostructure, one can perform first-principle calculations to obtain electronic properties such as band structure, transmission coefficient, DOS, and conductance. Matsumoto [163] used a tight-binding method to study various (6,6) CNT–graphene heterostructures, including CNT with open tip or capped, and CNT sandwiched between two graphene layers similar to that in Fig. 3.22b. The total energy minimization method is adopted to optimize the geometries using the tight-binding method. Although the (6,6) CNT is metallic, sizable direct bandgaps of 0.27 eV and

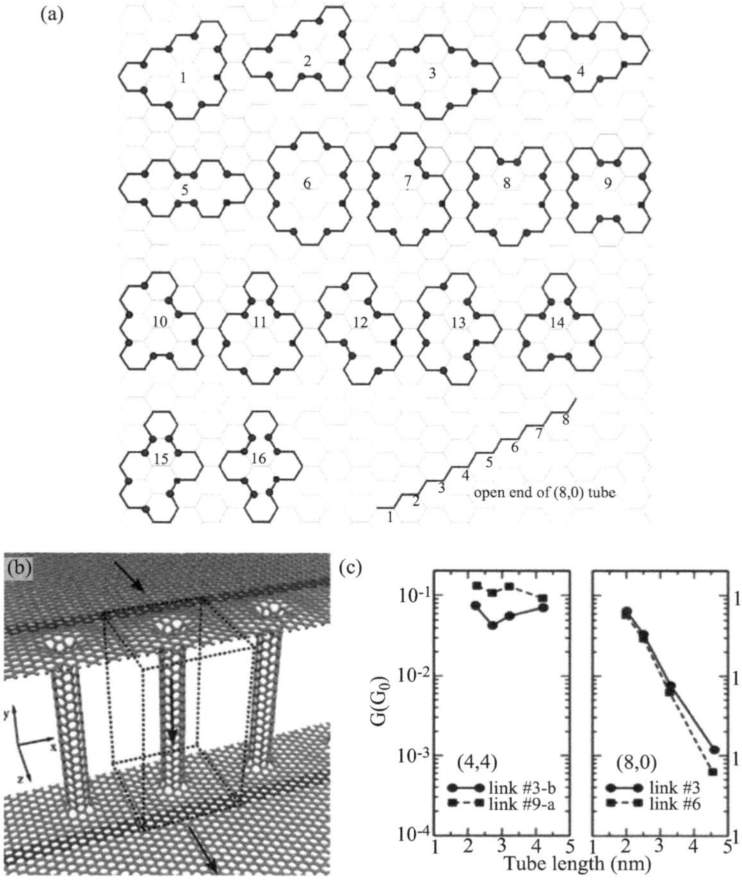

Figure 3.22 (a) Possible ways to seamlessly connecting (4,4) or (8,0) CNT with graphene. (b) Graphene–CNT–graphene heterostructure used to calculate the transmission coefficient, with the arrows showing the electron transport directions. (c) Deduced conductance of the heterostructure with different CNT type and tube length. Figure (a) reprinted from Ref. [152], Copyright 2007, with permission from Elsevier, and Figs. (b) and (c) reprinted with permission from Ref. [149], Copyright 2010, by the American Physical Society.

0.51 eV were predicted for the open and capped CNT–graphene heterostructures, respectively. An even larger bandgap was predicted for the sandwiched heterostructure. Another interesting structure proposed by Mao [164] also showed a similar effect that the metallic (5,5) CNT was transformed into a semiconductor with a bandgap of 0.2 eV. Strictly speaking, this structure is not a seamless CNT–graphene heterostructure, because the CNT is inserted into the graphene and the two ends of the CNT are connected to the graphene sheets through a hole on each. Nevertheless, a covalent bond is formed between each atom at the graphene hole edge and an atom on the CNT sidewall. This strong coupling results in a bandgap in the metallic (5,5) CNT. In contrast, the original bandgap of 0.65 eV vanishes for a semiconducting (8,0) CNT, because of the induced impurity states by the sp^3-like hybridization between the CNT and the holed graphene. Thus, one may conclude from the above theoretical study that the pristine CNT bandgap can be changed due to the strong covalent bond formed at the CNT–graphene interface.

Since the tight-binding method could not capture the junction-induced band offset between CNT and graphene [163], a first-principle calculation was performed to study the electronic transport properties of the (4,4) and (8,0) CNT–graphene heterostructures [148]. For the metallic (4,4) CNT–graphene heterostructures, two kinds of symmetrical connections containing six heptagonal rings at the interface (No. 3 and No. 9 in Fig. 3.22a) were adopted. The unit cell of the periodic 3D seamless heterostructure shown in Fig. 3.22b was constructed for electronic transport calculations using the nonequilibrium Green's function (NEGF) method, and the current flow is through the CNT–graphene junction and the CNT itself. The calculated transmission coefficient was between 0.01 and 1 and shows a weak dependence on the CNT length in the range of 2.2–4.2 nm, indicating a clear ballistic transport characteristic of the (4,4) metallic CNT–graphene heterostructure. The conductance deduced from the transmission curves also shows a similar weak dependence on CNT length for metallic CNTs, as shown in Fig. 3.22c. On the other hand, the conductance shows strong dependence on the contact structure, with the No. 9 contact structure exhibiting a higher transmission and conductance than the No. 3 case. In contrast, the conductance of the semiconducting (8,0) CNT–graphene heterostructure shows a strong dependence on the CNT length and weak dependence on the contact structure. Another interesting point for the (8,0) CNT–graphene heterostructure is that a relatively large conductance is predicted for the heterostructure with a small CNT length of 2 nm (Fig. 3.22c), showing the effect of tunneling. Although the study reveals some interesting electronic transport properties of the CNT–graphene heterostructure, their calculation cannot ascertain the exact contribution of the CNT–graphene junction to the total conductance.

To better understand the CNT–graphene contact properties, we have performed calculations on a two-point structure with the NEGF method. To extract the CNT–graphene contact resistance, graphene resistance, and CNT resistance, we calculate the I–V characteristics of the graphene–CNT–graphene heterostructure shown in Fig. 3.23a. Toward this end, we first compute the resistance of a graphene sheet for different lengths, which turns out to be 6.45 kΩ and independent of length, confirming ballistic transport. This result also serves as a validation of the calculation method. A typical I–V curve for the complete two-point structure is shown in Fig. 3.23b. The total resistance is found to be 91.5 kΩ for the heterostructure with a 2.44 nm long (8,0) CNT. The linear I–V behavior indicates ohmic conduction across the CNT–graphene junction and possibly along the CNT as well. Furthermore, we have also verified the previous study that semiconducting CNTs when contacted with graphene lead to metallic behavior. The DOS and the transmission coefficients of the (8,0) CNT–graphene heterostructure are shown in Figs. 3.23c and d, respectively. A finite DOS exists at the Fermi level (located at 0 eV), representative of the metallic nature of the CNT. The transmission coefficient shown in Fig. 3.23d also suggests that transmission indeed occurs at the Fermi level because of the available states.

Based on findings reported thus far, engineering the CNT–graphene interface remains a major challenge in forming a low-resistance heterostructure, as in CNT-metal contact in functionalizing CNT vias. Although the interface between CNTs and graphene can be refined to yield contact resistivity as low as $\sim 10^{-5}\ \Omega\cdot cm^2$ [165], this value is still high for acceptable chip operation. Further, CNTs must be grown on graphene at temperatures compatible to chip manufacturing, such as 550°C [166], 510°C [167], and 400°C [168], where contact resistance remains the critical challenge in its implementation.

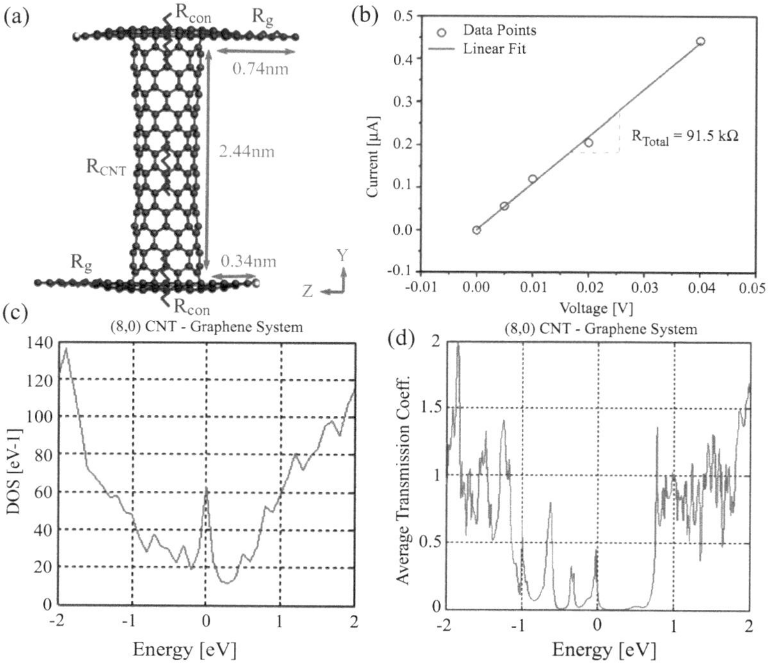

Figure 3.23 (a) Graphene–CNT–graphene heterostructure used for I–V calculations. (b) I–V curve of the graphene–CNT–graphene heterostructure. (c) DOS and (d) transmission coefficients of the (8,0) CNT–graphene system, showing finite DOS and transmission at the Fermi level (0 eV).

3.6 Summary and Conclusion

The electronic structures of graphene and CNT are briefly reviewed to lay the groundwork for understanding the electrical transport within these two nanocarbons in various configurations for potential on-chip interconnect applications. While conduction along the graphitic plane (1D in metallic CNT and 2D in graphene) is superior to conventional metals, high resistance across nanocarbon–metal contacts and across CNT–graphene interface remains a daunting challenge, both experimentally and theoretically. For CNT interconnect vias, the measured current-carrying capacities are generally two orders of magnitudes higher than those of current interconnect metals, Cu and W, while the measured via resistance is two orders higher. Such high resistance is attributed to large contact resistance, though imperfections in CNT itself could also play a role.

In principle, a 3D all-carbon heterostructure made up of CNT standing on a graphene layer would mitigate the contact resistance problem. In practice, the growth of vertically aligned CNTs on graphene without damaging the latter is a nontrivial task, and even simulating a seamless contact for first-principle calculations to shed some light on transport across the interface remains a challenge. Most of the CNT–graphene heterostructures have been synthesized by CVD methods, which are usually adapted from the CNT growth recipes with careful control of catalyst deposition and catalyst–substrate interactions. Thus far, the experimental findings, in conjunction with atomistic models used in theoretical calculations, are still a long way from being conclusive on the interfacial atomic arrangements. Therefore, novel techniques are needed to reveal detailed interfacial information. For example, the ratio of sp^2/sp^3 bonding can be extracted from the measured density of states, if an atomically clean CNT–graphene junction can be prepared for advanced TEM and STM analyses. Such an approach poses a great challenge in the experimental study of this heterostructure, while overcoming such challenge would yield enormous gain in understanding electronic transport in the heterostructure.

References

1. M. S. Dresselhaus, G. Dresselhaus, and P. C. Eklund, *Science of Fullerenes and Carbon Nanotubes: Their Properties and Applications* (New York: Academic), 1996.
2. M. S. Dresselhaus, G. Dresselhaus, and P. Avouris, *Carbon Nanotubes: Synthesis, Structure, Properties, and Applications* (Berlin: Springer), 2001.
3. S. Ijima, "Helical microtubules of graphitic carbon," *Nature,* **354**, pp. 56–58, 1991.
4. C. Dekker, "Carbon nanotubes as molecular quantum wires," *Physics Today,* **52**, pp. 22–28, 1999.
5. F. Kreupl, A. P. Graham, G. S. Duesberg, W. Steinhogl, M. Liebau, E. Unger, and W. Honlein, "Carbon nanotubes in interconnect applications," *Microelectronic Engineering,* **64**, pp. 399–408, 2002.

6. J. Li, Q. Ye, A. Cassell, H. T. Ng, R. Stevens, J. Han, and M. Meyyappan, "Bottom-up approach for carbon nanotube interconnects," *Applied Physics Letters,* **82**, pp. 2491–2493, 2003.
7. M. Nihei, A. Kawabata, D. Kondo, M. Horibe, S. Sato, and Y. Awano, "Electrical properties of carbon nanotube bundles for future via interconnects," *Japanese Journal of Applied Physics,* **44**, p. 1626, 2005.
8. L. Zhang, D. Austin, V. I. Merkulov, A. V. Meleshko, K. L. Klein, M. A. Guillorn, D. H. Lowndes, and M. L. Simpson, "Four-probe charge transport measurements on individual vertically aligned carbon nanofibers," *Applied Physics Letters,* **84**, pp. 3972–3974, 2004.
9. A. V. Melechko, V. I. Merkulov, T. E. McKnight, M. A. Guillorn, K. L. Klein, D. H. Lowndes, and M. L. Simpson, "Vertically aligned carbon nanofibers and related structures: controlled synthesis and directed assembly," *Journal of Applied Physics,* **97**, p. 041301, 2005.
10. Q. Ngo, T. Yamada, M. Suzuki, Y. Ominami, A. M. Cassell, L. Jun, M. Meyyappan, and C. Y. Yang, "Structural and electrical characterization of carbon nanofibers for interconnect via applications," *IEEE Transactions on Nanotechnology,* **6**, pp. 688–695, 2007.
11. A. K. Geim and K. S. Novoselov, "The rise of graphene," *Nature Materials,* **6**, pp. 183–191, 2007.
12. J. S. Moon, D. Curtis, M. Hu, D. Wong, C. Mcguire, P. Campbell, G. G. Jernigan, J. L. Tedesco, B. L. VanMil, R. L. Myers-Ward, C. R. Eddy, and D. K. Gaskill, "Epitaxial-graphene RF field-effect transistors on Si-face 6H-SiC substrates," *IEEE Electron Device Letters,* **30**, pp. 650–652, 2009.
13. S. Bae, S., H. Kim, Y. Lee, X. Xu, J.-S. Park, Y. Zheng, J. Balakrishnan, T. Lei, H. R. Kim, Y. I. Song, Y.-J. Kim, K. S. Kim, B. Özyilmaz, J.-H. Ahn, B. H. Hong, and S. Iijima, "Roll-to-roll production of 30-inch graphene films for transparent electrodes," *Nature Nanotechnology,* **5**, pp. 574–578, 2010.
14. J. S. Moon, D. Curtis, S. Bui, T. Marshall, D. Wheeler, I. Valles, S. Kim, E. Wang, X. Weng, and M. Fanton, "Top-gated graphene field-effect transistors using graphene on Si (111) wafers," *IEEE Electron Device Letters,* **31**, pp. 1193–1195, 2010.
15. A. Naeemi and J. D. Meindl, "Carbon nanotube interconnects," *Annual Review of Materials Research,* **39**, pp. 255–275, 2009.
16. E. Pop, D. Mann, J. Cao, Q. Wang, K. Goodson, and H. Dai, "Negative differential conductance and hot phonons in suspended nanotube molecular wires," *Physical Review Letters,* **95**, p. 155505, 2005.

17. H. Li, C. Xu, and K. Banerjee, "Carbon nanomaterials: the ideal interconnect technology for next-generation ICs," *IEEE Design & Test of Computers*, **27**, pp. 20–31, 2010.

18. M. F. L. De Volder, S. H. Tawfick, R. H. Baughman, and A. J. Hart, "Carbon nanotubes: present and future commercial applications," *Science*, **339**, pp. 535–539, 2013.

19. H. Kitsuki, T. Yamada, D. Fabris, J. R. Jameson, P. Wilhite, M. Suzuki, and C. Y. Yang, "Length dependence of current-induced breakdown in carbon nanofiber interconnects," *Applied Physics Letters*, **92**, p. 173110, 2008.

20. T. Saito, T. Yamada, D. Fabris, H. Kitsuki, P. Wilhite, M. Suzuki, and C. Y. Yang, "Improved contact for thermal and electrical transport in carbon nanofiber interconnects," *Applied Physics Letters*, **93**, p. 102108, 2008.

21. T. Yamada, T. Saito, D. Fabris, and C. Y. Yang, "Electrothermal analysis of breakdown in carbon nanofiber interconnects," *IEEE Electron Device Letters*, **30**, pp. 469–471, 2009.

22. M. Suzuki, T. Yamada, and C. Y. Yang, "Monte Carlo simulation of scanning electron microscopy bright contrast images of suspended carbon nanofibers," *Applied Physics Letters*, **90**, p. 083111, 2007.

23. T. Yamada, F. R. Madriz, and C. Y. Yang, "Inductance in one-dimensional nanostructures," *IEEE Transactions on Electron Devices*, **56**, pp. 1834–1839, 2009.

24. T. Yamada, H. Yabutani, T. Saito, and C. Y. Yang, "Temperature dependence of carbon nanofiber resistance," *Nanotechnology*, **21**, p. 265707, 2010.

25. ITRS 2011 International technology roadmap for semiconductors, www.itrs.net/Links/2012ITRS/ Home2012.htm (with 2012 updates).

26. T. Wang, K. Jeppson, N. Olofsson, E. E. Campbell, and J. Liu, "Through silicon vias filled with planarized carbon nanotube bundles," *Nanotechnology*, **20**, p. 485203, 2009.

27. J. U. Knickerbocker, C. S. Patel, P. S. Andry, C. K. Tsang, L. P. Buchwalter, E. J. Sprogis, H. Gan, R. R. Horton, R. J. Polastre, S. L. Wright, and J. M. Cotte, "3-D silicon integration and silicon packaging technology using silicon through-vias," *IEEE Journal of Solid-State Circuits*, **41**, pp. 1718–1725, 2006.

28. A. C. Neto, F. Guinea, N. M. Peres, K. S. Novoselov, and A. K. Geim, "The electronic properties of graphene," *Reviews of Modern Physics*, **81**, pp. 109–164, 2009.

29. S. Hermann, "Growth of carbon nanotubes on different support/catalyst systems for advanced interconnects in integrated circuits," Doctoral Thesis, Chemnitz University of Technology, 2011.

30. B. Partoens and F. M. Peeters, "From graphene to graphite: electronic structure around the K point," *Physical Review B*, **74**, p. 075404-4, 2006.

31. M. S. Dresselhaus, G. Dresselhaus, J. C. Charlier, and E. Hernandez, "Electronic, thermal and mechanical properties of carbon nanotubes," *Philosophical Transactions of the Royal Society A: Mathematical, Physical and Engineering Sciences,* **362**, pp. 2065–2098, 2004; R. Saito, G. Dresselhaus, and M. S. Dresselhaus, *Physical Properties of Carbon Nanotubes* (Imperial College Press), 1998.

32. J. J. Palacios, A. J. Pérez-Jiménez, E. Louis, E. SanFabián, and J. A. Vergés, "First-principles phase-coherent transport in metallic nanotubes with realistic contacts," *Physical Review Letters,* **90**, pp. 106801–106805, 2003.

33. R. A. Jishi, D. Inomata, K. Nakao, M. S. Dresselhaus, and G. Dresselhaus, "Electronic and lattice properties of carbon nanotubes," *Journal of the Physical Society of Japan,* **63**, pp. 2252–2260, 1994.

34. G. G. Samsonidze, A. R. Saito, D. A. Jorio, E. Pimenta, F. Souza, F. A. Grüneis, D. G. Dresselhaus, and M. Dresselhausa, "The concept of cutting lines in carbon nanotube science," *Journal of Nanoscience and Nanotechnology,* **3**, pp. 431–458, 2003.

35. K. Grove-Rasmussen, "Electronic transport in single wall carbon nanotubes, Ph.D. Thesis, University of Copenhagen, pp. 48–59, 2006.

36. J. W. Mintmire and C. T. White, "Universal density of states for carbon nanotubes," *Physical Review Letters,* **81**, pp. 2506–2509, 1998.

37. N. Wang, K. K. Fung, W. Y. Lu, and S. H. Yang, "Structural characterization of carbon nanotubes and nanoparticles by high-resolution electron-microscopy," *Chemical Physics Letters,* **229**, pp. 587–592, 1994.

38. M. Q. Liu and J. M. Cowley, "Growth-behavior and growth defects of carbon nanotubes," *Material Science and Engineering A,* **185**, pp. 131–140, 1994.

39. P. H. Tan, S. L. Zhang, K. T. Yue, F. M. Huang, Z. J. Shi, X. H. Zhou, and Z. N. Gu, "Comparative Raman study of carbon nanotubes prepared by DC arc discharge and catalytic methods," *Journal of Raman Spectroscopy,* **28**, pp. 369–372, 1997.

40. A. Naeemi and J. D. Meindl, "Compact physical models for multiwall carbon-nanotube interconnects," *IEEE Electron Device Letters,* **27**, pp. 338–340, 2006.

41. W. Wu, S. Krishnan, T. Yamada, X. Sun, P. Wilhite, R. Wu, K. Li, and C. Y. Yang, "Contact resistance in carbon nanostructure via interconnects," *Applied Physics Letters,* **94**, p. 163113, 2009.

42. A. Vyas, P. Wang, C. Zhou, M. Chan, and C. Y. Yang, "Contact resistance and reliability of 40 nm carbon nanotube vias," *IEEE International Interconnect Technology Conference,* pp. 203–205, 2016.

43. Y. H. Kim and H. S. Kim, "Anomalous length scaling of carbon nanotube-metal contact resistance: an ab initio study," *Applied Physics Letters,* **100**, p. 213113, 2012.

44. M. Karita, K. Asaka, H. Nakahara, and Y. Saito, "In situ TEM study on the improvement of contact resistance between a carbon nanotube and metal electrodes by local melting," *Surface and Interface Analysis,* **44**, pp. 674–677, 2012.

45. Y. A. Kim, S. Aoki, K. Fujisawa, Y. I. Ko, K. S. Yang, C. M. Yang, Y. C. Jung, T. Hayashi, M. Endo, M. Terrones, and M. S. Dresselhaus, "Defect-assisted heavily and substitutionally boron-doped thin multiwalled carbon nanotubes using high-temperature thermal diffusion," *Journal of Physical Chemistry C,* **118**, pp. 4454–4459, 2014.

46. Y. J. Wei, J. T. Wu, H. Q. Yin, X. H. Shi, R. G. Yang, and M. Dresselhaus, "The nature of strength enhancement and weakening by pentagon-heptagon defects in graphene," *Nature Materials,* **11**, pp. 759–763, 2012.

47. P. T. Araujo, M. Terrones, and M. S. Dresselhaus, "Defects and impurities in graphene-like materials," *Materials Today,* **15**, pp. 98–109, 2012.

48. M. Kalbac, Y. P. Hsieh, H. Farhat, L. Kavan, M. Hofmann, J. Kong, and M. S. Dresselhaus, "Defects in individual semiconducting single wall carbon nanotubes: Raman spectroscopic and in situ Raman spectroelectrochemical study," *Nano Letters,* **10**, pp. 4619–4626, 2010.

49. M. S. Dresselhaus, A. Jorio, A. G. Souza, and R. Saito, "Defect characterization in graphene and carbon nanotubes using Raman spectroscopy," *Philosophical Transactions of the Royal Society A: Mathematical Physical and Engineering Sciences,* **368**, pp. 5355–5377, 2010.

50. G. H. Zhu, H. Lee, Y. C. Lan, X. W. Wang, G. Joshi, D. Z. Wang, J. Yang, D. Vashaee, H. Guilbert, A. Pillitteri, M. S. Dresselhaus, G. Chen, and Z. F. Ren, "Increased phonon scattering by nanograins and point defects in nanostructured silicon with a low concentration of germanium," *Physical Review Letters,* **102**, p. 196803, 2009.

51. S. Vollebregt, A. N. Chiaramonti, R. Ishihara, H. Schellevis, and K. Beenakker, "Contact resistance of low-temperature carbon nanotube vertical interconnects," *12th IEEE Conference on Nanotechnology*, pp. 1–5, 2012.

52. M. H. van der Veen, B. Vereecke, C. Huyghebaert, D. J. Cott, M. Sugiura, Y. Kashiwagi, L. Teugels, R. Caluwaerts, N. Chiodarelli, P. M. Vereecken, G. P. Beyer, M. M. Heyns, S. De Gendt, and Z. Tokei, "Electrical characterization of CNT contacts with Cu Damascene top contact," *Microelectronic Engineering*, **106**, pp. 106–111, 2013.

53. N. Chiodarelli, A. Fournier, H. Okuno, and J. Dijon, "Carbon nanotubes horizontal interconnects with end-bonded contacts, diameters down to 50 nm and lengths up to 20 μm," *Carbon*, **60**, pp. 139–145, 2013.

54. N. Chiodarelli, A. Fournier, and J. Dijon, "Impact of the contact's geometry on the line resistivity of carbon nanotubes bundles for applications as horizontal interconnects," *Applied Physics Letters*, **103**, p. 053115, 2013.

55. N. Chiodarelli, S. Masahito, Y. Kashiwagi, Y. L. Li, K. Arstila, O. Richard, D. J. Cott, M. Heyns, S. De Gendt, G. Groeseneken, and P. M. Vereecken, "Measuring the electrical resistivity and contact resistance of vertical carbon nanotube bundles for application as interconnects," *Nanotechnology*, **22**, p. 085302, 2011.

56. N. Chiodarelli, K. Kellens, D. J. Cott, N. Peys, K. Arstila, M. Heyns, S. De Gendt, G. Groeseneken, and P. M. Vereecken, "Integration of vertical carbon nanotube bundles for interconnects," *Journal of the Electrochemical Society*, **157**, pp. K211–K217, 2010.

57. N. Chiodarelli, Y. L. Li, D. J. Cott, S. Mertens, N. Peys, M. Heyns, S. De Gendt, G. Groeseneken, and P. M. Vereecken, "Integration and electrical characterization of carbon nanotube via interconnects," *Microelectronic Engineering*, **88**, pp. 837–843, 2011.

58. C. Zhang, D. Cott, N. Chiodarelli, P. Vereecken, J. Robertson, and C. M. Whelan, "Growth of carbon nanotubes as horizontal interconnects," *Physica Status Solidi B-Basic Solid State Physics*, **245**, pp. 2308–2310, 2008.

59. Y. Chai, Z. Xiao, and P. C. Chan, "Horizontally aligned carbon nanotube bundles for interconnect application: diameter-dependent contact resistance and mean free path," *Nanotechnology*, **21**, p. 235705, 2010.

60. K. Hernadi, A. Fonseca, J. B. Nagy, D. Bernaerts, and A. A. Lucas, "Fe-catalyzed carbon nanotube formation," *Carbon*, **34**, pp. 1249–1257, 1996.

61. J. Tersoff, "Contact resistance of carbon nanotubes," *Applied Physics Letters,* **74**, pp. 2122–2124, 1999.

62. Y. Matsuda, W. Q. Deng, and W. A. Goddard, "Contact resistance for 'end-contacted' metal–graphene and metal–nanotube interfaces from quantum mechanics," *The Journal of Physical Chemistry C,* **114**, pp. 17845–17850, 2010.

63. S. C. Lim, J. H. Jang, D. J. Bae, G. H. Han, S. Lee, I. S. Yeo, and Y. H. Lee, "Contact resistance between metal and carbon nanotube interconnects: effect of work function and wettability," *Applied Physics Letters,* **95**, p. 264103, 2009.

64. F. Wakaya, K. Katayama, and K. Gamo, "Contact resistance of multiwall carbon nanotubes," *Microelectronic Engineering,* **67**, pp. 853–857, 2003.

65. H. H. Berger, "Models for contacts to planar devices," *Solid-State Electronics,* **15**, pp. 145–158, 1972.

66. M. Lundstrom and Z. Ren, "Essential physics of carrier transport in nanoscale MOSFETs," *IEEE Transactions on Electron Devices,* **49**, pp. 133–141, 2002.

67. F. Yang, X. Wang, D. Q. Zhang, J. Yang, D. Luo, Z. W. Xu, J. K. Wei, J. Q. Wang, Z. Xu, F. Peng, X. M. Li, R. M. Li, Y. L. Li, M. H. Li, X. D. Bai, F. Ding, and Y. Li, "Chirality-specific growth of single-walled carbon nanotubes on solid alloy catalysts," *Nature,* **510**, pp. 522–524, 2014.

68. P. Wilhite, A. A. Vyas, J. Tan, J. Tan, T. Yamada, P. Wang, J. Park, and C. Y. Yang, "Metal-nanocarbon contacts," *Semiconductor Science and Technology,* **29**, p. 054006, 2014.

69. L. Dong, S. Youkey, J. Bush, J. Jiao, V. M. Dubin, and R. V. Chebiam, "Effects of local Joule heating on the reduction of contact resistance between carbon nanotubes and metal electrodes," *Journal of Applied Physics.,* **101**, p. 024320, 2007.

70. T. Yamada, T. Saito, M. Suzuki, P. Wilhite, X. Sun, N. Akhavantafti, D. Fabris, and C. Y. Yang, "Tunneling between carbon nanofiber and gold electrodes," *Journal of Applied Physics,* **107**, p. 044304, 2010.

71. J. Svensson and E. E. Campbell, "Schottky barriers in carbon nanotube-metal contacts," *Journal of Applied Physics,* **110**, p. 111101, 2011.

72. T. Hertel, R. E. Walkup, and P. Avouris, "Deformation of carbon nanotubes by surface van der Waals forces," *Physical Review B,* **58**, p. 13870, 1998.

73. M. Knupfer, "Electronic properties of carbon nanostructures," *Surface Science Report,* **42**, pp. 1–74, 2001.

74. J. Zhao, A. Buldum, J. Han, and J. P. Lu, "Gas molecule adsorption in carbon nanotubes and nanotube bundles," *Nanotechnology,* **13**, pp. 195–200, 2002.

75. G. U. Sumanaskera, C. K. W. Adu, S. Fang, and P. C. Eklund, "Effects of gas adsorption and collisions on electrical transport in single-walled carbon nanotubes," *Physical Review Letters,* **85**, pp. 1096–1099, 2000.

76. O. Leenaerts, B. Partoens, and F. M. Peeters, "Adsorption of H_2O, NH_3, CO, NO_2 and NO on graphene: a first-principles study," *Physical Review B,* **77**, p. 125416, 2008.

77. R. Pati, Y. Zhang, S. K. Nayak, and P. M. Ajayan, "Effect of H_2O adsorption on electron transport in a carbon nanotube," *Applied Physics Letters,* **81**, pp. 2638–2640, 2002.

78. J. L. Blackburn, T. M. Barnes, M. C. Beard, Y. H. Kim, R. C. Tenent, T. J. McDonald, B. To, T. J. Coutts, and M. J. Heben, "Transparent conductive single-walled carbon nanotube networks with precisely tunable ratios of semiconducting and metallic nanotubes," *ACS Nano,* **2**, pp. 1266–1274, 2008.

79. A. Zahab, L. Spina, P. Poncharal, and C. Marliere, "Water-vapor effect on the electrical conductivity of a single-walled carbon nanotube mat," *Physical Review B,* **62**, pp. 10000–10003, 2000.

80. R. Ifuku, K. Nagashio, T. Nishimura, and A. Toriumi, "The density of states of graphene underneath a metal electrode and its correlation with contact resistivity," *Applied Physics Letters,* **103**, p. 033514, 2013.

81. S. Maeda, P. Wilhite, N. Kanzaki, T. Yamada, and C. Y. Yang, "Change in carbon nanofiber resistance from ambient to vacuum," *AIP Advances,* **1**, p. 022102, 2011.

82. P. G. Collins, K. Bradley, M. Ishigami, and A. Zettl, "Extreme oxygen sensitivity of electronic properties of carbon nanotubes," *Science,* **287**, pp. 1801–1804, 2000.

83. C. Lan, D. N. Zakharov, and R. G. Reifenberger, "Determining the optimal contact length for a metal/multiwalled carbon nanotube interconnect," *Applied Physics Letters,* **92**, p. 213112, 2008.

84. C. Lan, P. Srisungsitthisunti, P. B. Amama, T. S. Fisher, X. Xu, and R. G. Reifenberger, "Measurement of metal/carbon nanotube contact resistance by adjusting contact length using laser ablation," *Nanotechnology,* **19**, p. 125703, 2008.

85. F. Madriz, T. Yamada, X. Sun, J. G. Nickel, and C. Y. Yang, "Frequency-independent RC circuit model for one-dimensional carbon nanostructures," *IEEE Electron Device Letters,* **31**, pp. 263–265, 2010.

86. A. A. Vyas, F. Madriz, N. Kanzaki, P. Wilhite, X. Sun, T. Yamada, and C. Y. Yang, "Carbon nanofiber interconnect RF characteristics improvement with deposited tungsten contacts," *Journal of Nanoscience and Nanotechnology,* **14**, pp. 2683–2686, 2014.

87. S. Kim, D. Kulkarni, K. Rykaczewski, M. Henry, V. Tsukruk, and A. Fedorov, "Fabrication of an ultra-low-resistance ohmic contact to MWCNT-metal interconnect using graphitic carbon by electron beam induced deposition (EBID)," *IEEE Transactions on Nanotechnology,* **11**, pp. 1223–1230, 2012.

88. Y. Ominami, Q. Ngo, M. Suzuki, A. J. Austin, C. Y. Yang, A. M. Cassell, and J. Li, "Interface characteristics of vertically aligned carbon nanofibers for interconnect applications," *Applied Physics Letters.,* **89**, p. 263114, 2006.

89. R. Martel, V. Derycke, C. Lavoie, J. Appenzeller, K. K. Chan, J. Tersoff, and P. Avouris, "Ambipolar electrical transport in semiconducting single-wall carbon nanotubes," *Physical Review Letters.,* **87**, p. 256805, 2001.

90. M. S. Wang, D. Golberg, and Y. Bando, "Superstrong low-resistant carbon nanotube–carbide-metal nanocontacts," *Advanced Materials,* **22**. pp. 5350–5355, 2010.

91. L. Wang, I. Meric, P.Y. Huang, Q. Gao, Y. Gao, H. Tran, T. Taniguchi, K. Watanabe, L. M. Campos, D. A. Muller, J. Guo, P. Kim, J. Hone, K. L. Shepard, and C. R. Dean, "One-dimensional electrical contact to a two-dimensional material," *Science,* **342**, pp. 614–617, 2013.

92. *International Technology Roadmap for Semiconductors*, 2013 edition, available at www.itrs2.net.

93. C. J. Zhou, A. A. Vyas, P. Wilhite, P. Wang, M. S. Chan, and C. Y. Yang, "Resistance determination for sub-100-nm carbon nanotube vias," *IEEE Electron Device Letters,* **36**, pp. 71–73, 2015.

94. A. Vyas, "Carbon nanotube interconnects for end-of-roadmap semiconductor technology nodes," Ph.D. Thesis, Santa Clara University, 2016.

95. A. Powell, "Behaviour of the platinum metals at high temperatures," *Platinum Metals Review,* **2**, pp. 95–98, 1958.

96. A. P. Graham, G. S. Duesberg, R. V. Seidel, M. Liebau, E. Unger, W. Pamler, F. Kreupl, and W. Hoenlein, "Carbon nanotubes for microelectronics?," *Small,* **1**, pp. 382–390, 2005.

97. A. P. Graham, G. S. Duesberg, W. Hoenlein, F. Kreupl, M. Liebau, R. Martin, B. Rajasekharan, W. Pamler, R. Seidel, W. Steinhoegl, and E. Unger, "How do carbon nanotubes fit into the semiconductor roadmap?," *Applied Physics A: Materials Science & Processing,* **80**, pp. 1141–1151, 2005.

98. S. Vollebregt, F. D. Tichelaar, H. Schellevis, C. I. M. Beenakker, and R. Ishihara, "Carbon nanotube vertical interconnects fabricated at temperatures as low as 350 degrees C," *Carbon,* **71**, pp. 249–256, 2014.

99. S. Vollebregt, S. Banerjee, K. Beenakker, and R. Ishihara, "Thermal conductivity of low temperature grown vertical carbon nanotube bundles measured using the three-omega method," *Applied Physics Letters,* **102**, p. 191909, 2013.

100. S. Vollebregt, S. Banerjee, K. Beenakker, and R. Ishihara, "Size-dependent effects on the temperature coefficient of resistance of carbon nanotube vias," *IEEE Transactions on Electron Devices,* **60**, pp. 4085–4089, 2013.

101. S. Vollebregt, A. N. Chiaramonti, R. Ishihara, H. Schellevis, and K. Beenakker, "Contact resistance of low-temperature carbon nanotube vertical interconnects," *Nanotechnology (IEEE-NANO), 12th IEEE Conference,* pp. 1–5, 2012.

102. S. Vollebregt, J. Derakhshandeh, R. Ishihara, M. Y. Wu, and C. I. M. Beenakker, "Growth of high-density self-aligned carbon nanotubes and nanofibers using palladium catalyst," *Journal of Electronic Materials,* **39**, pp. 371–375, 2010.

103. Y. Matsuda, W. Q. Deng, and W. A. Goddard, "Contact resistance properties between nanotubes and various metals from quantum mechanics," *The Journal of Physical Chemistry C,* **111**, pp. 11113–11116, 2007.

104. A. A. Vyas, C. Zhou, Y. Chai, P. Wang, and C. Y. Yang, "Effect of improved contact on reliability of sub-60 nm carbon nanotube vias," *Nanotechnology,* **27**, p. 375202, 2016.

105. M. S. Dresselhaus, A. Jorio, and R. Saito, "Characterizing graphene, graphite, and carbon nanotubes by raman spectroscopy," *Annual Review of Condensed Matter Physics,* **1**, pp. 89–108, 2010.

106. M. S. Dresselhaus, A. Jorio, M. Hofmann, G. Dresselhaus, and R. Saito, "Perspectives on carbon nanotubes and graphene Raman spectroscopy," *Nano Letters,* **10**, pp. 751–758, 2010.

107. M. S. Dresselhaus, G. Dresselhaus, and M. Hofmann, "Raman spectroscopy as a probe of graphene and carbon nanotubes," *Philosophical Transactions of the Royal Society A: Mathematical Physical and Engineering Sciences,* **366**, pp. 231–236, 2008.

108. M. S. Dresselhaus, G. Dresselhaus, and A. Jorio, "Raman spectroscopy of carbon nanotubes in 1997 and 2007," *Journal of Physical Chemistry C,* **111**, pp. 17887–17893, 2007.

109. M. S. Dresselhaus, G. Dresselhaus, and M. Hofmann, "The big picture of Raman scattering in carbon nanotubes," *Vibrational Spectroscopy*, **45**, pp. 71–81, 2007.

110. M. S. Dresselhaus, G. Dresselhaus, R. Saito, and A. Jorio, "Raman spectroscopy of carbon nanotubes," *Physics Reports*, **409**, pp. 47–99, 2005.

111. M. S. Dresselhaus, G. Dresselhaus, A. M. Rao, A. Jorio, A. G. Souza, G. G. Samsonidze, and R. Saito, "Resonance Raman scattering on one-dimensional systems," *Indian Journal of Physics and Proceedings of the Indian Association for the Cultivation of Science-Part B*, **77B**, pp. 75–99, 2003.

112. M. S. Dresselhaus, G. Dresselhaus, A. Jorio, A. G. Souza, and R. Saito, "Raman spectroscopy on isolated single wall carbon nanotubes," *Carbon*, **40**, pp. 2043–2061, 2002.

113. S. Vollebregt, R. Ishihara, F. D. Tichelaar, Y. Hou, and C. I. M. Beenakker, "Influence of the growth temperature on the first and second-order Raman band ratios and widths of carbon nanotubes and fibers," *Carbon*, **50**, pp. 3542–3554, 2012.

114. Q. Ngo, D. Petranovic, S. Krishnan, A. M. Cassell, Q. Ye, J. Li, M. Meyyappan, and C. Y. Yang, "Electron transport through metal-multiwall carbon nanotube interfaces," *IEEE Transactions on Nanotechnology*, **3**, pp. 311–317, 2004.

115. Z. F. Zhao, Y. S. Zhou, C. Q. Zhang, and Z. Y. Wang, "Thermoset composites functionalized with carbon nanofiber sheets for EMI shielding," *Journal of Applied Polymer Science*, **132**, p. 41873, 2015.

116. X. P. Zhang, D. Liu, L. B. Li, and T. Y. You, "Direct electrochemistry of glucose oxidase on novel free-standing nitrogen-doped carbon nanospheres@carbon nanofibers composite film," *Scientific Reports*, **5**, p. 9885, 2015.

117. L. C. Zeng, Y. Jiang, J. Xu, M. Wang, W. H. Li, and Y. Yu, "Flexible copper-stabilized sulfur-carbon nanofibers with excellent electrochemical performance for Li-S batteries," *Nanoscale*, **7**, pp. 10940–10949, 2015.

118. S. Yu, J. W. Jung, and I. D. Kim, "Single layers of WS2 nanoplates embedded in nitrogen-doped carbon nanofibers as anode materials for lithium-ion batteries," *Nanoscale*, **7**, pp. 11945–11950, 2015.

119. Y. Yaakob, M. Z. M. Yusop, C. Takahashi, M. S. Rosmi, G. Kalita, and M. Tanemura, "In situ transmission electron microscopy of Ag-incorporated carbon nanofibers: the effect of Ag nanoparticle size on graphene formation," *RSC Advances*, **5**, pp. 5647–5651, 2015.

120. S. B. Xu, A. Akchurin, T. Liu, W. Wood, X. W. Tangpong, I. S. Akhatov, and W. H. Zhong, "Thermal properties of carbon nanofiber reinforced high-density polyethylene nanocomposites," *Journal of Composite Materials,* **49**, pp. 795–805, 2015.

121. W. Wang, Y. Sun, B. Liu, S. G. Wang, and M. H. Cao, "Porous carbon nanofiber webs derived from bacterial cellulose as an anode for high performance lithium ion batteries," *Carbon,* **91**, pp. 56–65, 2015.

122. M. S. Wang, W. L. Song, J. Wang, and L. Z. Fan, "Highly uniform silicon nanoparticle/porous carbon nanofiber hybrids towards free-standing high-performance anodes for lithium-ion batteries," *Carbon,* **82**, pp. 337–345, 2015.

123. E. Tsuji, T. Yamasaki, Y. Aoki, S. G. Park, K. Shimizu, and H. Habazaki, "Highly durable platelet carbon nanofiber-supported platinum catalysts for the oxygen reduction reaction," *Carbon,* **87**, pp. 1–9, 2015.

124. X. Tang, Z. Y. Xie, Q. Z. Huang, G. F. Chen, M. Hou, and B. L. Yi, "Mass-transport-controlled, large-area, uniform deposition of carbon nanofibers and their application in gas diffusion layers of fuel cells," *Nanoscale,* **7**, pp. 7971–7979, 2015.

125. K. X. Tang, Y. P. Li, H. B. Cao, F. Duan, J. Zhang, Y. Zhang, and Y. Wang, 'Integrated electrospun carbon nanofibers with vanadium and single-walled carbon nanotubes through covalent bonds for high-performance supercapacitors," *RSC Advances,* **5**, pp. 40163–40172, 2015.

126. A. Simon, M. Seyring, S. Kamnitz, H. Richter, I. Voigt, M. Rettenmayr, and U. Ritter, "Carbon nanotubes and carbon nanofibers fabricated on tubular porous Al_2O_3 substrates," *Carbon,* **90**, pp. 25–33, 2015.

127. S. Vollebregt, A. N. Chiaramonti, J. van der Cingel, K. Beenakker, and R. Ishihara, "Towards the integration of carbon nanotubes as vias in monolithic three-dimensional integrated circuits," *Japanese Journal of Applied Physics,* **52**, p. 04CB02, 2013.

128. K. C. Huang, D. B. Janes, K. J. Webb, and M. R. Melloch, "A transfer length model for contact resistance of two-layer systems with arbitrary interlayer coupling under the contacts," *IEEE Transactions on Electron Devices,* **43**, pp. 676–684, 1996.

129. P. Wilhite, A. Vyas, J. Tan, C. Y. Yang, P. Wang, J. Park, H. Ai, and M. Narasimhan, "Electrical and structural analysis of CNT–metal contacts in via interconnects," *Proceedings of the Seventh International Conference on Quantum, Nano, and Micro Technologies*, 2013, pp. 41–43, 2013.

130. M. Katagiri, Y. Yamazaki, M. Wada, M. Kitamura, N. Sakuma, M. Suzuki, S. Sato, M. Nihei, A. Kajita, T. Sakai, and Y. Awano, "Improvement in electrical properties of carbon nanotube via interconnects," *Japanese Journal of Applied Physics,* **50**, p. 05EF01, 2011.

131. Y. Awano, S. Sato, M. Nihei, T. Sakai, Y. Ohno, and T. Mizutani, "Carbon nanotubes for VLSI: interconnect and transistor applications," *Proceedings of the IEEE,* **98**, pp. 2015–2031, 2010.

132. A. P. Graham, G. S. Duesberg, R. V. Seidel, M. Liebau, E. Unger, W. Pamler, F. Kreupl, and W. Hoenlein, "Carbon nanotubes for microelectronics?," *Small,* **1**, pp. 382–390, 2005.

133. C. Adelmann, L. G. Wen, A. P. Peter, Y. K. Siew, K. Croes, J. Swerts, M. Popovici, K. Sankaran, G. Pourtois, and S. Van Elshocht, "Alternative metals for advanced interconnects," *IEEE International Interconnect Technology Conference*, pp. 173–176, 2014.

134. B. Q. Wei, R. Vajtai, and P. M. Ajayan, "Reliability and current carrying capacity of carbon nanotubes," *Applied Physics Letters,* **79**, pp. 1172–1174, 2001.

135. P. Wilhite, H. S. Uh, N. Kanzaki, P. Wang, A. Vyas, S. Maeda, T. Yamada, and C. Y. Yang, "Electron-beam and ion-beam-induced deposited tungsten contacts for carbon nanofiber interconnects," *Nanotechnology,* **25**, p. 375702, 2014.

136. Z. L. Wang, J. M. Petroski, T. C. Green, and M. A. El-Sayed, "Shape transformation and surface melting of cubic and tetrahedral platinum nanocrystals," *The Journal of Physical Chemistry B,* **102**, pp. 6145–6151, 1998.

137. M. Katagiri, Y. Yamazaki, N. Sakuma, M. Suzuki, T. Sakai, M. Wada, N. Nakamura, N. Matsunaga, S. Sato, and M. Nihei, "Fabrication of 70-nm-diameter carbon nanotube via interconnects by remote plasma-enhanced chemical vapor deposition and their electrical properties," *Interconnect Technology Conference, IEEE International*, pp. 44–46, 2009.

138. A. A. Vyas, C. Zhou, P. Wilhite, P. Wang, and C. Y. Yang, "Electrical properties of carbon nanotube via interconnects for 30 nm linewidth and beyond," *Microelectronics Reliability,* **61**, pp. 35–42, 2016.

139. G. S. Duesberg, A. P. Graham, M. Liebau, R. Seidel, E. Unger, F. Kreupl, and W. Hoenlein, "Growth of isolated carbon nanotubes with lithographically defined diameter and location," *Nano Letters,* **3**, pp. 257–259, 2003.

140. M. Katagiri, M. Wada, B. Ito, Y. Yamazaki, M. Suzuki, M. Kitamura, T. Saito, A. Isobayashi, A. Sakata, N. Sakuma, A. Kajita, and T. Sakai, "Fabrication and characterization of planarized carbon nanotube via interconnects," *Japanese Journal of Applied Physics,* **51**, p. 05ED02, 2012.

141. M. Wada, B. Ito, T. Saito, D. Nishide, T. Ishikura, A. Isobayashi, M. Katagiri, Y. Yamazaki, T. Matsumoto, and M. Kitamura, "Selective carbon nanotube growth in via structure using novel arrangement of catalytic metal," *IEEE International Interconnect Technology Conference,* 2014.

142. Q. Ngo, A. M. Cassell, A. J. Austin, J. Li, S. Krishnan, M. Meyyappan, and C. Y. Yang, "Characteristics of aligned carbon nanofibers for interconnect via applications," *IEEE Electron Device Letters,* **27**, pp. 221–224, 2006.

143. S. Hermann, S. Schulze, R. Ecke, A. Liebig, P. Schaefer, D. R. T. Zahn, M. Albrecht, M. Hietschold, S. E. Schulz, and T. Gessner, "Growth of carbon nanotube forests between a bi-metallic catalyst layer and a SiO_2 substrate to form a self-assembled carbon-metal heterostructure," *Carbon,* **50**, pp. 4765–4772, 2012.

144. S. Hermann, B. Pahl, R. Ecke, S. E. Schulz, and T. Gessner, "Carbon nanotubes for nanoscale low temperature flip chip connections," *Microelectronic Engineering,* **87**, pp. 438–442, 2010.

145. S. Hermann, R. Ecke, S. Schulz, and T. Gessner, "Controlling the formation of nanoparticles for definite growth of carbon nanotubes for interconnect applications," *Microelectronic Engineering,* **85**, pp. 1979–1983, 2008.

146. M. H. van der Veen, Y. Barbarin, Y. Kashiwagi, and Z. Tokei, "Electron mean-free path for CNT in vertical interconnects approaches Cu," *Interconnect Technology Conference/Advanced Metallization Conference (IITC/AMC), IEEE International,* pp. 181–184, 2014.

147. S. Vollebregt, S. Banerjee, A. N. Chiaramonti, F. D. Tichelaar, K. Beenakker, and R. Ishihara, "Dominant thermal boundary resistance in multi-walled carbon nanotube bundles fabricated at low temperature," *Journal of Applied Physics,* **116**, p. 023514, 2014.

148. W. Du, Z. Ahmed, Q. Wang, C. Yu, Z. Feng, G. Li, M. Zhang, C. Zhou, R. Senegor, and C. Y. Yang, "Structures, properties, and applications of CNT-graphene heterostructures," *2D Materials,* **6**, p. 042005, 2019.

149. F. D. Novaes, R. Rurali, and P. Ordejón, "Electronic transport between graphene layers covalently connected by carbon nanotubes," *ACS Nano,* **4**, pp. 7596–7602, 2010.

150. C. J. Zhou, R. Senegor, Z. Baron, Y. Chen, S. Raju, A. A. Vyas, M. Chan, Y. Chai, and C. Y. Yang, "Synthesis and interface characterization of CNTs on graphene," *Nanotechnology*, **28**, pp. 054007–054016, 2017.

151. D. H. Lee, J. E. Kim, T. H. Han, J. W. Hwang, S. Jeon, S. Y. Choi, S. H. Hong, W. J. Lee, R. S. Ruoff, and S. O. Kim, "Versatile carbon hybrid films composed of vertical carbon nanotubes grown on mechanically compliant graphene films," *Advanced Materials,* **22**, pp. 1247–1252, 2010.

152. D. Baowan, B. J. Cox, and J. M. Hill, "Two least squares analyses of bond lengths and bond angles for the joining of carbon nanotubes to graphenes," *Carbon,* **45**, pp. 2972–2980, 2007.

153. D. Baowan, B. J. Cox, N. Thamwattana, and J. M. Hill, "Two minimisation approximations for joining carbon nanostructures," *IUTAM Symposium on Modelling Nanomaterials and Nanosystems*, pp. 109–121.

154. H. W. Kroto, "The stability of the fullerenes Cn, with n = 24, 28, 32, 36, 50, 60 and 70," *Nature,* **329**, pp. 529–531, 1987.

155. H. Terrones and A. L. Mackay, "The geometry of hypothetical curved graphite structures," *Carbon,* **30**, pp. 1251–1260, 1992.

156. H. Terrones and M. Terrones, "Curved nanostructured materials," *New Journal of Physics,* **5**, p. 126, 2003.

157. T. W. Ebbesen and T. Takada, "Topological and SP3 defect structures in nanotubes," *Carbon,* **33**, pp. 973–978, 1995.

158. G. K. Dimitrakakis, E. Tylianakis, and G. E. Froudakis, "Pillared graphene: A new 3-D network nanostructure for enhanced hydrogen storage," *Nano Letters,* **8**, pp. 3166–3170, 2008.

159. N. Yang, D. Yang, L. Chen, D. Liu, M. Cai, and X. Fan, "A first-principle theoretical study of mechanical and electronic properties in graphene single-walled carbon nanotube junctions," *Materials,* **10**, p. 1300, 2017.

160. J. Taylor, H. Guo, and J. Wang, "Ab initio modeling of quantum transport properties of molecular electronic devices," *Physical Review B,* **63**, p. 245407, 2001.

161. J. M. Soler, E. Artacho, J. D. Gale, A. García, J. Junquera, P. Ordejón, and D. Sánchez-Portal, "The SIESTA method for ab initio order-N materials simulation," *Journal of Physics: Condensed Matter,* **14**, pp. 2745–2779, 2002.

162. M. D. Hanwell, D. E. Curtis, D. C. Lonie, T. Vandermeersch, E. Zurek, and G. R. Hutchison, "Avogadro: an advanced semantic chemical editor,

visualization, and analysis platform," *Journal of Cheminformatics,* **4**, p. 17, 2012.

163. T. Matsumoto and S. Saito, "Geometric and electronic structure of new carbon-network materials: nanotube array on graphite sheet," *Journal of the Physical Society of Japan,* **71**, pp. 2765–2770, 2002.

164. Y. Mao and J. Zhong, "The computational design of junctions by carbon nanotube insertion into a graphene matrix," *New Journal of Physics,* **11**, p. 093002, 2009.

165. R. Ramos, A. Fournier, M. Fayolle, J. Dijon, C. P. Murray, and J. McKenna, "Nanocarbon interconnects combining vertical CNT interconnects and horizontal graphene lines," *2016 IEEE International Interconnect Technology Conference/Advanced Metallization Conference (IITC/AMC),* pp. 48–50, 2016.

166. K. Ghosh, N. Ranjan, Y. K. Verma, and C. S. Tan, "Graphene–CNT hetero-structure for next generation interconnects," *RSC Advances,* **6**, pp. 53054–53061, 2016.

167. D. Kondo, D., S. Sato, and Y. Awano, "Self-organization of novel carbon composite structure: graphene multi-layers combined perpendicularly with aligned carbon nanotubes," *Applied Physics Express,* **1**, p. 074003, 2008.

168. V. Jousseaume, J. Cuzzocrea, N. Bernier, and V. T. Renard, "Few graphene layers/carbon nanotube composites grown at complementary-metal-oxide-semiconductor compatible temperature," *Applied Physics Letters,* **98**, p. 123103, 2011.

Chapter 4

Carbon Nanotube Transistors

Min Zhang, Chunhui Du, Qiuyue Huang, Zhiqiang Liao, Yanyan Deng, Weihong Huang, and Xiaofang Wang
School of Electronic and Computer Engineering,
Peking University Shenzhen Graduate School,
Shenzhen 518055, China
zhang.min@pku.edu.cn

4.1 Introduction

In recent years, new technologies such as carbon nanotube (CNT) computer [1, 2], capacitive sensor interface circuit [3], wearable sensor [4, 5], electronic paper [6], RF tag [7], and flexible electronics [8–11] have radically changed people's perspectives toward modern electronics. Reviewing the history, the advancement of transistors relies largely on the material science and process technology. The development of micro-/nanoelectronics expands from silicon-based electronics [12, 13] to metal oxide [14, 15], organic [16–18], and low dimensional material (graphene [19, 20], CNT [10, 21, 22], and MoS_2 [23, 24]) based electronics. In the regime of nanoelectronics, several critical barriers prevent the further scaling down of Si metal-

Nanocarbon Electronics
Edited by Changjian Zhou, Min Zhang, and Cary Y. Yang

ISBN 978-981-4877-11-4 (Hardcover), 978-1-003-04308-9 (eBook)
www.jennystanford.com

oxide-semiconductor field-effect transistor (MOSFET), including the accompanying short channel effect [25], the fabrication difficulty for transistors with sub-10 nm channel, the increasing junction and gate-oxide leakage current [26], the parasitic capacitance among interconnects, and the heat production [27, 28]. These factors have blocked the way to further minimize effective power consumption and improve operation frequency. Electronics is approaching major intergeneration innovations since silicon transistor scaling no longer yields historical energy-efficiency benefits, speeding research toward beyond-silicon nanotechnologies.

For planar display and other non-silicon platforms, the conventional amorphous silicon (α-silicon) based electronics cannot satisfy the requirement of high frame rate due to the low carrier mobility [29] and poor stability [30, 31], and the application of polysilicon is limited by the processing cost and large-area uniformity. The commercialized indium gallium zinc oxide (IGZO) has a carrier mobility that is one order of magnitude higher than that of α-silicon [32], satisfying the requirement of medium quality display. However, it suffers from the instability to environment [33] and the exhaustion of rare-earth element. Moreover, both silicon and metal oxide have limitation for flexible electronics due to their brittle nature. Organic materials are widely used in flexible electronics, and all the components, including substrate, channel, electrodes, and dielectric, can be formed by polymers [18, 34]. The critical problems of organic material are the low mobility and the mobility degradation at a high gate voltage [35, 36]. Two-dimensional (2D) materials such as single- or few-layered graphene, black phosphorus [37, 38], and transition metal dichalcogenides (TMDs) [39] share common merits on quasi-transparency, in-planar electron transport, and mechanical flexibility, demonstrating a great potential for transparent and flexible nanoelectronics. However, many problems still exist ahead to be addressed, such as instability, harsh synthetic conditions, and low on/off current ratio (I_{on}/I_{off}).

As a promising material for transistors, CNT has attracted numerous research interests due to their excellent electrical properties with mean free path up to several micrometers, excellent thermal conductivity, compatibility with high-κ dielectrics, as well as mechanical flexibility [40–44]. The semiconducting or metallic conduction induced by different chirality makes CNTs applicable as both channel and electrode materials. It is reported that semiconducting single-walled CNTs (SWCNTs) can exhibit a field-effect mobility of 10^4 cm^2 V^{-1}s^{-1}, while metallic SWCNTs have extremely high current-carrying capability of 10^9 A cm^{-2} [40, 42, 45]. Using semiconducting enriched SWCNTs as channel material, the CNT transistors can reach a field-effect mobility higher than that of silicon by more than two magnitudes and an I_{on}/I_{off} over 10^6 [46], which are more suitable to drive the high-speed circuit. The metallic or unsorted CNTs are also promising to replace the existing indium tin oxide (ITO) electrode and metal electrode [47, 48]. Besides the superior electrical property, the mechanical flexibility in nature also makes CNTs the most promising candidate for flexible electronics.

Over the past decades, tremendous progress has been made in CNT transistor and its application for digital integrated circuits (ICs), advanced displays, as well as sensors. In particular, the thickness of high-κ dielectric was reduced to 2–3 nm, producing a subthreshold swing as low as 60 mV/decade at room temperature [49]. A quasi-ballistic CNT transistor based on dense CNT array was also fabricated, with a gate length of only 5 nm [50]. Standalone circuit elements such as half-adder sum generators and D-latches based on CNT transistors were realized [51]. Additionally, CNT computer [1, 2] and capacitive sensor interface circuit [3] also proved the feasibility of CNT transistors. In particular, Cao et al. have recently reduced all components of CNT transistors to 40 nm, which broke the prediction of ITRS [52]. Ghazali et al. fabricated tunnel barriers and single electron transistors (SETs) using suspended multi-walled carbon nanotubes (MWCNTs) [53]. The tunnel barriers were formed by irradiating MWCNTs with focused Ga ion beam, and the SETs were fabricated by forming the double barriers in the suspended MWCNTs [53]. In the display area, the CNT transistors also received numerous research efforts. Unlike conventional silicon-based electronics, the processing and integration of CNT transistors can be realized in a low-temperature, low-cost and ecofriendly way on a large scale. Solution-processed methods such as spin-coating, ink-jet printing, roll-to-roll (R2R) printing, and transfer printing are widely investigated [11]. These methods are applied at near-ambient temperature and have no limitation on substrate size, matching the manufacturing requirements for the next-generation displays.

Though significant advances have been made, many critical issues remain to be addressed before the practical application of CNT transistors. In this chapter, we will review the basics of CNT transistor, shed light on the critical issues that affect the performance of device, and update the state of the art for CNT transistors. Then, the flexible and stretchable CNT transistors will also be reviewed in detail, including the material selection and the updated fabrication techniques. Among these discussions, the latest advances on CNT transistors will be exhibited. Finally, the applications of CNT transistors in integrated circuits, display electronics, and bio-detection will be discussed.

4.2 CNT Fabrication and Separation

Traditional methods to synthesize CNTs include arc discharge [54], laser ablation [55], and chemical vapor deposition (CVD). Among them, CVD using a compound of ferritin or metal oxide as catalyst is the most widely adopted method because of the controllable synthesis and high yield [56]. For thermal CVD, the synthesis process is conducted at high temperature (generally over 800°C) under delicate control over the catalyst thickness and gas source, during which the continuous catalyst film will be broken into nano-sized islands to catalyze the vertical growth of CNTs through a tip or base growth mechanism [57]. The catalyst thickness and the size of catalyst are the keys to determine the density and the diameter of the CNTs. The plasma-enhanced CVD (PECVD) method is also widely used since it was developed in 1992 [58].

The synthesized CNTs contain various chirality, which means metallic and semiconducting SWCNTs coexist in the system. Assuming that all possible chirality is grown with equal probability, one-third of the grown SWCNTs are metallic SWCNTs and two-thirds are semiconducting SWCNTs. The existence of metallic tubes has great influence on the I_{on}/I_{off} especially for short channel CNT transistors and CNT array transistors. The existence of metallic CNTs may form the percolation path and drastically degrade the switching characteristics, which greatly impedes the application of CNT transistors. Therefore, a high-efficiency separation process is needed to obtain high-purity semiconducting tubes or remove the effect of metallic tubes.

In general, the separation or sorting methods of SWCNTs can be divided into two categories. One is achieved in the synthesis process, by improving process conditions or catalyst systems to obtain higher-purity SWCNTs (or MWCNTs), with a purity above 95% [59]. However, in the products of the selective synthesis, impurities appear to be unavoidable. Therefore, combining the controlled synthesis with the post-separation technology can further increase the purity of SWCNTs.

Density-gradient ultracentrifugation (DGU) is commonly used to obtain semiconducting enriched SWCNTs, with a high purity over 98%. The method utilizes the differences in the buoyant densities of different SWCNTs to separate metallic and semiconducting nanotubes. In this method, CNTs are surrounded by surfactants and spatially separated under the effect of centrifugal force, as shown in Fig. 4.1 [60]. However, this method usually separates a small amount of CNTs and requires multiple separation steps. Therefore, Ghosh et al. proposed a method with nonlinear DGU [61]. What is more, the commercially available highest purity of semiconducting SWCNTs now has reached 99.9%, according to the website data of Nano Integris. In 2014, Lee and his group experimentally demonstrated high-yield, high-performance CNT transistors by using a highly separated 99.9% semiconducting SWCNT solution [62].

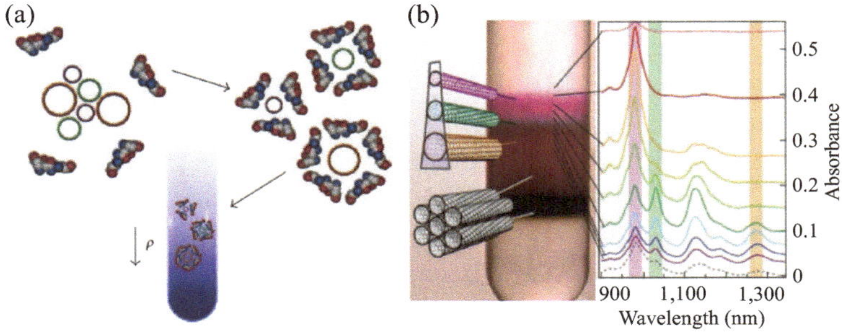

Figure 4.1 (a) Schematic of SWCNTs surrounded by surfactant during their sorting. (b) Layering of SWCNTs after separation and the corresponding optical absorbance spectra. Reprinted with permission from Ref. [60], Copyright 2006, Nature Publishing Group.

Separation and purification of carbon nanotubes by chromatographic techniques have wide applications, including gel permeation chromatography (GPC) and ion exchange chromatography (IEC) [63, 64]. These methods introduce gels or DNA molecules. The underlying principle is to selectively entangle and adsorb SWCNTs of different chiralities and diameters, and then separate them by the corresponding stationary phase.

The GPC separation process is very simple, and it facilitates quantitative separation. Tanaka et al. developed an agarose gel adsorption extrusion method and found that the interaction between the gel and semiconducting SWCNTs (S-SWCNTs) was stronger than that of metallic SWCNTs (M-SWCNTs) [65]. Based on this, M-SWCNTs and S-SWCNTs were successfully separated. Later on, Liu et al. improved the experimental process using this gel as the medium and devised a single-surfactant multicolumn gel chromatography method, in which several gel columns were connected vertically in series to achieve large-scale chirality separation [66]. Generally, the diameter of carbon nanotubes prepared by GPC is less than 1 nm. In 2017, Yang et al. used NaOH to regulate the adsorption of surfactant molecules on the surface of carbon nanotubes and polysaccharide gels of different structures to achieve large-diameter CNTs (diameter >1.2 nm) [67].

IEC can separate the CNTs according to the diameter or chirality, which is considered to be the most promising separation technology [64]. In the case of sorting SWCNTs, two armchair species, (6,6) and (7,7), were successfully separated from a synthetic mixture with the application of a series of DNA, as shown in Fig. 4.2 [68]. Due to the strong selective binding between DNA and SWCNTs, this method has high selectivity for the separation of carbon nanotube diameter and chirality. On the other hand, the separation cost is too high, which is not conducive to achieving large-scale separation. In addition, DNA molecules adsorbed on SWCNTs are difficult to be cleared and may affect the subsequent application.

For an as-fabricated transistor with mixed CNT channel, the electrical breakdown is widely used to eliminate the effect of metallic tubes and improve $I_{\mathrm{on}}/I_{\mathrm{off}}$ [69]. Briefly, a high electrical field is applied between the source and drain electrodes, while a bias voltage is applied to the gate electrode to turn off the semiconducting CNTs. Then, when high current flows through the metallic CNTs, the massive Joule heat produced will break the metallic CNTs where defects are most concentrated.

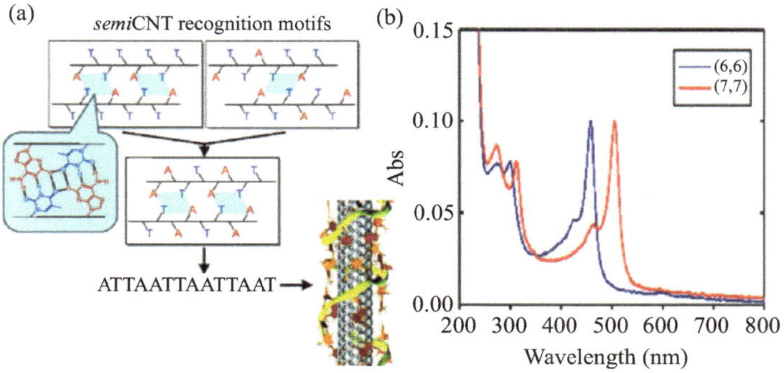

Figure 4.2 (a) Sequence motif variation to derive a new sequence pattern for metallic tube recognition. (b) Optical absorption spectra and data analysis. Reprinted with permission from Ref. [68], Copyright 2011, American Chemical Society.

Besides the methods mentioned above, methods such as electrophoresis [70, 71] and self-sorting of SWCNTs via selective surface interaction [72–74] are also developed to selectively remove metallic CNTs. These methods have yielded some progresses in the laboratory but are still unpractical for wide applications. More reliable and low-cost methods are expected.

4.3 Progress of CNT Transistors

The first CNT transistor was reported by Tans in 1998, which has a global back-gate structure consisting of silicon substrate, silicon oxide dielectric, SWCNT active channel, and platinum electrodes [75]. The fabricated CNT transistors displayed an on-state metal–SWCNT contact resistance of 1 MΩ, a maximum on-current of about 20 nA, and a maximum gain of 0.35. Besides, the device exhibited strong bipolar conduction behavior. In the following years, CNT transistors have demonstrated significant improvement on electrical performance, structural configuration, and mechanism investigation. Generally speaking, the performance of a CNT transistor is mainly determined by the property of dielectric and gate configuration, the contact between electrodes and SWCNTs, as well as chemical and physical state of SWCNTs. Taking a panoramic view of the CNT transistors, the existing issues can mainly be classified into four aspects: gate control, Schottky barrier (SB), conduction behavior, and SWCNT channel.

4.3.1 Gate Control

The CNT transistor functions as a gate-controlled electronic switch for both SB CNT transistor and MOS CNT transistor. To be specific, the gate voltage modulates the width and height of energy barriers for metal/CNT contact. At the same time, it modulates the CNT channel. Both effects tune the possibility of tunneling and thermionic emission. A strong gate control will lead to small subthreshold swing (SS), large transconductance (g_m), and high on-state current (I_{on}).

First, the gate capacitance is an important index of gate control. In a standard MOSFET model (working in saturation mode), the drain current I_D can be approximately defined by [76]

$$I_D = \mu C_{ox} \frac{W}{L} \frac{(V_G - V_{th})^2}{2} \tag{4.1}$$

where W is the width of transistor channel, L is the channel length, μ is the channel carrier mobility (assumed constant here), C_{ox} is the capacitance density associated with the gate dielectric when the underlying channel is in the inversion state, V_G is the gate voltage applied to the transistor, V_{DS} is the drain voltage, and V_{th} is the threshold voltage. For a given transistor, the drain current is proportional to gate capacitance when the state of channel is inversion and, more specifically, proportional to the permittivity constant and inversely proportional to the thickness of dielectric. Dielectrics with high permittivity constant and high insulating property are preferred to cut off the leakage current and yield better performance.

The subthreshold swing characterizes the speed to switch the device from the off-state to the on-state. It can be expressed using [77]

$$SS = \left(\frac{k_B T}{e}\right) \ln(10) \left(1 + \frac{C_{IT}}{C_{ox}}\right), \tag{4.2}$$

where C_{IT} is the capacitance resulted from interface traps, C_{ox} is the gate capacitance, k_B is the Boltzmann constant, T is temperature, and e is the elementary charge of an electron. The value of SS is determined by C_{IT}/C_{ox}. If C_{IT} is very small compared to C_{ox}, SS approaches the limit of 60 mV/decade at room temperature for Si MOSFET. The capacitance generated by interfacial charge traps is very common in solution-processed transistor.

Fortunately, the SWCNT has perfect compatibility with high-κ dielectrics. To be specific, the SWCNTs and the high-κ dielectrics are bonded only by weak van der Waals force on their interface, so the intrinsic chemical and physical natures of the nanotubes are preserved well after the deposition of high-κ dielectrics by atomic layer deposition (ALD). The top-gate CNT transistors reported by Javey et al. used an 8 nm ZrO_2 thin film as gate dielectric by ALD [78]. The SS was reduced to 60–80 mV/decade for p-type CNT transistors and 90–100 mV/decade for n-type CNT transistors. And a high carrier mobility of up to 3000 $cm^2\ V^{-1}\ s^{-1}$ was observed. Additionally, the transconductance of these transistors reached ~6000 S m^{-1} and the complementary inverters exhibited a voltage gain of up to 60. Liyanage et al. investigated the effects of atomic layer deposited high-κ dielectrics (Al_2O_3 and HfO_2) on SWCNTs using Raman spectroscopy [79]. The Raman peak shifts of the G-peak and the 2D (G′) peaks, which is the first overtone of the D-peak, suggest that doping and stress induce effects on the CNTs by the surrounding high-κ oxide environment.

It is necessary to continue reducing the thickness of dielectrics for sub-10 nm technology node. Since there is no dangling bond on the surface of CNT, about 8 nm high-κ dielectric is needed to fully cover an SWCNT [78]. In order to realize conformal deposition, the surface of SWCNT needs to be modified. Farmer et al. reported that reactive diazonium species can covalently functionalize the sidewalls of SWCNTs [80]. However, the covalent functionalization would influence the transport property of carriers. As shown in Fig. 4.3, Lu et al. reported that poly-T DNA can enhance the nucleation of high-κ dielectrics on the SWCNT sidewalls by noncovalent interaction, and only 2–3 nm high-κ dielectric is needed to cover a single SWCNT [49]. The SS of the fabricated CNT transistor was reduced to the ultimate value of 60 mV/decade, and the high maximum transconductance of 5000 S m^{-1} was obtained.

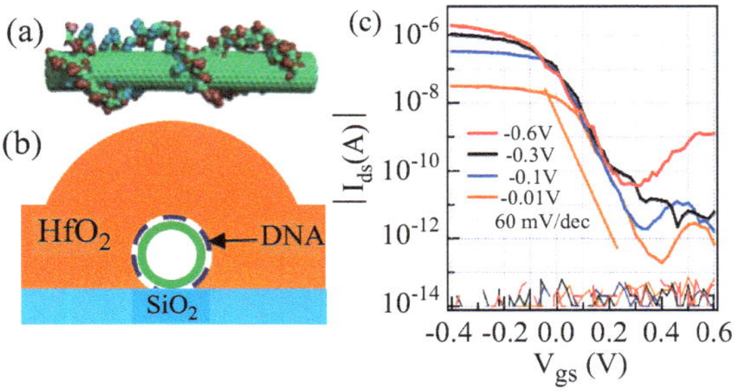

Figure 4.3 The schematic diagram of (a) an SWCNT is modified by poly-T DNA top-gate CNT transistor and (b) the top-gate structure, (c) the transfer characteristic of top-gate CNT transistors with 2–3 nm dielectrics. Reprinted with permission from Ref. [49], Copyright 2006, American Chemical Society.

Besides, gate design is significant to suppress the leakage current of a CNT transistor. Experimental measurements of CNT field-effect transistors (CNFETs) often exhibit substantial off-state leakage current (I_{off}). Recently, Srimani et al. experimentally revealed the significant impact of gate-induced drain leakage (GIDL) on CNFETs and designed asymmetric gates for mitigating the off-state leakage current [81].

4.3.2 Schottky Barrier

An SB exists at the CNT–metal contact and presents a major obstacle, which impedes the realization of ballistic transport CNT transistors. SB is formed between metal and SWCNTs when source/drain metal electrodes directly contact with SWCNTs, and CNT transistor fabricated in this way is referred to as SB CNT transistor. The gate electrode tunes the height and width of the SB to control the tunneling possibility of carriers [82]. It has been reported that the SB is sensitive to work function of the contact metal [83]. SB CNT transistors using metal (Au, Ti, Ni, etc.) as source/drain electrodes suffer from the presence of a large SB [84–86], which severely limits the on-state conductance. When using metal Pd (5.1 eV) [87], Sc (3.3 eV), and Y (3.3 eV) [88, 89] as source/drain, the work function will form a small or even a negative SB. Javey et al. reported a single-tube ballistic CNT transistor, which exhibited a high on-state conductance of up to 0.65 G_0 at room temperature and a high saturation current of up to 25 μA by using Pd as source/drain electrodes [41]. The on-current conductance of the CNT transistor changes sensitively with the work function. H_2 has been used to decrease the work function of Pd [90].

Besides the work function of metal electrode, the wettability of metal on SWCNT also affects the formation of metal/SWCNT interface [91–94]. Perebeinos et al. found that SWCNT arrays with very small pitch suffered from contact problems [95]. More seriously, the metal dewetted from the substrate and rode on top of the CNTs. Brady et al. reported a quasi-ballistic CNT transistor using SWCNT array with a density of 47 tube/μm as conduction channel in 2016 [96]. They conducted extensive solvent rinsing and annealing treatments at 400°C in vacuum to remove solvent residues and to insulate side chains of the polyfluorene wrappers [97–98]. By this way, the dewetting between Pd/SWCNTs interface was overcome. Consequently, the on-state current density and conductance per SWCNT were up to 900 μA/μm and 0.46 G_0, respectively.

Even though the SB has been suppressed by selecting proper contact metal, the SB will still inevitably result in bipolar conduction behavior. For a CNT transistor with a zero SB for holes, it still exhibits bipolar behavior when the thickness of dielectric is reduced. Two reasons account for this phenomenon: (1) The tunneling barrier is very thin after the gate-oxide thickness is greatly reduced. (2) Furthermore, the small effective mass of electrons in CNTs facilitates the quantum-mechanical tunneling [99].

In contrast to the direct contact between source/drain and SWCNTs in SB CNT transistor, chemical or electrical doping at the source/drain extension area can form SB-free CNT transistors [100, 101]. The chemical or electrical doping process eliminates the SB between the source/SWCNTs, and CNT transistors fabricated by this way are referred to as metal-oxide-semiconductor (MOS) CNT transistors. This method suppresses the bipolar conduction induced by SB, extends the channel length scaling limit because the density of metal induced gap states will be significantly reduced, delivers more on-current since it operates like an SB CNT transistor with a negative SB height [102], and allows faster operation since the parasitic capacitance between the source and gate electrode can be eliminated by self-aligned structure [87, 103]. As demonstrated by Zhang et al., the n-type CNT transistors with self-aligned structure yielded a very small gate delay time of 0.86 ps [104]. In addition, smaller leakage current can be obtained because the leakage current is controlled by the full bandgap of CNTs and band-to-band tunneling. Javey et al. exposed an SB CNT transistor to K vapor in vacuum, and the treated FET exhibited a high on-current of 8 μA at V_{ds}= 0.5 V and a transconductance of 20 μS [105]. However, there are no functional groups on the surface of SWCNTs, so it is difficult to form stable chemical doping [49].

As shown in Fig. 4.4, Shiomi et al. reported a laterally integrated heterojunction of graphene and a CNT network film with individual gate electrode to tune the band alignment corresponding to the Fermi-level shift of graphene in contact with the semiconducting CNT network film [106]. Multiple gate structures were designed to apply a gate bias to CNTs and graphene separately. With the graphene junction, the I_{on} was improved. This method is based on the thermionic emission theory: the barrier height between graphene and CNTs for both the conduction and valence band sides varies from 70 to 85 meV, with a linear change as a function of the applied gate voltage to graphene.

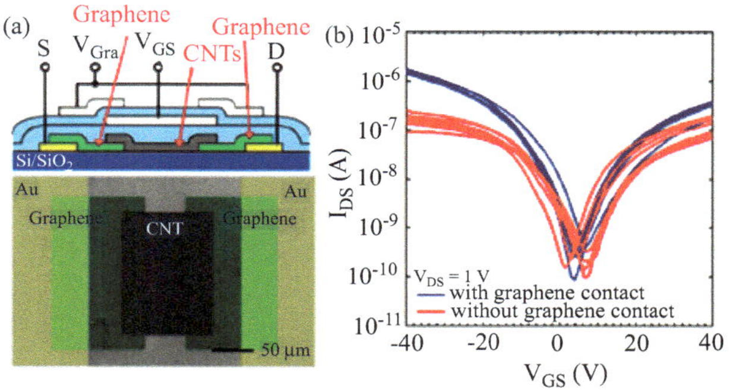

Figure 4.4 (a) Schematic of the graphene/CNT junction transistor (b) I_{DS}–V_{GS} curves with and without a graphene junction. Reprinted with permission from Ref. [106], Copyright 2019, American Chemical Society.

4.3.3 Unipolar Conduction Behavior

CMOS technology is the most common circuit paradigm for modern ICs because of its low static power dissipation and high noise margin [107]. Design and fabrication of p- and n-type CNT transistors with matched performance are vital to develop CNT-transistor-based CMOS technology. The intrinsic CNTs have symmetric E-k relation for both electron and hole [108, 109], which means the potential to fabricate complementary CNT transistors with symmetric performance. However, the fabricated CNT transistors exhibit predominant p-type conduction behavior due to the adsorption of oxygen molecules [110]. Therefore, in order to fabricate stable complementary CNT transistors, it is important to review the physical mechanism and research progress of conduction behavior.

Some gases have a great influence on the conduction behavior of CNT transistors. Collins et al. experimentally verified that the electronic properties of SWCNTs were extremely sensitive to the chemical environment [111]. Oxidizing gases such as O_2 [111] and NO_2 [112] adsorbed on SWCNTs are equivalent to negative fix charges stored near the source and drain contacts, which will decrease the barrier height for holes and lead to p-type dominant behavior of CNT transistors. In contrast, H_2O and NH_3 molecules [113, 114] prefer to contribute their electrons to substrates, resulting in n-type conduction behavior. The problem is that the polarity modulation through gas adsorption is not stable, since gas molecules can be easily desorbed via annealing [115–117].

Besides, it is reported that using a high work function metal (such as Pd) [41, 87] and a low work function metal (Sc, Y, and Ca) [88, 89] as source/drain electrodes can selectively inject holes or electrons into the intrinsic SWCNTs, thus achieving high-performance p/n-type CNT transistors. High-performance CNT transistors with an on-current conductance up to 0.65 G_0 and an on-current of about 25 μA per tube have been reported [41].

It is worth mentioning that the unipolar SB CNT transistors can also be realized by tuning the thickness of oxide. When the thickness of dielectric is reduced, SB CNT transistors with zero SB exhibit bipolar characteristics [118]. The bipolar characteristics become pronounced when the thickness is reduced from 500 nm to 7 nm [41, 87, 88, 104].

Chemical or electrical doping can also suppress the bipolar conduction [119, 120]. Lau et al. demonstrated an electrostatic doping technique. The key to this technique is utilizing ALD to encapsulate CNTs with nonstoichiometric oxides. It was shown that ALD achieved precise control of oxide stoichiometry, which translated to direct control on the amount of CNT doping [121].

4.3.4 SWCNT Channel

Though it is more feasible to use a transistor based on a single SWCNT to investigate the physical mechanism of CNT, SWCNT random networks and arrays are adopted in practical applications more commonly. It is commonly accepted that the SWCNT random networks form conducting paths through tube–tube contact, and tunneling effect [122, 123] also contributes as a minor factor. Figure 4.5 shows the effect of semiconducting purity and channel length on electrical parameters, including on-current density, I_{on}/I_{off}, transconductance, and mobility of CNT transistor [77]. As can be found from the graph, the ratio of semiconducting to metallic nanotubes leads to a trade-off between the on-current, I_{on}/I_{off}, and device mobility. The higher semiconducting purity means higher I_{on}/I_{off} but lower on-current and mobility. The CNT transistor demonstrated a device mobility of 67 $cm^2\ V^{-1}s^{-1}$ and an I_{on}/I_{off} of 10^3 by using 95% semiconducting nanotubes. When the purity of semiconducting nanotubes was increased to 98%, the mobility of CNT transistor was decreased to 31 $cm^2\ V^{-1}s^{-1}$ while the I_{on}/I_{off} was increased to 3×10^5. As channel length decreased, the device mobility of transistors fabricated with 95% semiconducting CNTs increased while transistors fabricated with 98% semiconducting CNTs displayed opposite trend.

Besides the purity of semiconducting nanotubes and channel length, the density of CNT network also affects the device performance. By controlling the rotation speed during spin-coating and the adopted volume of the CNT solution, Lemieux et al. obtained different densities of CNT networks from 3 tube/μm^2 to 22 tube/μm^2. The AFM images are shown in Fig. 4.6a–d [124]. At a lower nanotube density, the on-current and I_{on}/I_{off} will increase with the increase in CNT density. However, once the density reaches a certain level where the metallic nanotubes form pathways to bridge the source and drain electrodes, the I_{on}/I_{off} drops drastically, as shown in Fig. 4.6e.

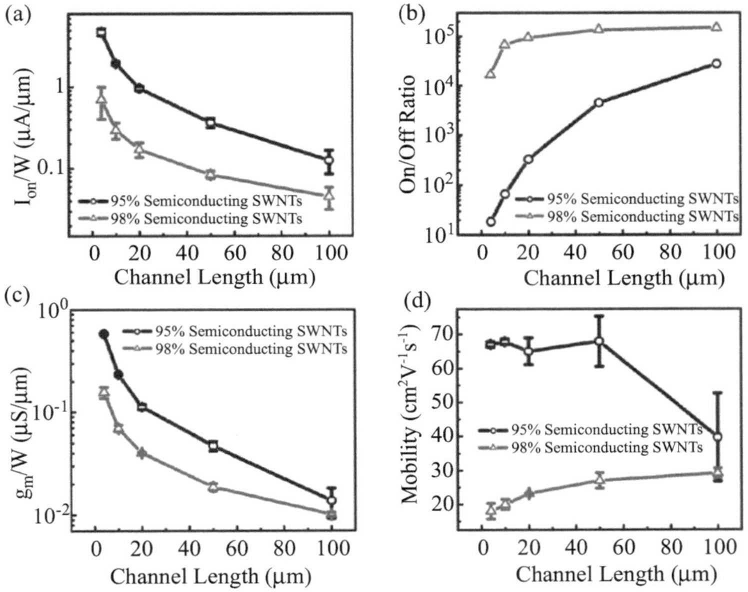

Figure 4.5 Effect of two different semiconducting nanotubes purity of 95% (black trace) and 98% (red trace) on the electrical performance. (a) Current density (I_{on}/W), (b) average I_{on}/I_{off}, (c) normalized maximum transconductance (g_m/W), and (d) device mobility. These data are based on 175 CNT transistors. Reprinted with permission from Ref. [77], Copyright 2010, American Chemical Society.

However, it is still not practical for either CNT random network transistor or single-tube CNT transistor to be used in high-speed circuits due to the low on-state current. Besides, problems such as variation among CNT transistors, low yield, and drive current make single SWCNT-based CNT transistors still not capable of working as the basis of IC technology. In order to improve CNT transistor performance, densely and horizontally aligned arrays of pure SWCNTs are favored for large-scale ICs [125, 126]. In this array channel, the CNT semiconducting purity is critical for transistor performance. Besides, methods for fabricating SWCNT arrays, such as CVD growth on crystal substrate, physical transfer, shear force, Langmuir–Schaefer/Blodgett, and AC dielectrophoresis, have been proposed.

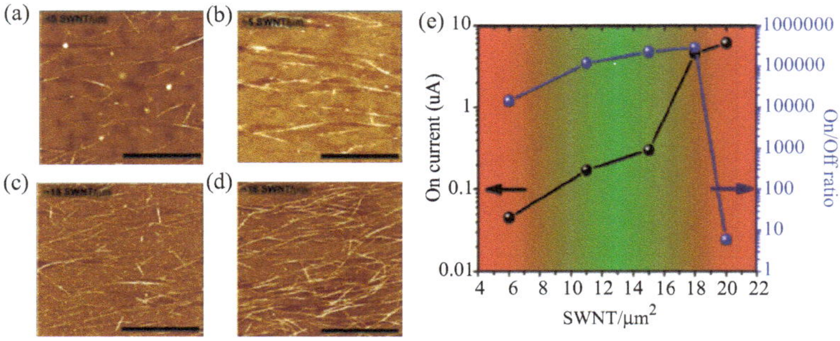

Figure 4.6 Typical AFM images (2×2 μm, z-scale=10 nm) of the different density of CNT network: (a) smaller than 5 tubes/μm; (b) about 5 tubes/μm; (c) about 15 tubes/μm; and (d) about 18 tubes/μm. The scale bars are all 1 μm. (e) The plot of the on-current and on–off ratio of CNTs as a function of CNT network density. Reprinted with permission from Ref. [124], Copyright 2009, American Chemical Society.

In the early stage, SWCNT arrays can be generated by CVD-guided growth on substrates such as mis-cut sapphire [127, 128], and α-plane and γ-plane of sapphire [129, 130]. Earlier in 2005, Kocabas et al. utilized the horizontally grown CNT arrays to demonstrate a high device mobility of 125 $cm^2\ V^{-1}s^{-1}$ and "per tube" mobility of ~9000 $cm^2\ V^{-1}s^{-1}$ [128]. It was reported that more than 99.9% of the SWCNTs lied along the direction of $[2\bar{1}\bar{1}0]$ of the quartz substrate [131]. Besides, both n-/p-type CNT transistors revealed the mobility and scaled g_m approaching 1000 $cm^2\ V^{-1}s^{-1}$ and 3000 S m^{-1}, respectively. Moreover, Hong et al. demonstrated that multiple CVD growth increased the density of highly aligned SWCNT on quartz, and the densities vary from 20 to 30 SWCNTs/μm with a peak value approaching 45 SWCNTs/μm [132]. However, the temperature required for CVD growth reaches 900°C or above [132, 133]. Besides, catalyst metallic particles used for growth will remain in CNT transistors.

The physical transfer method is also used to fabricate SWCNT arrays. The transfer process is schematically shown in Fig. 4.7 [134].

This method works on the principle that van der Waals interaction between SWCNT layer and the receiving surface is greater than the van der Waals interaction and bond strength of the nanotubes to the growth substrate [135, 136]. Combined with H_2/H_2O vapor etching [136], which releases chemical bonds between catalyst and the nanotubes, the patterned SWCNT can be easily lifted away while the catalyst remains on substrate. The key enablers include bilayer, Au, and polyimide (or polyvinyl alcohol, adhesive tapes) [137, 138], which effectively encapsulate CNTs and form adhesive force to peel-off. The density, alignment, and other features of SWCNTs are preserved by the transfer method. Besides, bilayer CNT array superstructure by this method is sufficiently adherent to the substrate to allow for device fabrication without substantially altering the coverage or arrangements of the SWCNT arrays. Shulaker et al. improved the transfer method by using sacrificial layer [137]. Four-time transfer performance was conducted to increase the density of carbon nanotube linearly from 2 to 8 SWCNTs/μm time after time, resulting in an increase in drain/source current density from 4.3 to 17.4 μA/μm. Although the alignment of CNT array by the physical transfer method is nearly perfect, the density of CNT is still too low. Patil et al. estimated that the density must be comparable to 250 CNTs/μm for CNT transistor to surpass Si CMOS FET [139]. Moreover, the physical transfer requires manual placement.

The Langmuir–Blodgett (LB) [140, 141] and Langmuir–Schaefer (LS) [126] techniques are also widely used to fabricate CNT array with well-controlled alignment. One of the advantages is that this method can use purified SWCNT solution to fabricate SWCNT arrays, resulting in high purity. These methods can be explained using a "logs-on-a-river" analogy. After dispersing nanotubes on the surface of water, they spread out to form an SWCNT thin film due to surface tension. An external compressive force is then applied to align them in a certain direction. The difference between the LB method and the LS method is that the former uses a vertical transfer method, while the latter uses a horizontal transfer method. Compared with the LB method, the LS method produces films with better alignment and higher yield because of the reduced disturbance of the CNT film on water. Using the LS method, Cao et al. [126] reported the successful assembly of semiconducting SWCNT array and achieved a density of more than 500 tubes/μm, a drive current density of more than

120 μA μm^{-1}, a transconductance larger than 40 μS μm^{-1} and I_{on}/I_{off} over 10^3.

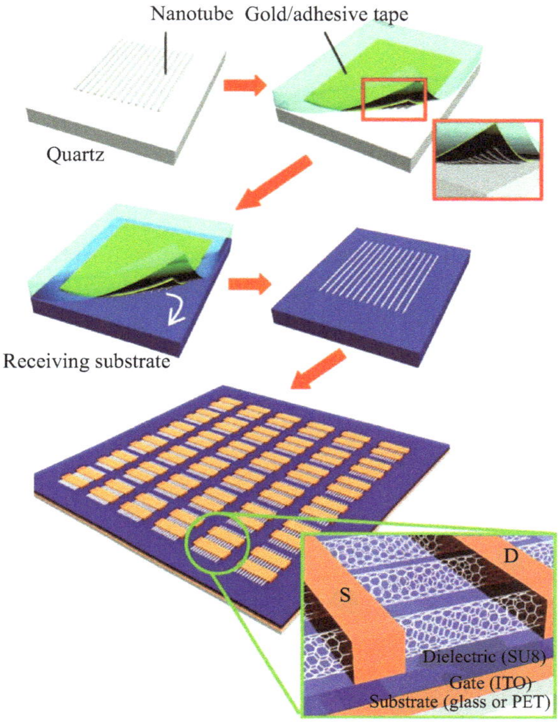

Figure 4.7 Schematic diagram of a CVD-grown SWCNT film transfer from quartz substrate to glass or PET. Reprinted with permission from Ref. [134], Copyright 2009, American Chemical Society.

Recently, Jin et al. [142] reported pure SWCNT array with thermocapillary effect in thin organic coating. As shown in Fig. 4.8, the organic coats all the CNTs, acting as a thermocapillary resist. While a large voltage applied, a large current goes through metallic SWCNTs and produces Joule heating that drives thermocapillary resist away to form open trenches. Then the reactive ion etching physically eliminates the metallic SWCNTs. The purity of SWCNTs was increased up to 99.9%, the I_{on}/I_{off} of CNT transistor is about 10^4, and mobility reached 1000 $cm^2\ V^{-1}s^{-1}$.

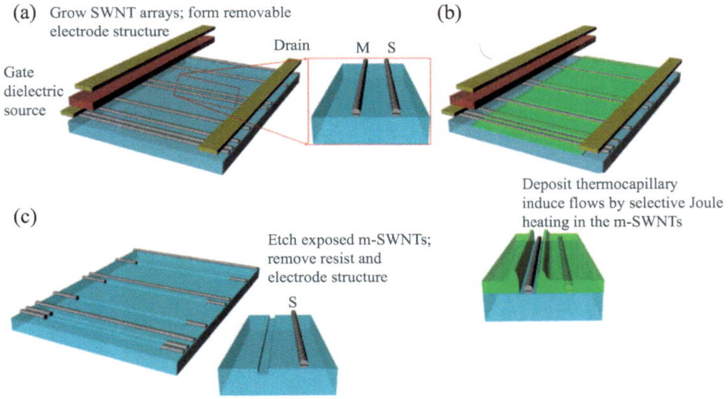

Figure 4.8 Process for exploiting thermocapillary effects in the purification of arrays of SWCNTs. Reprinted from Ref. [142], Copyright 2013, by permission from Springer Nature.

The conjugated polymers such as regioregular poly (3-alkylthiophene)s, dithiafulvalene, and thiophene copolymers can selectively disperse SWCNTs [143–145] and fabricate densely aligned CNT arrays [146], as shown in Fig. 4.9a. By this method, Cao et al. reported their 99% purity SWCNT arrays by the Langmuir–Schaefer method, with a high density of more than 500 CNTs/μm. As shown in Fig. 4.9b [126], the current density of CNT transistors based on this array is more than 120 μA μm^{-1}. By using a large excess of the conjugated polymer poly [(9,9-dioctylfluorenyl-2,7-diyl)-alt-co-(6,60-(2,20-bipyridine))] (PFO-BPy) in toluene to selectively wrap the semiconducting species [147] and then fabricating the SWCNT arrays with a density of 50 tubes/μm by floating evaporative self-assembly method, Brady et al. reported the first quasi-ballistic CNT array transistors [96]. After removing the organic solvent residue and surfactant clinging to the surface of SWCNTs by annealing in vacuum under 300 to 400°C, the CNT array transistor exhibited a high conductance of 0.46 G_0 per CNT, and the conductance of the arrays reached 1.7 mS μm^{-1}.

Recently, by using the vacuum-filtration technique, He et al. obtained a uniform, wafer-scale (> cm^2) SWCNT film with controllable thickness (several 100 nm). The SWCNT film has high-level alignment and cross-sectional packing density ($\approx 10^6$ μm^{-2}), satisfying the requirement of the optoelectronic applications due to the controllable infrared (IR) light absorption [148]. Besides solution-based SWCNT aligning techniques, Tune et al. introduced a versatile post-fabrication dry shearing treatment to bring in SWCNT alignment on the film surface, along with roughness and uniformity improvement [149].

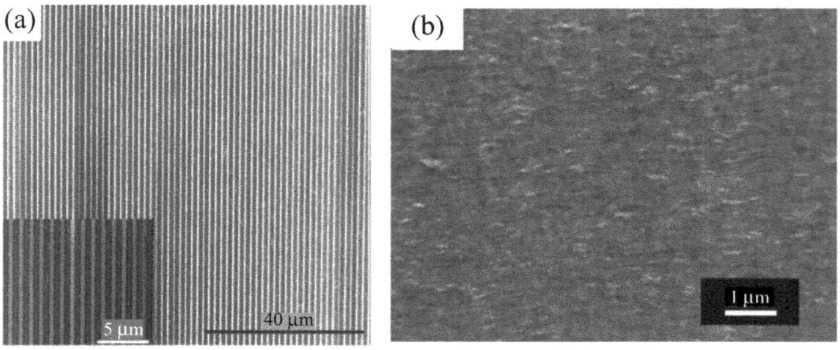

Figure 4.9 (a) Large-area SEM image of the shear-aligned SWCNTs on 500 nm wide solvent-wetting regions. The inset is a zoomed-in image of the same area. (b) SEM images of aligned nanotube arrays transferred onto solid substrates. Figure (a) reprinted from Ref. [146], Copyright 2015, with permission from John Wiley and Sons, and Fig. (b) reprinted from Ref. [126], Copyright 2013, by permission from Springer Nature.

Although many methods have been proposed to fabricate dense and pure SWCNT arrays, they are still limited to small-area production, which greatly impedes the real application. More practical methods are being explored to realize the CNT array on large scale in a more efficient way.

4.4 Flexible and Stretchable CNT Transistors

Owing to the superior mechanical flexibility and electrical properties, CNT has a great potential to be applied to wearable, human–machine interface, display, prosthetics systems, and other consumer electronics. Transistors on rigid and soft substrates share many common critical issues, which have been thoroughly discussed in the previous section. In this section, we will mainly discuss the state-of-the-art advances on flexible and stretchable CNT transistors.

4.4.1 Fabrication Technology

Fabricating CNT transistors on flexible electronics is different from that on rigid substrate in many aspects. For example, flexible substrate made from polymer cannot bear high temperature; the conventional deposition methods such as ALD and sputtering cannot be used to deposit CNTs and other organic materials, and the polymer cannot sustain the traditional plasma etching as well. Over decades, various fabrication methods have been proposed to realize flexible electronics. Here we introduce the most widely used printing techniques, including ink-jet printing, R2R printing, and transfer printing. These methods have their own specialties in solving the issues mentioned above.

Ink-jet printing was first used to fabricate organic light-emitting diode (OLED) by Hebner et al. in 1998 [150]. After that, it is intensively investigated to fabricate flexible electronics. In 2011, Chen at al. successfully fabricated fully printed SWCNT circuits for OLED control and display [151]. The printed transistors adopted 98% semiconducting SWCNTs as the channel and printed Ag as electrodes, achieving a high carrier mobility of 10–30 $cm^2\ V^{-1}s^{-1}$ and I_{on}/I_{off} of 10^4–10^7 (Fig. 4.10). Generally, ink-jet printing consists of a printer and the ink made by metal or semiconducting materials. The ink requires a low viscosity of about 10 mN s^{-1} to be smoothly ejected from the nozzles and wet the substrate. However, the low viscosity causes the expansion of deposited drops and affects the resolution of printed pattern. The surface energy of substrate also affects the spreading of ink. The low surface energy can help to prevent the spreading of ink, while it decreases the wettability and adhesion. The typical volume of a droplet ranges from 1 to 100 pL [152] corresponding to a diameter of around 10 μm. Many efforts are taken to push the limit, such as modifying the surface energy of substrate and the drive voltage applied on actuator.

To date, R2R processing has been applied in numerous fields such as flexible electronic devices, flexible solar panels, and printed thin-film batteries, due to advantages in low-cost and large-scale production. R2R is a family of manufacturing techniques involving continuous processing of electronics on flexible substrate, as it is characterized by transferring materials between two moving rolls of plates. The manufacturing process can be described as additive or

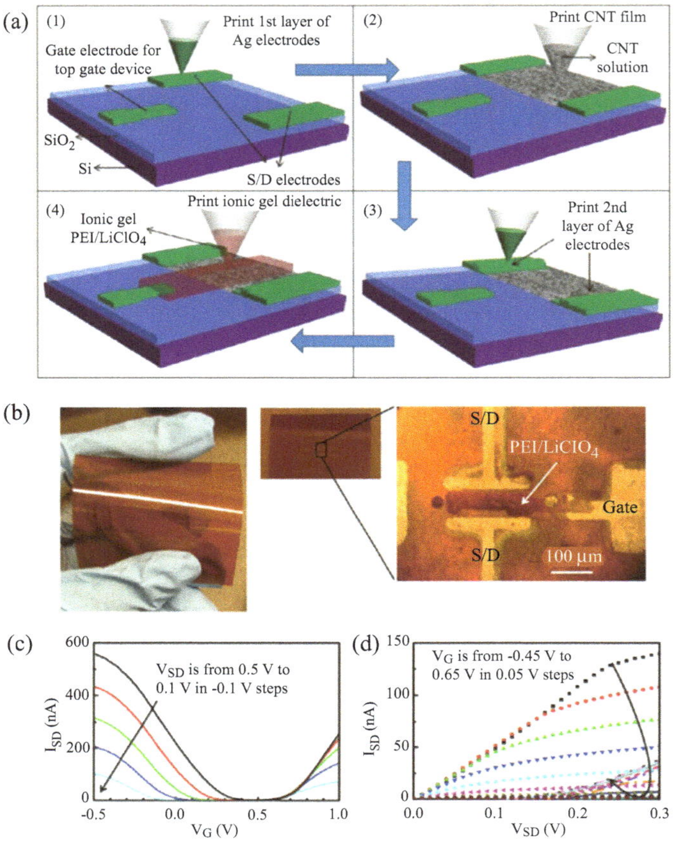

Figure 4.10 (a) Schematic diagram illustrating the fully printed back-gated and top-gated SWCNT fabrication process. (b) Optical images of top-gated SWCNT sprinted on Kapton. (c) Transfer (I_D–V_G) characteristics of one typical top-gated SWCNT on Kapton with bipolar behavior. (d) Output (I_D–V_D) characteristics of the same device. Reprinted with permission from Ref. [151], Copyright 2011, American Chemical Society.

subtractive printing on a roll of plastic or metal foil in a continuous manner. Gravure printing, flexographic printing, and reverse-offset printing are the most popular R2R techniques [153–155]. The process of gravure printing can be divided into three steps. First, the cell engraved on the cliché is filled with inks, then the excess ink is wiped out from the surface of the cliché, and finally the ink remaining in the engraved cells is transferred to the substrate (Fig. 4.11). Therefore, the ink-transfer ratio is mainly determined by the capillary effect, viscosity of ink, and surface energy of receiver substrate. Compared with gravure printing, the flexographic printing uses flexible photopolymer printing plates wrapping around rotating cylinders. Only the raised area in the pattern cylinder is inked, and the pattern is transferred to the substrate. It is essentially a modern version of letterpress, which can be used for printing on a large variety of substrates at high speed. The reverse-offset printing is a commonly

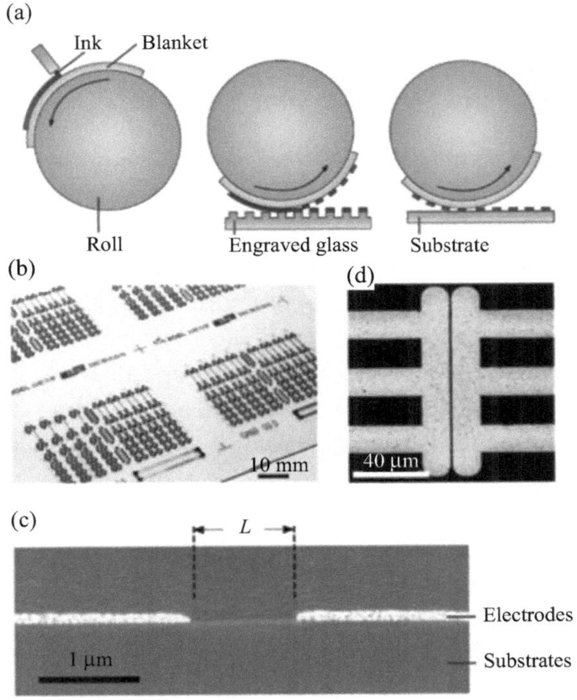

Figure 4.11 Reverse-offset printed conductive electrodes. (a) Schematic illustration of the reverse-offset printing method. (b) Top-view image of reverse-offset printed electrodes formed on 120 × 120 mm glass substrates. (c) Photograph of printed electrodes with a width/gap of 15/0.6 μm. (d) Cross-sectional SEM image of the printed electrodes. Reprinted from Ref. [155], Copyright 2015, with permission from WILEY-VCH Verlag GmbH & Co. KGaA, Weinheim.

used technique, in which subtractive and additive processes are both employed. Briefly, the ink-coated blanket (usually made of rubber) is engraved by glass to form the pattern, and the ink remaining on the blanket is then transferred to the receiver substrate. In 2015, Fukuda et al. reported that the reverse-offset printing method could pattern devices with channel lengths shorter than 1.0 μm [155]. In 2018, Zschieschang et al. reported that the stencil lithography method could pattern organic thin film transistors (TFTs) with a channel length of 300 nm on Si substrates and 125 μm on PEN substrates [156], demonstrating its potential for large-area production in flexible electronics.

4.4.2 Flexible CNT Thin-Film Transistors

Realization of flexible CNT transistors with high mechanical flexibility, durability, and environmental stability essentially requires each component to own these properties. Wang et al. used Au, Pd, and Ti as source/drain/gate electrode on PI (polyimide) substrate to achieve a bend curvature of about 1.27 mm [157]. More recently, Cao et al. fabricated ultra-flexible CNTs using 1.4 μm poly(ethylene terephthalate) (PET) film as substrate, 5 nm poly-L-lysine functionalized SiO_2 as dielectric, and Ti/Pd as source/drain [158]. They reported a bending radius down to 40 μm. In addition, the resulted transistors showed mobility of 12.04 $cm^2\ V^{-1}s^{-1}$, high I_{on}/I_{off} of about 10^6, ultralight weight (<3 g m^{-2}), and good mechanical robustness, which could accommodate severe crumpling and 67% compressive strain, as shown in Fig. 4.12. Recently, Zhang et al. achieved high-mobility flexible CNT p-type and n-type transistors with mobility up to 80.26 $cm^2\ V^{-1}s^{-1}$ and 97.09 $cm^2\ V^{-1}s^{-1}$, respectively, together with an I_{on}/I_{off} of more than 10^5 [159]. They chose Pd and Sc metal as source/drain electrodes to form p-type and n-type transistors, respectively. Because of the perfect match between these metals and the semiconducting CNTs, the contacts performed well and contributed to the high mobility. As shown in Fig. 4.13b–d, the performance of p-type and n-type transistors has high symmetry and uniformity, which can endure 2 mm bending radius with I_{on} only decreased by 6.5% after 100 bending cycles.

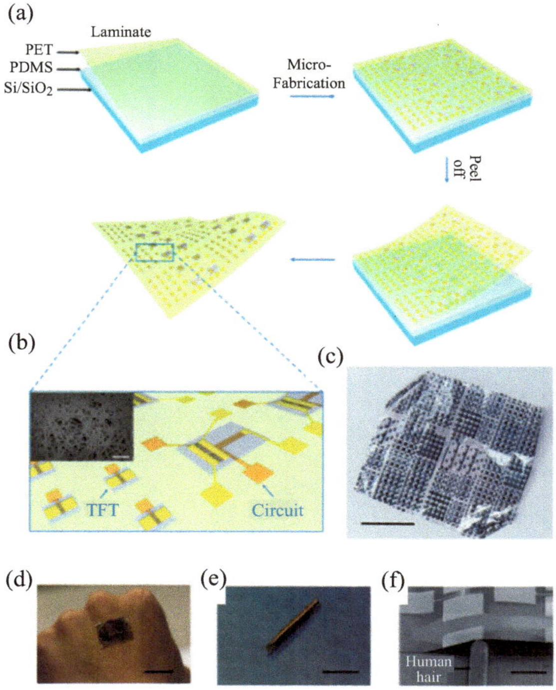

Figure 4.12 Imperceptible carbon nanotube macroelectronics. (a) Schematic diagrams showing fabrication procedure of carbon nanotube electronic foil. (b) Zoom-in schematic diagram of fabricated electronic foil, showing TFTs and circuits. Inset is an SEM image of a typical SWCNT thin film. Scale bar is 500 nm. (c) Photograph of as-fabricated imperceptible nanotube macroelectronics. Scale bar is 1 cm. (d, e) Ultrathin carbon nanotube electronic foil laminated onto human skin (d) and in a rolled-up state (e). Scale bar: 2 cm in (d) and 1 cm in (e). (f) SEM image of the ultrathin SWCNT electronic foil laminated onto human hair sitting on a substrate. Scale bar is 150 μm. Reprinted with permission from Ref. [158], Copyright 2016, American Chemical Society.

Although metal electrodes are most widely used in flexible electronics, they have very limited bend radius and endurance for tensile strain. A promising candidate material to be used as a conductor in flexible and stretchable electronics is the liquid metal alloy. Andrews et al. used eutectic GaIn (EGaIn) liquid metal for source and drain contacts to printed CNT channels. But the surface adhesion between EGaIn and CNTs was poor, which limited the device performance [160]. One more drawback is that metals lack transparency. ITO electrode is another widely used material for transparent and flexible electronics due to its duality of transparency and electrical conductivity. Like most metal oxides, the ITO electrode also suffers from the fatal limit of brittleness.

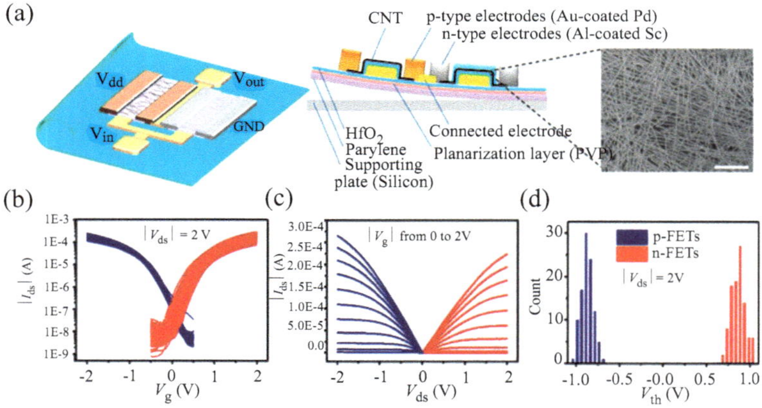

Figure 4.13 (a) Schematic layout and cross-sectional diagram of the flexible CNT-based CMOS structure. (b) Transfer characteristics of 100 p-type and 100 n-type flexible CNT transistors. (c) Typical pair of output characteristics of the p-type and n-type flexible CNT transistors. (d) Statistical distribution of the threshold voltage of the 100 p-type and 100 n-type flexible CNT transistors. Reprinted with permission from Ref. [159], Copyright 2018, American Chemical Society.

Metallic or unsorted CNT is an ideal candidate to work as electrode in flexible and stretchable electronics owing to its high conductivity and flexibility. All-carbon-nanotube (ACNT) transistors applying CNT as channel and electrode simultaneously have attracted numerous interests in recent decades. Xia et al. reported that the ACNT transistors exhibited a high stability at a large strain of 22.5%, where PDMS was used as both substrate and gate dielectric [47, 48]. As shown in Fig. 4.14, Aikawa et al. realized the deformable and transparent ACNT transistors using SWCNTs as channel and electrode, and poly(vinyl alcohol) (PVA) as substrate and dielectric.

They achieved a small bend radius of 1.0 mm and transparency more than 80% [47]. Chen et al. proposed a high-throughput method to fabricate high-performance ACNT transistor with an I_{on}/I_{off} of more than 10^5, a mobility of 33 $cm^2\ V^{-1}s^{-1}$, a bending radius of 3.7 mm, and a small hysteresis. The ACNT transistors also exhibited good uniformity [161]. Recently, Du et al. proposed photolithography-incorporated filtration (PIF) platform. As shown in Fig. 4.15, the fabricated ACNT transistors show an ultralow operating voltage of less than −1 V, an extraordinary subthreshold swing of 170 mV dec^{-1}, a low threshold voltage of −0.3 V, and a high carrier mobility up to 105 $cm^2\ V^{-1}s^{-1}$. Moreover, the device shows a small bending radius of 1 mm and a transistor transparency of 94% [162].

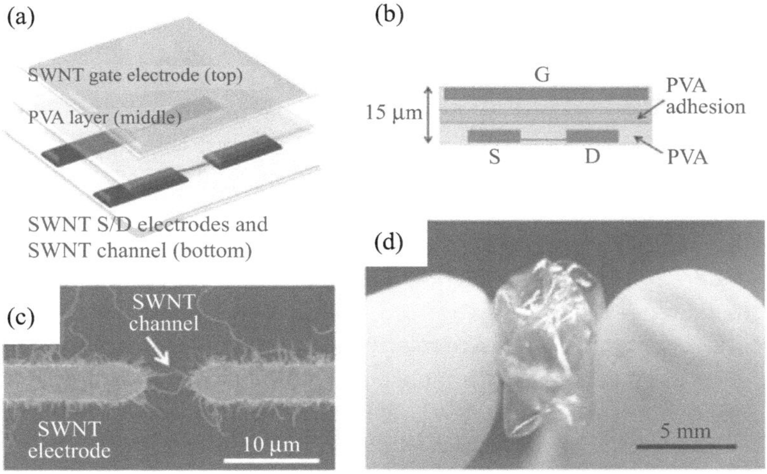

Figure 4.14 (a) Schematic of the layered structure of fabricated ACNT-FETs. All electrodes (source, drain, and gate) and channels consist only of SWCNTs. The substrate and dielectrics layers are composed of flexible polymer. The top layer is the PVA-coated SCWNT global gate electrode; the middle layer is the PVA adhesion layer acting as dielectrics; and the bottom layer contains PVA-coated SWCNT source/drain electrodes and the SCWNT channel. (b) Schematic cross-sectional diagram of the device. (c) SEM image of the channel region. Both the channel length and width are 5 μm. (d) Photograph of a crumpled, yet functional, ACNT-FET device. Reprinted from Ref. [47], Copyright 2012, with the permission of AIP Publishing.

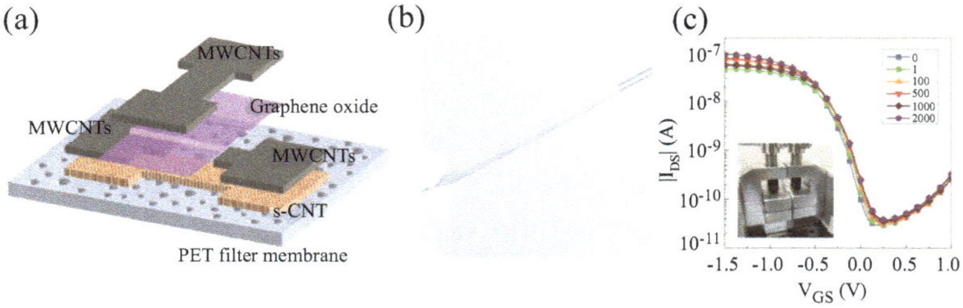

Figure 4.15 (a) Schematic of the flexible and transparent ACNT transistor on the PET filter membrane. (b) Photograph showing the extreme bendability of the fabricated ACNT transistors. (c) Transfer characteristics of an ACNT transistor after different bending cycles at a bending radius of 1 mm. Reprinted from Ref. [162], Copyright 2019, the Royal Society of Chemistry.

In general, dielectrics are often directly responsible for device failure. Metal oxides such as Al_2O_3, HfO_2, and Y_2O_3 are most widely studied high-κ materials. A uniform, ultrathin, and high-quality film can be obtained by ALD or sputtering techniques to ensure a high gate capacitance and a low leakage current. The fragile nature, however, impedes their wide application in flexible electronics. Organic materials such as PMMA [163], PDMS [164], PI [165], PVA [47], SEBS [166], and Epoxy (SU-8) [134] are also explored. These materials have much higher bendability and stretchability than metal oxides and are easier to be processed by spin coating. One drawback is that they have relatively lower dielectric constants. Ion gel is another kind of printable novel dielectric material, developed by Frisbie et al. in 2008 [167]. This material has been widely used to fabricate flexible electronics with printing techniques.

4.4.3 Stretchable CNT Thin-Film Transistors

Stretchable electronics have many novel applications that distinguish them from traditional silicon electronics [168], especially in electronic skin (e-skin) [166] and implantable health care equipment [169]. Owing to the extreme aspect ratio, CNTs, with the modulus up to 1 TPa [170], become the ideal material for stretchable electronics, which have been explored as stretchable conductors [166], supercapacitors [171], and sensors [172, 173] successfully in recent years. However, the realization of stretchable TFTs based on

CNTs is much more difficult. Typical inorganic gate oxides are fragile and easily induce degradation on the device performance after bending or stretching, while polymer dielectrics usually have large leakage current due to their nature [174]. Here, we briefly present the main progress on stretchable CNT TFTs in recent years.

In 2013, Chae et al. used wrinkled Al_2O_3 as the dielectric layer, SWCNT film as the channel, graphene as the electrodes to fabricate a transparent stretchable thin-film transistor, as shown in Fig. 4.16 [175]. Figures 4.16a–c show the device fabrication process. There existed air gaps in wrinkled Al_2O_3, so that the gate capacitance was equivalent to the Al_2O_3 capacitor in series with the air capacitor. This structure achieved a gate leakage current of 10^{-13} A, which was an order of magnitude lower than devices with the flat Al_2O_3 dielectric. The device displayed a high carrier mobility of 40 $cm^2\ V^{-1}\ s^{-1}$ (parallel-plate model), I_{on}/I_{off} of 10^5, and an operating voltage less than 1 V. After horizontal and vertical biaxial stretching for 1000 times at 10% strain, the I_{on}, mobility, and I_{on}/I_{off} of the device did not change significantly, as shown in Figs. 4.16d and e, demonstrating a promising solution for designing flexible, bendable, twistable, and stretchable electronic devices.

However, the structure-engineering approaches used in bulked materials to fabricate stretchable electronic devices are very complicated and costly and are difficult to control, which results in a low device yield [154]. The devices usually cannot withstand too much strain. Therefore, using intrinsically stretchable components is a more effective method to develop stretchable electronics.

Ion gel is one of the intrinsically stretchable materials with high ionic conductivity, specific large capacitance [176, 177], which can be used as dielectric layers in stretchable electronic devices. In 2014, Xu et al. fabricated stretchable CNT TFTs by transfer and pre-strain approach using ion gel as the dielectric layer, SWCNT film as the channel, Au/Cr (25 nm/4 nm) metal films as source, drain, and side-gate electrodes, as shown in Figs. 4.17a–d [178]. The main reason to use the pre-stretching is that the electrode is made of a bulk metal film. The device could operate in a large tensile strain range from 0 to 50%. Owing to the large capacitance and stretchability of the ion gel gate dielectric, an $I_{on}/I_{off} > 10^4$, a mobility of 10 $cm^2\ V^{-1}s^{-1}$, and a low operating voltage of < 2 V were achieved and maintained during repeated mechanical stretch cycle, as shown in Figs. 4.17e and f.

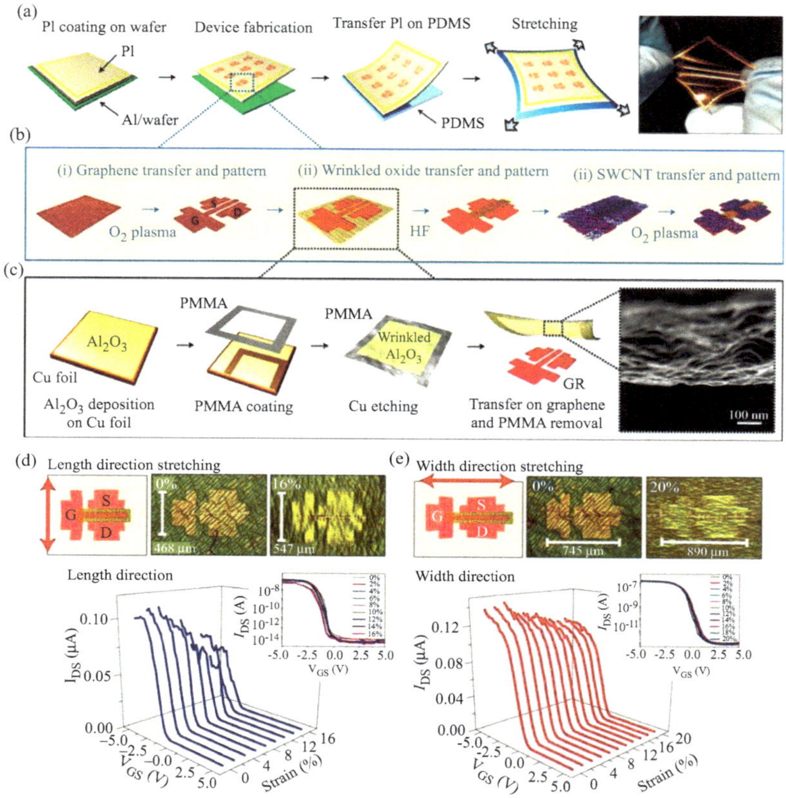

Figure 4.16 Stretchable CNT TFTs enabled by wrinkled Al_2O_3 gate dielectrics. (a) Schematic of whole fabrication procedure. (b) Schematic of TFT fabrication process. (c) Schematic of wrinkled Al_2O_3 layer fabrication and transfer process (d). (e) Transfer characteristics (V_{DS} = 1 V) of devices with tensile strain applied along the length direction (d) and width direction (e). Reprinted from Ref. [175], Copyright 2013, by permission from Springer Nature.

Utilization of intrinsically stretchable conducting materials not only simplifies the manufacturing process but also improves stretchability. However, the ion gel is sensitive to environment and displays slow response to external electrical field [179], which is not suitable for high-speed applications.

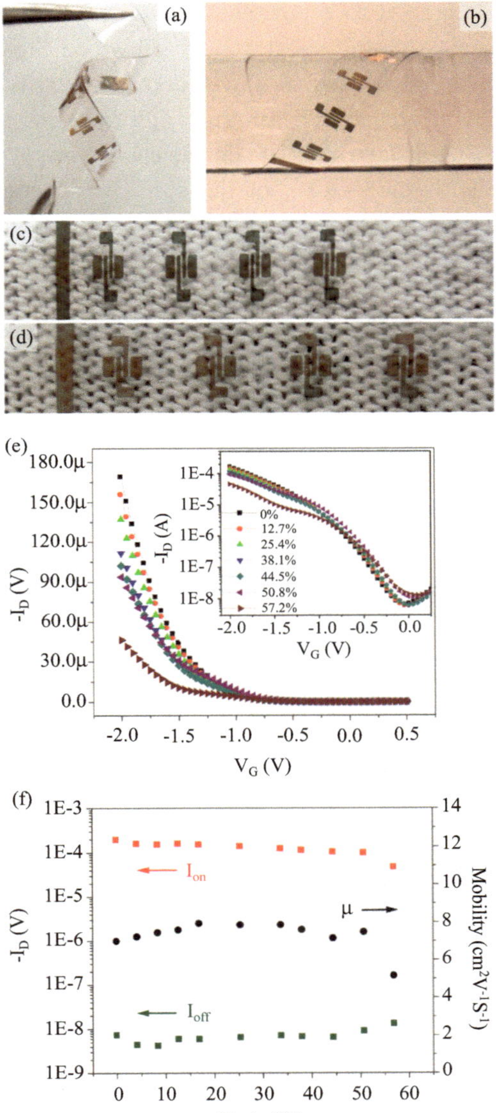

Figure 4.17 Stretchable CNT TFTs enabled by ion gel gate dielectrics. (a) Twisted by tweezers. (b) Wrapped around a glass rod. (c) and (d) Adhered to a textile and stretched to 30%. (e) Transfer characteristics under different strains. (f) Device performance as a function of applied strains. Reprinted with permission from Ref. [178], Copyright 2014, American Chemical Society.

In 2015, Liang et al. reported stretchable and transparent CNT TFTs on an elastomeric substrate, as shown in Fig. 4.18a. Spray-coated AgNWs were used as the source and drain electrodes to achieve high stretchability. Another AgNWs/elastomer film was laminated as gate electrode. The device characteristics had a strong connection with the density of the spin-coated enriched-SWCNT channels, which could be controlled by the amount of SWCNT ink drop casted on the substrates. In order to achieve high carrier mobility, the spin-coated polymeric PU-co-polyethylene glycol gate dielectric with high stretchability and compatibility with the source/drain/gate composite electrode was applied. The TFT exhibited a mobility of 30 $cm^2\ V^{-1}s^{-1}$, I_{on}/I_{off} of $10^3{\sim}10^4$, which could drive an OLED in the full brightness range [180]. Due to the application of all stretchable components, the TFT could be stretched up to 50% while retaining a high mobility, as shown in Figs. 4.18b and c.

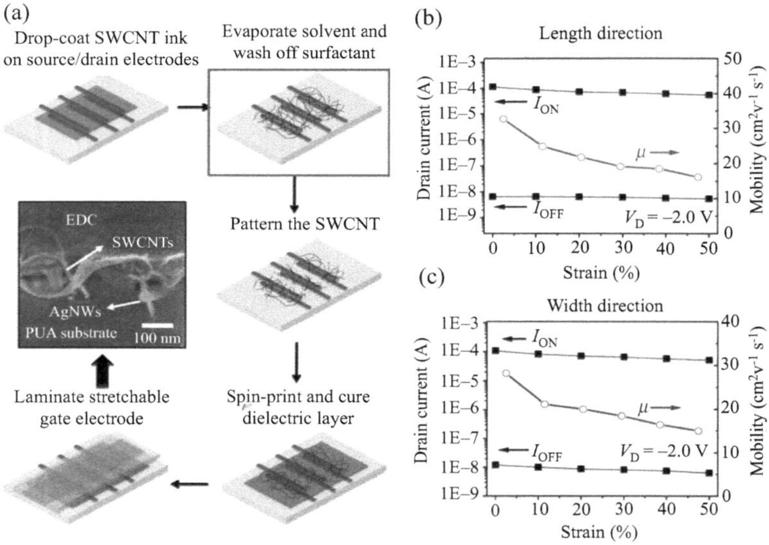

Figure 4.18 Stretchable CNT TFTs enabled by print method. (a) Schematic of fabrication process. (b, c) On, off current and mobility as a function of applied stains (b) along channel length direction, and (c) along channel width direction. Reprinted with permission from Ref. [180], Copyright 2015, Macmillan Publishers Limited.

Large difference exists in mechanical and chemical properties for different components, making it challenging to achieve a stable interface that can survive long-term operations under large strains [181, 182]. Thus, it is better to use the same material for different layers.

Chortos et al. reported the realization of all-carbon stretchable TFTs to withstand large tensile strains in 2015, which could be stretched up to 100%, as shown in Fig. 4.19 [183]. The unsorted CNT networks were spray coated and patterned as source, drain, and gate electrodes. And semiconducting CNT networks were used as channel. Thermoplastic polyurethane (TPU) was used for both dielectrics and substrates. The devices displayed an average low mobility of 0.18 ± 0.03 $cm^2\ V^{-1}s^{-1}$ due to the low ion conductivity of TPU. However, ionic dielectrics can enable a high density of semiconducting CNTs in the channel while retaining a favorable I_{on}/I_{off} of 10^4, compared to electrostatic dielectrics, and the devices could also survive 1000 stretching cycles between 0 and 100%, as shown in Figs 4.19e–g.

Diameter distribution exists in CNT TFT devices, which may affect device performance since the diameter is related to the bandgap for SWCNTs [184]. In 2017, Chortos et al. investigated the limit factors in stretchable all-carbon transistors, as shown in Fig. 4.20 [179]. Different SWCNTs, including high-pressure carbon monoxide (HiPCO) (d = 0.8–1.2 nm), plasma discharge (PD) (d = 1.0–1.5 nm), arc-discharged (AD) (d = 1.2–1.7 nm), and Tuball (d = 1.4–2.2 nm) SWCNTs, were used as channel, respectively. SEBS and unsorted CNTs were used as substrates and electrodes, respectively. The SEBS dielectric aims to minimize hysteresis and bias stress effects. Figures 4.20c–f indicate devices based on large-bandgap SWCNTs exhibited trap-limited behavior caused by the low capacitance of the dielectric, while devices based on smaller-bandgap SWCNTs were more limited by contact resistance. Devices with large-bandgap SWCNTs showed more strain dependence in the electrical characteristics, whereas contact-limited devices were not obvious, as shown in Fig. 4.20b.

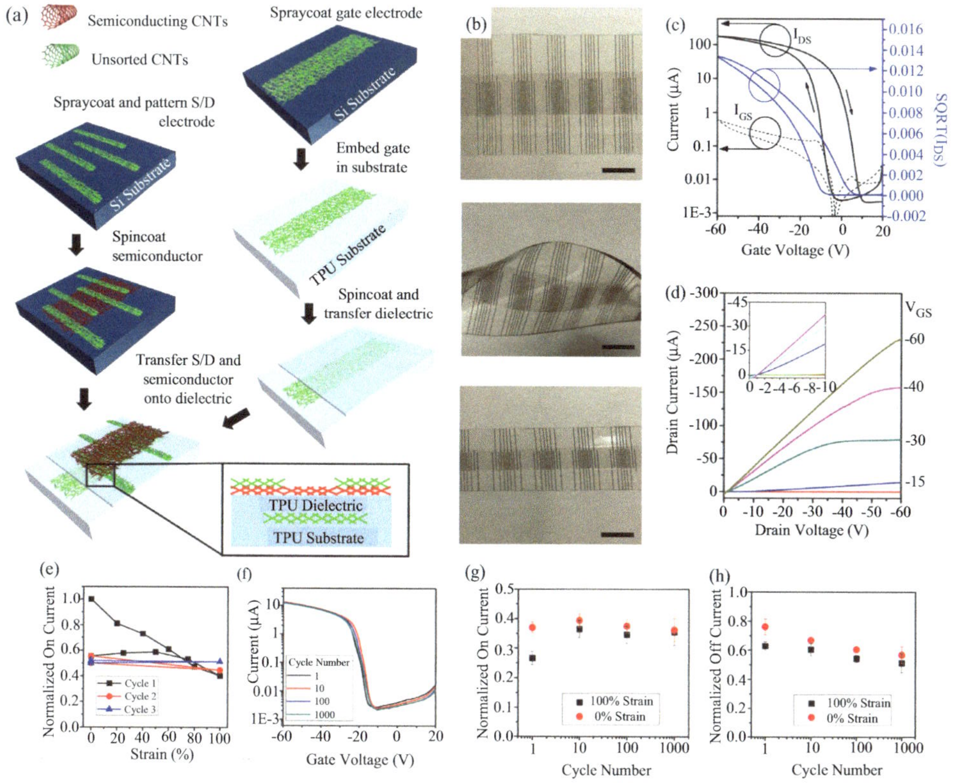

Figure 4.19 Stretchable all-carbon TFTs enabled by TPU gate dielectrics. (a) Schematic of all fabrication procedure. (b) Images of a device unstrained, twisted, and stretched to 100%. (c) Typical transfer characteristic curve, V_D= −60 V. (d) Output characteristics (W = 5 mm, L = 50 μm), inset displays gate voltage from 0 to 10 V. (e) Normalized on-current as a function of applied strains. (f) Transfer characteristics for a device that was cycled to 100%. (g) Normalized on-current and (h) normalized off current during cycling from 0 to 100%. Reprinted from Ref. [183], Copyright 2016, with permission from WILEY-VCH Verlag GmbH & Co. KGaA, Weinheim.

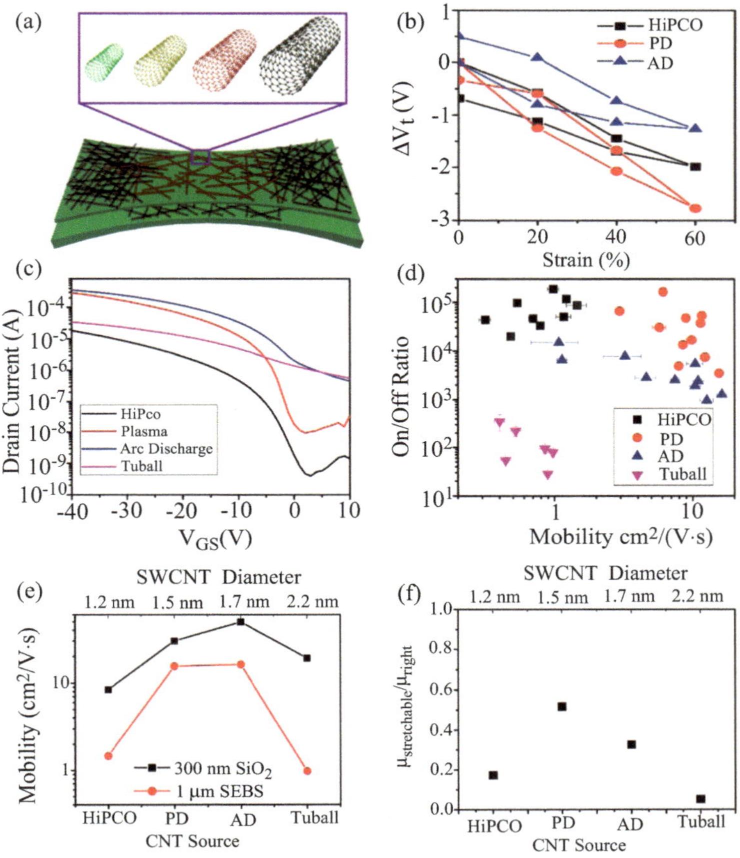

Figure 4.20 Stretchable all-carbon TFTs enabled by four different semi-SWCNT sources. (a) Schematic of SWCNTs with different diameters. (b) Shift in threshold voltage with strain. (c) Transfer curves for devices with each type of SWCNT (W = 4 mm, L = 200 μm, V_D = −40 V). (d) Mobility versus I_{on}/I_{off} for each type of SWCNTs. (e) Mobility versus tube diameter for rigid and stretchable devices. (f) Relative mobility for stretchable devices compared to rigid devices. Reprinted with permission from Ref. [179], Copyright 2017, American Chemical Society.

More functionalities such as washability [185], self-healing ability [186], self-powering [187], etc. have been developed to satisfy practical application requirements. In 2019, Khatib et al. reported all-carbon stretchable transistors with repeatable intrinsic and autonomic self-healing ability, benefiting from the intrinsic self-healing disulfide-containing poly(urea urethane) (PUU), as shown in Fig. 4.22 [188]. A good compromise between the mechanical property and self-healing ability was found by adjusting the proportion of the two substances making up the polymer. The device exhibited an average mobility of 10 $cm^2\ V^{-1}s^{-1}$ and could be stretched up to 50% without significant electrical degradation during 200 cycles. The gate electrode was cut by a sharp blade, and the transistor was recovered after 1 h with negligible change in performance, as shown in Figs. 4.21d and e. The recovery speed was slow when the S/D electrodes were cut, and the device performance could not be restored to the original, as shown in Figs. 4.21f and g. The channel had the worst recovery efficiency, as shown in Fig. 4.21i.

In order to avoid the influence of stretching on the sensors, strain suppression with dynamic differential circuits is developed. In 2018, Zhu et al. reported stretchable temperature-sensing circuits based on CNT TFTs, which could suppress the effects from stretching, as shown in Fig. 4.22 [189]. The circuits used unsorted SWCNT networks as the source, drain electrodes and sorted SWCNTs networks as the channel. The hydrogenated elastomer SEBS substrate by a transfer and lamination method, nonpolar SEBS dielectric layer, and modified PEDOT:PSS via and interconnects were applied to achieve good stretchability, as shown in Figs. 4.22a–c. Figure 4.22d shows good mechanical property of the fabricated devices. Because all parts of the devices applied had a similar Young's modulus and high stretchability, the sensor exhibited stable electrical output and nearly identical temperature response characteristics at 0, 20%, 40%, and 60% tensile strain, as shown in Fig. 4.22g. The stretchable devices demonstrated in this work showed robust and accurate temperature-sensing capability.

To summarize, significant progress has been achieved in CNT-based stretchable electronics, but several challenges remain to be overcome. At present, the device characteristics exhibit low carrier mobility, high operating voltage, large size, low uniformity, low stability, and device degradation under large stresses, which do not match requirements in practical applications. Gate dielectric layers used currently often result in low gate control efficiency [179].

Therefore, stretchable dielectric materials with high dielectric constant and excellent stretchability should be further researched. Furthermore, a reliable integration platform needs to be further explored.

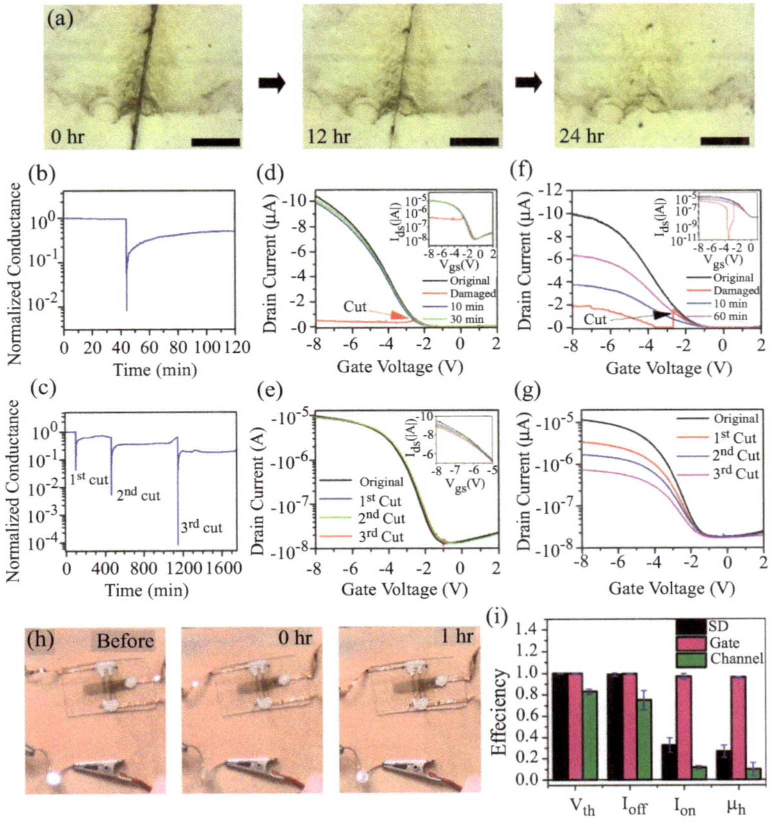

Figure 4.21 Self-healing ability of the transistor. (a) Images of PUU film self-healing process. (b) Conductivity self-healing process in CNT-PUU films. (c) Recovery process after several damage cycles. (d) The healing process of the gate and (e) the healing ability of the gate after several cuts. (f) The healing process of the S/D electrodes and (g) the healing ability of the S/D electrodes after several cuts. (h) The image of the device self-healing process. (i) Self-healing efficiencies after different parts were damaged. Reprinted from Ref. [188], Copyright 2019, with permission from WILEY-VCH Verlag GmbH & Co. KGaA, Weinheim.

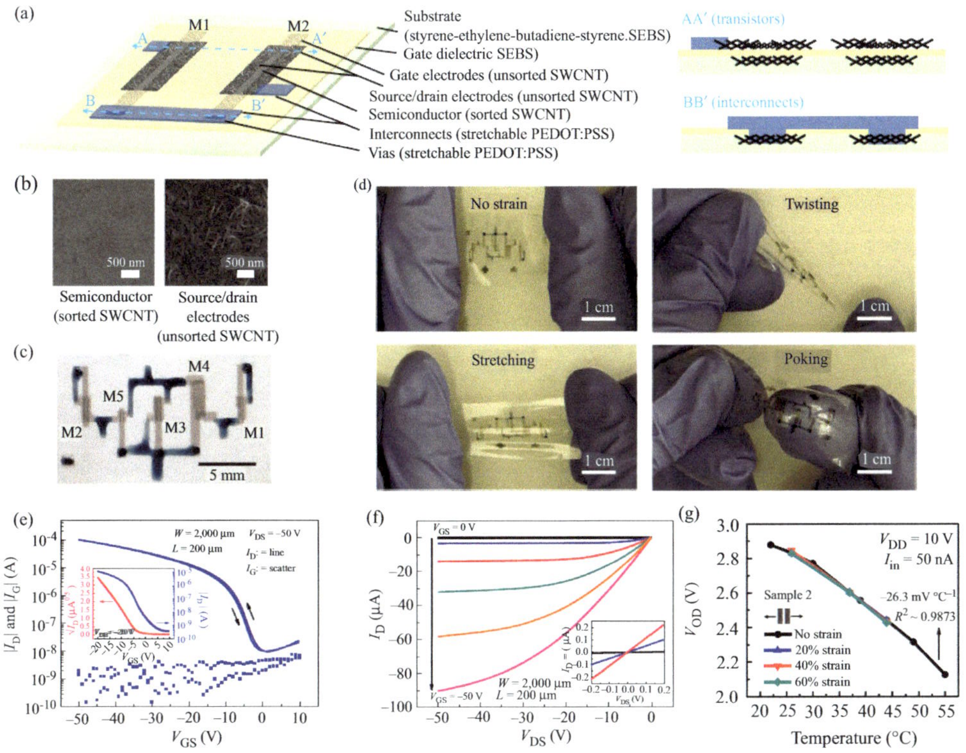

Figure 4.22 Stretchable CNT TFT circuits and temperature sensors. (a) Schematic of devices. (b) SEM images of sorted SWCNT network channels and unsorted SWCNT network electrodes. (c) Image of the circuit for temperature sensor. (d) Image of a temperature sensor under different operations. (e) Transfer characteristics of a CNT TFT. (f) Output characteristics of a CNT TFT. (g) Temperature-sensing performance of a sample with uniaxial 0, 20, 40, and 60% strains. Reprinted from Ref. [189], Copyright 2018, by permission from Springer Nature.

4.5 Applications of CNT Transistors

High-performance CNT transistors with high on-current, mobility, and I_{on}/I_{off}, as well as excellent mechanical characteristics, provide a wide range of applications. Although some issues still remain to be addressed as mentioned above, a lot of experimental demonstrations have shown that CNT-based technologies have promising potential for low-energy and high-performance circuit implementations. In this section, we will review the applications of CNT transistors in ICs, display circuits, as well as sensors.

4.5.1 Integrated Circuits

The extremely high intrinsic carrier mobility and current-carrying capacity make SWCNT a promising candidate for next-generation circuits. Many groups have fabricated integrated circuits based on p-type CNT TFT platforms, especially for high performance, flexible and transparent systems. Wang et al. successfully demonstrated high-performance logic gates, including inverters, NAND, NOR gates, with superior bending stabilities [157]. As shown in Fig. 4.23a, it is desirable to have a narrow and steep transition region to maximize the noise tolerance and the voltage gain [11]. Figure 4.23b presents the inverter gain and noise margin as a function of the supply voltage V_{DD}, showing a respectable voltage gain of 30 at V_{DD} = 5 V and a noise margin close to 0.4 V_{DD}, indicating that the circuit is highly immune to environment-induced electrical noise. Chen et al. fabricated top-gated p-type CNT transistors with high yield, uniform performance, small threshold voltage variation and hysteresis, and realized ICs with medium scale and complicated functions such as a 4-bit adder and a 2-bit multiplier for the first time. Particularly, the CNT ICs were powered by a single low voltage of 2 V [190].

In addition, researchers have spared no effort to convert p-type CNT TFTs to n-type TFTs, aiming to realize complementary CNT transistors. In the early years, Dercyke et al. fabricated the first inverter with a single SWCNT. The gain of the inverter was more than 1, with a large working voltage of 2 V [191]. Javey et al. improved the gain of the inverter to 60 using ZrO_2 as gate dielectrics [78]. In 2003, Javey et al. demonstrated an inverter with a gain of 3.5 and a switching voltage of 1 V [192]. What is worth mentioning is that they fabricated the n-type carbon nanotube transistors by tuning the contacts rather than chemical doping or annealing at high temperature [191, 192]. With the same design scheme, Liang et al. demonstrated the first all-carbon nanotube inverter with an ultralow switching voltage of 0.2 V and a high gain of 9.5 [193]. An n-type CNFET was fabricated by using Sc as source/drain electrode, and complementary CNFETs were interconnected by multiple-walled CNTs. To avoid the effect from the variation of SWCNT diameter, Zhang et al. demonstrated an inverter fabricated on a single SWCNT, which produced a voltage gain as high as 160 [89]. In addition, the end-bonded scheme has been introduced by Tang et al. to fabricate scalable inverter [194]. The gain of inverter was 5, and switching voltage was 0.5 V. This scheme only needs a lower annealing temperature (400–600°C) than 900°C, which is typical for the formation of metal carbides [195]. Recently, the same group successfully fabricated flexible CMOS integrated circuits based on carbon nanotubes, exhibiting great performance. The fabricated flexible ring oscillators achieved a stage delay of only 5.7 ns, due to the reliable n-type doping process, together with highly purified CNTs and high-density CNT deposition [196].

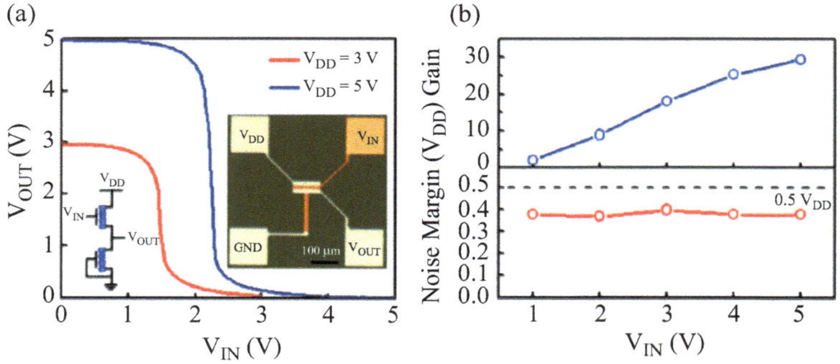

Figure 4.23 CNT integrated circuits. (a) Inverter voltage transfer characteristics (VTC) measured with V_{DD} of 3 or 5 V. (b) Inverter gain (blue line) and noise margin (red line) as a function of supply voltage V_{DD}. Reprinted with permission from Ref. [157], Copyright 2012, American Chemical Society.

Figure 4.24 shows the early demonstration of NOR, OR, NAND, and AND logic gates [197]. The working voltages are rather high, with a value of −1.5 V for OR, AND, and −2 V for NAND. Moreover, the voltages are −4 V for state "1" and 0 V for state "0", respectively.

The n-type CNFETs used in logic gates were obtained by applying a −40 V gate voltage and a 20 V source–drain bias to desorb oxygen molecules on CNT channel. The yield of n-type CNFETs by the so-called "electrical n-doping" is as low as 50%. Additionally, the obtained n-type CNFETs were not stable in air. Wang et al. demonstrated NAND and NOR logic gates, where the n-type CNFET was fabricated by hydrazine doping. The input voltages used for each logic were 2 V for state "1" and −2 V for state "0". Besides, the working voltage was 1 V [198]. Zhang et al. demonstrated the AND and OR logic gates with a working voltage of 1 V. The input voltages used for each logic were 1 V for state "1" and 0 V for state "0" [89].

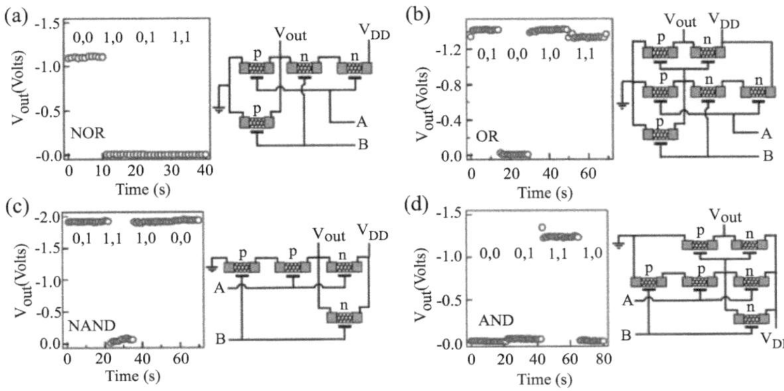

Figure 4.24 Output characteristics and structures of complementary (a) NOR, (b) OR, (c) NAND, (d) AND logic gates. Reprinted with permission from Ref. [197], Copyright 2002, American Chemical Society.

In order to eliminate metallic CNTs and fabricate high-performance CNFET ICs, Patil et al. proposed a method called very large-scale integration compatible metallic CNT removal (VMR) to fabricate the multiple-CNT FETs, which exhibited high I_{on}/I_{off} of 10^3–10^5 [51]. A CNFET half-adder sum generator and a D-latch were demonstrated with p-type CNFETs. The key step is the fabrication of the interdigitated Pd layout after the transfer of CNTs. Electrical breakdown was performed by applying a 5–12 V voltage to CNTs before carrying out the next fabrication process. The purity of obtained SWCNT can reach 99.99%. The bias voltage (V_{bias}) and working voltage (V_{DD}) for half-adder sum generator are 4

V and 3 V, respectively. The V_{bias} and V_{DD} for D-latch are 5 V and 3 V, respectively. The high working voltage and bias voltage are required because the Joule heating degrades the conductance of the remaining semiconducting SWCNTs. Instead of using CVD-synthesized SWCNTs, some researchers use sorted high-purity semiconducting enriched SWCNTs as the channel material, because the solution-processed high-purity SWCNTs offer high electrical performance and ambient compatibility simultaneously.

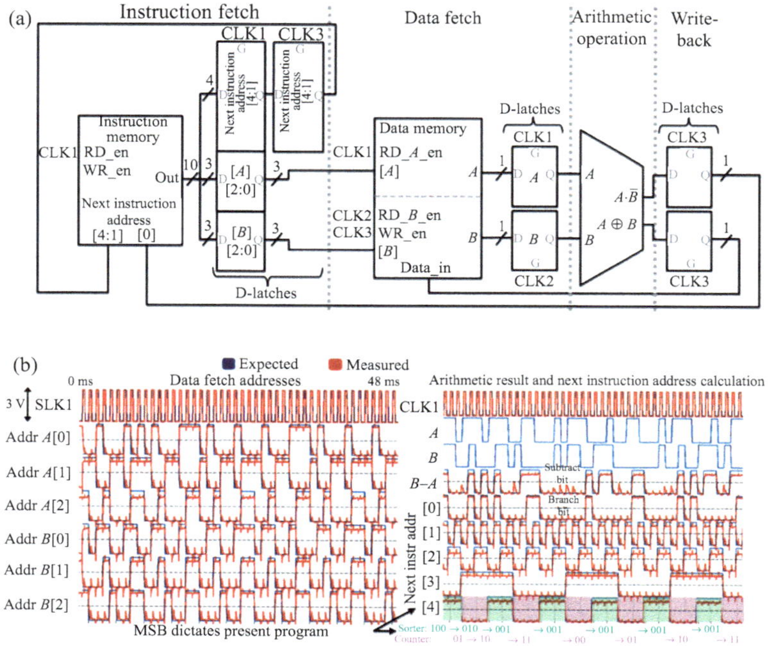

Figure 4.25 (a) Four circuits corresponding with four tasks of the CNT computer. (b) The measured outputs from the CNT computer. Reprinted from Ref. [1], Copyright 2013, by permission from Springer Nature.

Nevertheless, there is still a large gap between simple CNT circuits and a CNT computer. It is not until 2013 that Shulaker et al. demonstrated a programmable machine with entirely CNT transistors [1]. As shown in Fig. 4.25a, the computer performs four tasks, including instruction fetch, data fetch, arithmetic operation, and write back with the instruction and data memories implemented off-chip. The instruction fetch supplies instruction memory (10 D-latches) with the address to read. Data fetch supplies the data memory with the addresses given by the instruction to read. Arithmetic operation performs the computation (subtraction and comparison with zero) on the two data values supplied by the data-fetch unit. Write-back task writes back the result in the data memory at the address of the second data address. As shown in Fig. 4.25b, the CNT computer is a synchronous digital system, which not only can perform counting and integer-sorting simultaneously, but also can run 20 different commercial MIPS (million instructions per second) instruction sets. Although the structure of the computer is compact and the performance demonstrated is simple compared with silicon-based personal computer, the demonstration proves that the imperfections can be overcome and the CNFET technology is feasible for computer.

In 2019, Hills et al. further overcame the intrinsic CNT defects and variations to demonstrate a beyond-silicon microprocessor named RV16X-NANO, built entirely from CNFETs [2], as shown in Fig. 4.26.

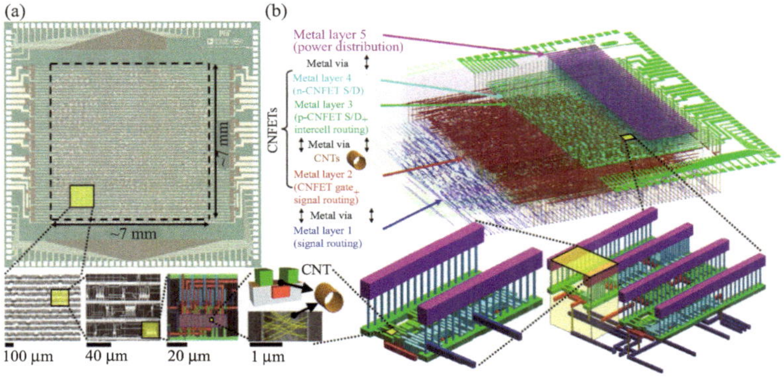

Figure 4.26 RV16X-NANO. (a) Fabricated RV16X-NANO chip. The die area is 6.912 mm × 6.912 mm, with input/output pads placed around the periphery. SEM images with increasing magnification are shown below. RV16X-NANO is fabricated entirely from CNFET CMOS, in a wafer-scalable, VLSI-compatible, and silicon-CMOS compatible fashion. (b) Three-dimensional to-scale rendered schematic of the RV16X-NANO physical layout. RV16X-NANO leverages a new three-dimensional physical architecture in which the CNFETs are physically located in the middle of the stack, with metal routing both above and below. Reprinted from Ref. [2], Copyright 2019, by permission from Springer Nature.

Figure 4.27 shows the architecture and design of RV16X-NANO. The 16-bit microprocessor is based on the RISC-V instruction set, runs standard 32-bit instructions on 16-bit data and addresses, comprises more than 14,000 complementary metal-oxide-semiconductor CNFETs, and operates as modern microprocessors do today. RV16X-NANO was designed and fabricated using industry-standard design flows and processes. Besides, they proposed a manufacturing methodology for carbon nanotubes, a set of combined processing and design techniques for overcoming nanoscale imperfections at macroscopic scales across full wafer substrates. This work further experimentally validates a promising path toward practical beyond-silicon electronic systems [2].

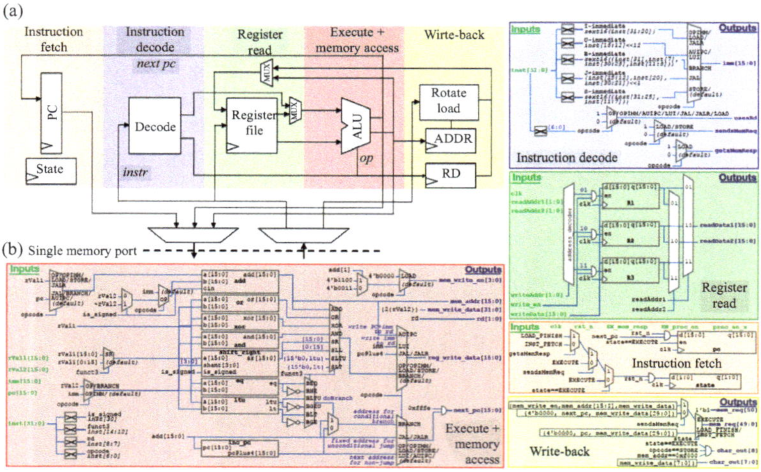

Figure 4.27 Architecture and design of RV16X-NANO. (a) Block diagram showing the organization of RV16X-NANO, including the instruction fetch, instruction decode, register read, execute + memory access, and write-back stages. (b) Schematics describing the high-level register transfer level description of each stage, including inputs, outputs, and signal connections. Reprinted from Ref. [2], Copyright 2019, by permission from Springer Nature.

A capacitive sensor interface circuit is another example, which shows the feasibility of CNFET technology [3]. As shown in Fig. 4.28a, Shulaker et al. presented the image of the live demonstration of the CNFET-controlled handshaking robot. The sub-system consists of two 9-stage ring oscillators and D-latch, in

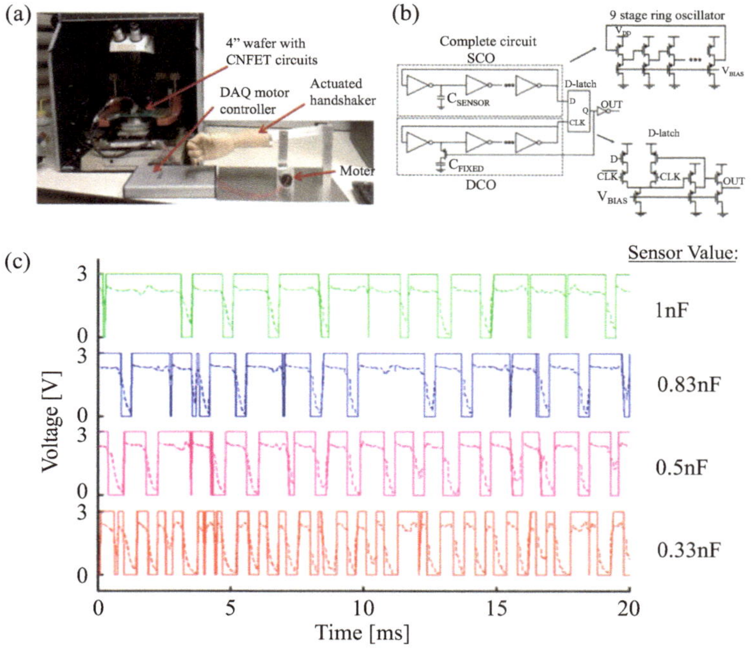

Figure 4.28 (a) Image of the live demonstration of the CNFET-controlled handshaking robot. (b) Circuit implementation of the capacitive sensor to digital interface circuit. (c) Experimentally measured (dotted lines) and digitized (solid lines) over-sampled outputs of the full CNFET sensor interface circuit (over time), for different values of the capacitive sensor. Reprinted with permission from Ref. [3], Copyright 2014, IEEE.

which the CNFETs are p-type and fabricated by the VMR method. The system consists of modulation and demodulation modules: (1) Modulation: the sensor-controlled oscillator (SCO) generates a frequency-modulated oscillating signal dependent on the capacitive sensor value. The change of sensor value will influence the frequency of one inverter delay cell of the SCO. Based on the principle mentioned above, the sensor information can be incorporated in the frequency of the SCO. (2) Demodulation: The digital bang-bang phase-locked loop structure, consisting of a single-bit phase detector (D-latch) and a single-bit digitally controlled oscillator (DCO), demodulates the sensor-modulated signal, digitizing the sensor information. The CNT-transistor-based sub-system demonstrated that the mismatch between the measured and the expected outputs was only 5.5%

(root-mean-square error) at 2013. Besides, the total interface circuit consumed only 336 μW from a 3 V supply. The capacitive sensor to digital interface circuit experimentally demonstrated that the VRM could overcome the inherent CNT imperfections and the CNT transistor is practical for low-power applications.

4.5.2 Flexible Display Circuits

CNT TFTs are also attractive in the application of display area, such as driving the organic light-emitting diode (OLED) [8, 199] and active matrix (AM) OLED [200, 201] due to the high on-current, high mobility, and high I_{on}/I_{off}. In 2009, Wang et al. achieved solution-based and wafer-scaled CNT transistor with high electrical performance of I_{on}/I_{off} (>10^4), current density (~10 μA/μm), mobility (~52 $cm^2\ V^{-1}\ s^{-1}$), as well as yield (>98%) [199]. As a proof of concept, they further demonstrated an OLED circuit controlled by the CNT, realizing the modulation of the light intensity, as shown in Fig. 4.29. Zou et al. [200] and Zhang et al. [201] also demonstrated the CNT driver circuits for AMOLED display. Besides that, efforts have been focused on integrating CNTs into the flexible and transparent displays. In particular, by combining the advantages of printed electronics and CNTs, Chen et al. successfully fabricated fully printed CNT circuits for OLED control and display for the first time [151]. Moreover, Cao et al. have fabricated fully screen-printed top-gated CNT TFTs on both rigid and flexible substrates [202]. All these works have further qualified the application potential of CNTs in the next-generation advanced displays.

Furthermore, numerous researchers have been exploring a scalable, low-cost approach toward e-skin applications using CNT TFTs. As early as 2013, Wang et al. reported the first user-interactive e-skin, which not only spatially mapped the applied pressure but also provided an instantaneous visual response by a built-in active-matrix OLED display with red, green, and blue pixels controlled by CNT TFTs [203]. This system consisted of three components—thin-film transistor, pressure sensor, and OLED arrays—monolithically integrated over large areas on a single plastic substrate (Fig. 4.30). When a certain pressure was applied, the OLED would partially emit light, and the intensity was determined by the pressure. In 2018, Nela et al. demonstrated a large-area high-performance flexible pressure sensor with carbon nanotube active matrix. This electronic skin possessed 16 × 16 pixels spanning over a 4 inch area, indicating the ability to build high-speed flexible complementary integrated circuits using CNT TFTs [204].

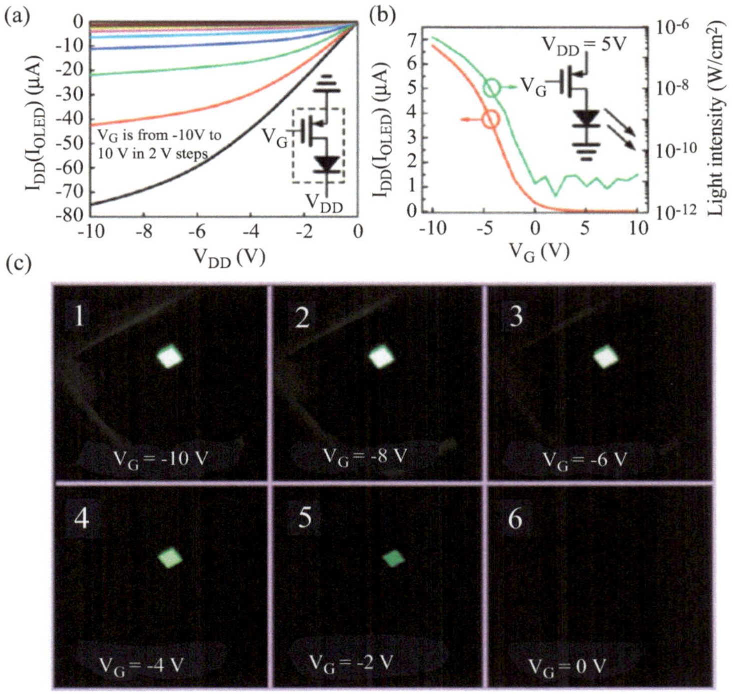

Figure 4.29 OLED control circuit by CNT. (a) Characteristics of the OLED control circuit where the current flow through the OLED is measured by sweeping the V_{DD} and input voltage V_G. V_G are ranged from −10 to 10 V in 2 V steps. (b) Plot of the current through the OLED (red line) and OLED light intensity (green line) versus V_G with V_{DD} = 5 V. Inset: The circuit diagram of an OLED driven by a CNT. (c) Photographs of the OLED driven by CNT under different inputs showing the on and off of the OLED. Reprinted with permission from Ref. [199], Copyright 2009, American Chemical Society.

4.5.3 Biosensors

Besides the superb electrical properties, CNT also possesses unique one-dimensional (1D) structure. Every atom of a CNT is located on the tube surface, thus trivial environment changes could result in drastic changes in its electrical properties. In addition, the size compatibility, all-carbon composition, and high aspect ratio

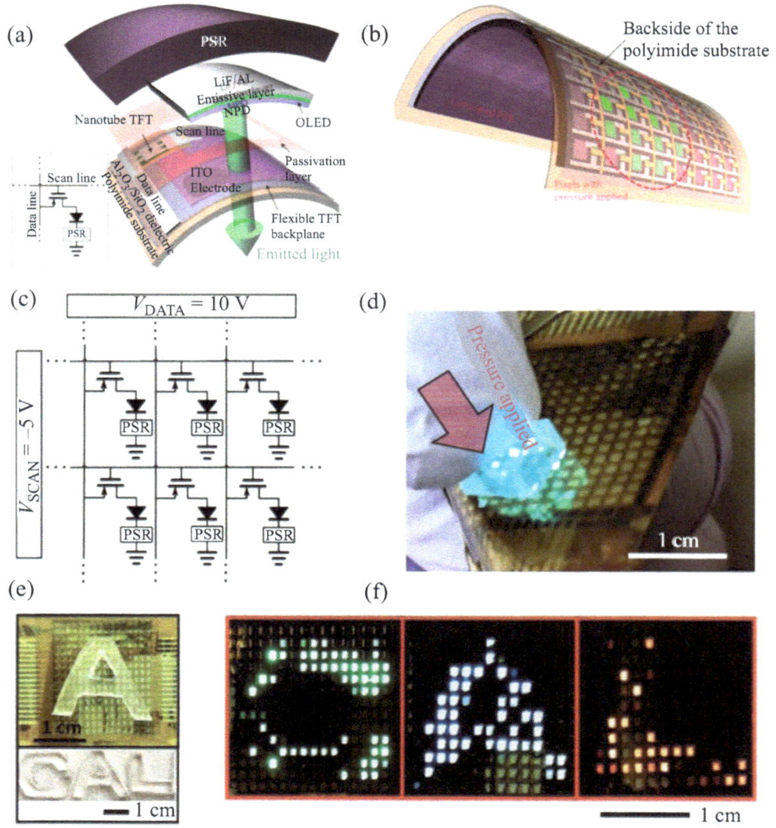

Figure 4.30 (a) Schematic layout of a single pixel, consisting of a nanotube TFT, an OLED, and a pressure sensor integrated vertically on a polyimide substrate. (b) Schematic diagram of an array of pixels (16 × 16) functioning as an interactive e-skin. (c) Circuit schematic of the e-skin matrix. (d) Photograph of a fabricated device (16 × 16 pixels), showing that light is locally emitted where the surface is touched. (e) PDMS slabs with C, A, and L shapes are prepared and used to apply pressure onto the sensor array. (f) Green, blue, and red color interactive e-skins are used to spatially map and display the pressure applied with C- (left), A- (center), and L-shaped (right) PDMS slabs, respectively. Reprinted from Ref. [203], Copyright 2013, by permission from Springer Nature.

of CNT have made it possible to achieve immediate detection of biological species at significantly lower concentration compared with existing sensors. Therefore, CNTs have received considerable attention in biosensor applications and show great potential for realizing direct, label-free, real-time electrical detection of biomolecules with ultrahigh sensitivity and excellent selectivity [205–209]. Noted that pristine CNTs have very limited selectivity when interacting with analyte, so functionalization of CNTs is very important, including noncovalent functionalization of CNTs with small molecule units, wrapping of CNTs with polymers, CNTs decorated with metal nanoparticles and covalent functionalization.

CNT-based electrochemical DNA sensors were originally pursued as a result of their high sensitivity, selectivity, and reproducibility. These early sensors relied on the immobilization of single-stranded DNA (ssDNA) on the electrode and then caused changes in electrical current triggered by hybridization of the complementary sequence [210]. For example, He et al. demonstrated aligned CNT–DNA sensors, with a high sensitivity and selectivity for probing complementary DNA and target DNA chains of specific sequences [211]. Besides, some groups have reported the application of CNT TFTs to detect the change in electrical signal, such as the conductance or resistance, induced by DNA analyte. For instance, Dong et al. have achieved a reliable detection of DNA at a concentration as low as ~100 fM by using gold-nano-enhanced CNTs [212]. Besides the detection of DNA [213, 214], CNT-based biosensors have been widely used to detect other biomolecules [215, 216], such as proteins, glucose, and neurotransmitters [217, 218].

4.6 Summary

CNTs are promising candidates in micro-/nanoscale electronic applications. Over the past three decades, the physical mechanisms of CNT transistors have been well studied. Meanwhile, materials science and new processing technology are also developed to realize the large-area production of CNT transistors in facile, ecofriendly, and less energy-consuming ways. Significant progress has been achieved for the applications of CNT transistors in integrated circuits, display electronics, as well as various sensors.

Despite these advances, many challenges remain to be addressed before large-scale applications of CNT transistors. Specifically, in materials preparation, the presence of metallic SWCNTs and the diameter variations continue to be bottlenecks for further progress.

For device considerations, critical issues involve alignment and density control of SWCNT channel, metal-CNT contact, metallic and semiconducting CNT separation, stable channel doping techniques, large-area fabrication process for SWCNT arrays, reliable high-κ dielectric and CNT-dielectric interface for flexible transistor, improvement on reliability, and feature size as well as scaling accuracy. In circuits, key challenges lie in mismatched performance of n-/p-type, scalability to sub-10 nm fabrication techniques, sensitivity to variations in device performance, and compatibility of p-/n-type fabrication processes. Further advances in nanocarbon science and CNT technology development will hopefully overcome these material, device, and circuit challenges in the coming years.

References

1. M. M. Shulaker, G. Hills, N. Patil, H. Wei, H. Y. Chen, H. S. Wong, and S. Mitra, "Carbon nanotube computer," *Nature,* **501**, 7468, pp. 526–530, 2013.
2. G. Hills, C. Lau, A. Wright, S. Fuller, M. D. Bishop, T. Srimani, P. Kanhaiya, R. Ho, A. Amer, Y. Stein, D. Murphy, Arvind, A. Chandrakasan, and M. M. Shulaker, "Modern microprocessor built from complementary carbon nanotube transistors," *Nature,* **572**, 7771, pp. 595–602, 2019.
3. M. M. Shulaker, J. Van Rethy, G. Hills, H. Wei, H. Y. Chen, G. Gielen, H. S. P. Wong, and S. Mitra, "Sensor-to-digital interface built entirely with carbon nanotube FETs," *IEEE Journal of Solid-State Circuits,* **49**, 1, pp. 190–201, 2014.
4. M. M. Rodgers, V. M. Pai, and R. S. Conroy, "Recent advances in wearable sensors for health monitoring," *IEEE Sensors Journal,* **15**, 6, pp. 3119–3126, 2015.
5. S. C. Mukhopadhyay, "Wearable sensors for human activity monitoring: A review," *IEEE Sensors Journal,* **15**, 3, pp. 1321–1330, 2015.
6. J. A. Rogers, Z. Bao, K. Baldwin, A. Dodabalapur, B. Crone, V. R. Raju, V. Kuck, H. Katz, K. Amundson, J. Ewing, and P. Drzaic, "Paper-like electronic displays: Large-area rubber-stamped plastic sheets of electronics and microencapsulated electrophoretic inks," *Proceedings of the National Academy of Sciences of the United States of America,* **98**, 9, pp. 4835–4840, 2001.
7. M. Jung, J. Kim, J. Noh, N. Lim, C. Lim, G. Lee, J. Kim, H. Kang, K. Jung, A. D. Leonard, J. M. Tour, and G. Cho, "All-printed and roll-to-roll-printable

13.56-MHz-operated 1-bit RF tag on plastic foils," *IEEE Transactions on Electron Devices,* **57**, 3, pp. 571–580, 2010.

8. J. Zhang, C. Wang, and C. Zhou, "Rigid/flexible transparent electronics based on separated carbon nanotube thin-film transistors and their application in display electronics," *ACS Nano,* **6**, 8, pp. 7412–7419, 2012.
9. W. Xu, Z. Liu, J. Zhao, W. Xu, W. Gu, X. Zhang, L. Qian, and Z. Cui, "Flexible logic circuits based on top-gate thin film transistors with printed semiconductor carbon nanotubes and top electrodes," *Nanoscale,* **6**, 24, pp. 14891–14897, 2014.
10. D. Sun, C. Liu, W. Ren, and H. Cheng, "A review of carbon nanotube- and graphene-based flexible thin-film transistors," *Small,* **9**, 8, pp. 1188–1205, 2013.
11. S. Park, M. Vosguerichian, and Z. Bao, "A review of fabrication and applications of carbon nanotube film-based flexible electronics," *Nanoscale,* **5**, 5, pp. 1727–1752, 2013.
12. L. Venema, "Silicon electronics and beyond," *Nature,* **479**, 7373, pp. 309–309, 2011.
13. R. A. Street, "Amorphous-silicon electronics," *Mrs Bulletin,* **17**, 11, pp. 70–76, 1992.
14. L. Petti, N. Munzenrieder, C. Vogt, H. Faber, L. Buthe, G. Cantarella, F. Bottacchi, T. D. Anthopoulos, and G. Troster, "Metal oxide semiconductor thin-film transistors for flexible electronics," *Applied Physics Reviews,* **3**, 2, p. 021303, 2016.
15. P. Nguyen, H. T. Ng, T. Yamada, M. K. Smith, J. Li, J. Han, and M. Meyyappan, "Direct integration of metal oxide nanowire in vertical field-effect transistor," *Nano Letters,* **4**, 4, pp. 651–657, 2004.
16. U. Zschieschang, F. Ante, T. Yamamoto, K. Takimiya, H. Kuwabara, M. Ikeda, T. Sekitani, T. Someya, K. Kern, and H. Klauk, "Flexible low-voltage organic transistors and circuits based on a high-mobility organic semiconductor with good air stability," *Advanced Materials,* **22**, 9, pp. 982–985, 2010.
17. T. Sekitani, U. Zschieschang, H. Klauk, and T. Someya, "Flexible organic transistors and circuits with extreme bending stability," *Nature Materials,* **9**, 12, pp. 1015–1022, 2010.
18. A. Facchetti, "Semiconductors for organic transistors," *Materials Today,* **10**, 3, pp. 28–37, 2007.
19. C. Yan, J. H. Cho, and J. H. Ahn, "Graphene-based flexible and stretchable thin film transistors," *Nanoscale,* **4**, 16, pp. 4870–4882, 2012.

20. N. O. Weiss, H. Zhou, L. Liao, Y. Liu, S. Jiang, Y. Huang, and X. Duan, "Graphene: An emerging electronic material," *Advanced Materials,* **24**, 43, pp. 5782–5825, 2012.

21. L. Hu, D. S. Hecht, and G. Gruner, "Carbon nanotube thin films: Fabrication, properties, and applications," *Chemical Reviews,* **110**, 10, pp. 5790–5844, 2010.

22. L. Cai and C. Wang, "Carbon nanotube flexible and stretchable electronics," *Nanoscale Research Letters,* **10**, 1, p. 320, 2015.

23. Y. Yoon, K. Ganapathi, and S. Salahuddin, "How good can monolayer MoS_2 transistors be?" *Nano Letters,* **11**, 9, pp. 3768–3773, 2011.

24. Z. Yin, H. Li, H. Li, L. Jiang, Y. Shi, Y. Sun, G. Lu, Q. Zhang, X. Chen, and H. Zhang, "Single-layer MoS_2 phototransistors," *ACS Nano,* **6**, 1, pp. 74–80, 2017.

25. Y. J. Wang, T. Al-Ameri, X. S. Wang, V. P. Georgiev, E. Towie, S. M. Amoroso, A. R. Brown, B. J. Cheng, D. Reid, C. Riddet, L. Shifren, S. Sinha, G. Yeric, R. Aitken, X. Y. Liu, J. F. Kang, and A. Asenov, "Simulation study of the impact of quantum confinement on the electrostatically driven performance of n-type nanowire transistors," *IEEE Transactions on Electron Devices,* **62**, 10, pp. 3229–3236, 2015.

26. J. Han, V. Moroz, A. Kucherov, D. Maheshwari, V. Abramzon, Z. Or-Bach, Y. Nishi, and Y. Widjaja, "A CMOS-compatible boosted transistor having >2× drive current and low leakage current," *European Solid-State Device Research Conference,* pp. 214–217, 2016.

27. A. P. Graham, G. S. Duesberg, W. Hoenlein, F. Kreupl, M. Liebau, R. Martin, B. Rajasekharan, W. Pamler, R. Seidel, W. Steinhoegl, and E. Unger, "How do carbon nanotubes fit into the semiconductor roadmap?" *Applied Physics A,* **80**, 6, pp. 1141–1151, 2005.

28. J. Ida, A. Ohtomo, and M. Yoshimaru, "Influence of low dielectric SiOF film on metal oxide semiconductor field effect transistor characteristics and its impact on circuit performance," *Japanese Journal of Applied Physics,* **37**, 11, pp. 5843–5848, 1998.

29. H. Miki, T. Tobita, T. Nakanishi, and K. Kariya, "The field-effect mobility and the current-voltage characteristics of amorphous-silicon thin-film transistors," *Japanese Journal of Applied Physics,* **30**, 11A, pp. 2740–2741, 1991.

30. C. S. Yang, L. L. Smith, C. B. Arthur, and G. N. Parsons, "Stability of low-temperature amorphous silicon thin film transistors formed on

glass and transparent plastic substrates," *Journal of Vacuum Science & Technology B,* **18**, 2, pp. 683–689, 2000.

31. A. T. Hatzopoulos, N. Arpatzanis, D. H. Tassis, C. A. Dimitriadis, F. Templier, M. Oudwan, and G. Kamarinos, "Stability of amorphous-silicon and nanocrystalline silicon thin-film transistors under DC and AC stress," *IEEE Electron Device Letters,* **28**, 9, pp. 803–805, 2007.

32. C. J. Chiu, S. P. Chang, and S. J. Chang, "High-performance a-IGZO thin-film transistor using Ta_2O_5 gate dielectric," *IEEE Electron Device Letters,* **31**, 11, pp. 1245–1247, 2010.

33. B. Ryu, H. K. Noh, E. A. Choi, and K. J. Chang, "O-vacancy as the origin of negative bias illumination stress instability in amorphous In-Ga-Zn-O thin film transistors," *Applied Physics Letters,* **97**, 2, p. 022108, 2010.

34. H. E. Katz, "Recent advances in semiconductor performance and printing processes for organic transistor-based electronics," *Chemistry of Materials,* **16**, 23, pp. 4748–4756, 2004.

35. O. Knopfmacher, M. L. Hammock, A. L. Appleton, G. Schwartz, J. Mei, T. Lei, J. Pei, and Z. Bao, "Highly stable organic polymer field-effect transistor sensor for selective detection in the marine environment," *Nature Communications,* **5**, 1, p. 2954, 2014.

36. M. R. Shijeesh, L. S. Vikas, M. K. Jayaraj, and J. Puigdollers, "Degradation study and calculation of density-of-states in PTCDI-C8 channel layer from the electrical characteristics of thin-film transistors," *Journal of Applied Physics,* **116**, 2, p. 024507, 2014.

37. Y. Du, H. Liu, Y. Deng, and P. D. Ye, "Device perspective for black phosphorus field-effect transistors: Contact resistance, ambipolar behavior, and scaling," *ACS Nano,* **8**, 10, pp. 10035–10042, 2014.

38. M. Buscema, D. J. Groenendijk, S. I. Blanter, G. A. Steele, H. S. van der Zant, and A. Castellanos-Gomez, "Fast and broadband photoresponse of few-layer black phosphorus field-effect transistors," *Nano Letters,* **14**, 6, pp. 3347–3352, 2014.

39. Q. H. Wang, K. Kalantar-Zadeh, A. Kis, J. N. Coleman, and M. S. Strano, "Electronics and optoelectronics of two-dimensional transition metal dichalcogenides," *Nature Nanotechnology,* **7**, 11, pp. 699–712, 2012.

40. X. Zhou, J.-Y. Park, S. Huang, J. Liu, and P. L. J. P. R. L. McEuen, "Band structure, phonon scattering, and the performance limit of single-walled carbon nanotube transistors," *Physical Review Letters,* **95**, 14, p. 146805, 2005.

41. A. Javey, J. Guo, Q. Wang, M. Lundstrom, and H. Dai, "Ballistic carbon nanotube field-effect transistors," *Nature*, **424**, 6949, pp. 654–657, 2003.

42. T. Durkop, S. A. Getty, E. Cobas, and M. S. Fuhrer, "Extraordinary mobility in semiconducting carbon nanotubes," *Nano Letters*, **4**, 1, pp. 35–39, 2004.

43. C. Wang, K. Takei, T. Takahashi, and A. Javey, "Carbon nanotube electronics - moving forward," *Chemical Society Reviews*, **42**, 7, pp. 2592–2609, 2013.

44. M. F. Yu, B. S. Files, S. Arepalli, and R. Ruoff, "Tensile loading of ropes of single wall carbon nanotubes and their mechanical properties," *Physical Review Letters*, **84**, 24, p. 5552, 2000.

45. B. Q. Wei, R. Vajtai, and P. M. Ajayan, "Reliability and current carrying capacity of carbon nanotubes," *Applied Physics Letters*, **79**, 8, pp. 1172–1174, 2001.

46. D. Sun, M. Y. Timmermans, Y. Tian, A. G. Nasibulin, E. I. Kauppinen, S. Kishimoto, T. Mizutani, and Y. Ohno, "Flexible high-performance carbon nanotube integrated circuits," *Nature Nanotechnology*, **6**, 3, pp. 156–161, 2011.

47. S. Aikawa, E. Einarsson, T. Thurakitseree, S. Chiashi, E. Nishikawa, and S. Maruyama, "Deformable transparent all-carbon-nanotube transistors," *Applied Physics Letters*, **100**, 6, p. 063502, 2012.

48. M. Xia, Z. Cheng, J. Han, and S. Zhang, "Extremely stretchable all-carbon-nanotube transistor on flexible and transparent substrates," *Applied Physics Letters*, **105**, 14, p. 143504, 2014.

49. Y. Lu, S. Bangsaruntip, X. Wang, L. Zhang, Y. Nishi, and H. Dai, "DNA functionalization of carbon nanotubes for ultrathin atomic layer deposition of high kappa dielectrics for nanotube transistors with 60 mV/decade switching," *Journal of the American Chemical Society*, **128**, 11, pp. 3518–3519, 2006.

50. C. Qiu, Z. Zhang, M. Xiao, Y. Yang, D. Zhong, and L. M. Peng, "Scaling carbon nanotube complementary transistors to 5-nm gate lengths," *Science*, **355**, 6322, pp. 271–276, 2017.

51. N. Patil, A. Lin, J. Zhang, H. Wei, K. Anderson, H. S. P. Wong, and S. Mitra, "Scalable carbon nanotube computational and storage circuits immune to metallic and mispositioned carbon nanotubes," *IEEE Transactions on Nanotechnology*, **10**, 4, pp. 744–750, 2011.

52. Q. Cao, J. Tersoff, D. B. Farmer, Y. Zhu, and S. J. Han, "Carbon nanotube transistors scaled to a 40-nanometer footprint," *Science,* **356**, 6345, pp. 1369–1372, 2017.

53. N. M. Ghazali, H. Tomizawa, N. Hagiwara, K. Suzuki, A. M. Hashim, T. Yamaguchi, S. Akita, and K. Ishibashi, "Fabrication of tunnel barriers and single electron transistors in suspended multi-wall carbon nanotubes," *AIP Advances,* **9**, 10, p. 105015, 2019.

54. S. Karthikeyan, P. Mahalingam, and M. Karthik, "Large scale synthesis of carbon nanotubes," *Nature,* **6**, 1, pp. 1–12, 2009.

55. T. Guo, P. Nikolaev, A. Thess, D. T. Colbert, and R. E. Smalley, "Catalytic growth of single-walled nanotubes by laser vaporization," *Chemical Physics Letters,* **243**, 1–2, pp. 49–54, 1995.

56. A. Huczko, "Synthesis of aligned carbon nanotubes," *Applied Physics A,* **74**, 5, pp. 617–638, 2002.

57. A. Maiti, C. J. Brabec, C. Roland, and J. Bernholc, "Theory of carbon nanotube growth," *Physical Review B Condensed Matter,* **52**, 20, pp. 14850–14858, 1995.

58. T. C. Maganas and A. L. Harrington, Intermittent film deposition method and system, US Patent, US5143745A, 1992.

59. W. Zhou, S. Zhan, L. Ding, and J. Liu, "General rules for selective growth of enriched semiconducting single walled carbon nanotubes with water vapor as in situ etchant," *Journal of the American Chemical Society,* **134**, 34, pp. 14019–14026, 2012.

60. M. S. Arnold, A. A. Green, J. F. Hulvat, S. I. Stupp, and M. C. Hersam, "Sorting carbon nanotubes by electronic structure using density differentiation," *Nature Nanotechnology,* **1**, 1, pp. 60–65, 2006.

61. S. Ghosh, S. M. Bachilo, and R. B. Weisman, "Advanced sorting of single-walled carbon nanotubes by nonlinear density-gradient ultracentrifugation," *Nature Nanotechnology,* **5**, 6, pp. 443–450, 2010.

62. D. Lee, M.-L. Seol, D.-I. Moon, P. Bennett, N. Yoder, J. Humes, J. Bokor, Y.-K. Choi, and S.-J. Choi, "High-performance thin-film transistors produced from highly separated solution-processed carbon nanotubes," *Applied Physics Letters,* **104**, 14, p. 143508, 2014.

63. D. Chattopadhyay, S. Kim, and F. Papadimitrakopoulos, "Length separation of zwitterion-functionalized single wall carbon nanotubes by GPC," *Journal of the American Chemical Society,* **124**, 5, pp. 728–729, 2002.

64. M. Zheng, A. Jagota, M. S. Strano, A. P. Santos, P. Barone, S. G. Chou, B. A. Diner, M. S. Dresselhaus, R. S. McLean, G. B. Onoa, G. G. Samsonidze, E. D. Semke, M. Usrey, and D. J. Walls, "Structure-based carbon nanotube sorting by sequence-dependent DNA assembly," *Science*, **302**, 5650, pp. 1545–1548, 2003.

65. T. Tanaka, H. Jin, Y. Miyata, S. Fujii, H. Suga, Y. Naitoh, T. Minari, T. Miyadera, K. Tsukagoshi, and H. Kataura, "Simple and scalable gel-based separation of metallic and semiconducting carbon nanotubes," *Nano Letters*, **9**, 4, pp. 1497–1500, 2009.

66. H. Liu, D. Nishide, T. Tanaka, and H. Kataura, "Large-scale single-chirality separation of single-wall carbon nanotubes by simple gel chromatography," *Nature Communications*, **2**, 1, p. 309, 2011.

67. D. Yang, J. Hu, H. Liu, S. Li, W. Su, Q. Li, N. Zhou, Y. Wang, W. Zhou, S. Xie, and H. Kataura, "Structure sorting of large-diameter carbon nanotubes by NaOH tuning the interactions between nanotubes and gel," *Advanced Functional Materials*, **27**, 40, p. 1700278, 2017.

68. X. Tu, A. R. Hight Walker, C. Y. Khripin, and M. Zheng, "Evolution of DNA sequences toward recognition of metallic armchair carbon nanotubes," *Journal of the American Chemical Society*, **133**, 33, pp. 12998–13001, 2011.

69. N. Patil, A. Lin, J. Zhang, H. Wei, K. Anderson, H.-S. P. Wong, and S. Mitra, "VMR: VLSI-compatible metallic carbon nanotube removal for imperfection-immune cascaded multi-stage digital logic circuits using carbon nanotube FETs," *IEEE International Electron Devices Meeting*, pp. 1–4, 2009.

70. S. K. Doorn, M. S. Strano, M. J. O'Connell, E. H. Haroz, K. L. Rialon, R. H. Hauge, and R. E. Smalley, "Capillary electrophoresis separations of bundled and individual carbon nanotubes," *Journal of Physical Chemistry B*, **107**, 25, pp. 6063–6069, 2003.

71. M. Lopez-Pastor, A. Dominguez-Vidal, M. J. Ayora-Canada, B. M. Simonet, B. Lendl, and M. Valcarcel, "Separation of single-walled carbon nanotubes by use of ionic liquid-aided capillary electrophoresis," *Analytical Chemistry*, **80**, 8, pp. 2672–2679, 2008.

72. D. Chattopadhyay, I. Galeska, and F. Papadimitrakopoulos, "A route for bulk separation of semiconducting from metallic single-wall carbon nanotubes," *Journal of the American Chemical Society*, **125**, 11, pp. 3370–3375, 2003.

73. A. Nish, J. Y. Hwang, J. Doig, and R. J. Nicholas, "Highly selective dispersion of single-walled carbon nanotubes using aromatic polymers," *Nature Nanotechnology*, **2**, 10, pp. 640–646, 2007.

74. L. An, Q. A. Fu, C. G. Lu, and J. Liu, "A simple chemical route to selectively eliminate metallic carbon nanotubes in nanotube network devices," *Journal of the American Chemical Society*, **126**, 34, pp. 10520–10521, 2004.

75. S. J. Tans, A. R. M. Verschueren, and C. Dekker, "Room-temperature transistor based on a single carbon nanotube," *Nature*, **393**, 6680, pp. 49–52, 1998.

76. S. M. Sze and K. K. Ng, *Physics of Semiconductor Devices*. John wiley & sons, 2006.

77. C. Wang, J. Zhang, and C. Zhou, "Macroelectronic integrated circuits using high-performance separated carbon nanotube thin-film transistors," *ACS Nano*, **4**, 12, pp. 7123–7132, 2010.

78. A. Javey, H. Kim, M. Brink, Q. Wang, A. Ural, J. Guo, P. McIntyre, P. McEuen, M. Lundstrom, and H. Dai, "High-kappa dielectrics for advanced carbon-nanotube transistors and logic gates," *Nature Materials*, **1**, 4, pp. 241–246, 2002.

79. L. S. Liyanage, D. J. Cott, A. Delabie, S. Van Elshocht, Z. Bao, and H. S. Wong, "Atomic layer deposition of high-k dielectrics on single-walled carbon nanotubes: A Raman study," *Nanotechnology*, **24**, 24, p. 245703, 2013.

80. D. B. Farmer and R. G. Gordon, "ALD of high-kappa dielectrics on suspended functionalized SWNTs," *Electrochemical and Solid State Letters*, **8**, 4, pp. G89–G91, 2005.

81. T. Srimani, G. Hills, X. Zhao, D. Antoniadis, J. A. D. Alamo, and M. M. Shulaker, "Asymmetric gating for reducing leakage current in carbon nanotube field-effect transistors," *Applied Physics Letters*, **115**, 6, p. 063107, 2019.

82. A. Javey, J. Guo, D. B. Farmer, Q. Wang, D. Wang, R. G. Gordon, M. Lundstrom, and H. Dai, "Carbon nanotube field-effect transistors with integrated ohmic contacts and high-k gate dielectrics," *Nano Letters*, **4**, 3, pp. 447–450, 2004.

83. S. Heinze, J. Tersoff, R. Martel, V. Derycke, J. Appenzeller, and P. Avouris, "Carbon nanotubes as Schottky barrier transistors," *Physical Review Letters*, **89**, 10, p. 106801, 2002.

84. J. Appenzeller, J. Knoch, V. Derycke, R. Martel, S. Wind, and P. Avouris, "Field-modulated carrier transport in carbon nanotube transistors," *Physical Review Letters*, **89**, 12, p. 126801, 2002.

85. V. Pachauri, K. Kern, and K. Balasubramanian, "Field-effect-based chemical sensing using nanowire-nanoparticle hybrids: The ion-

sensitive metal-semiconductor field-effect transistor," *Applied Physics Letters,* **102**, 2, p. 023501, 2013.

86. J. Xia, G. Dong, B. Tian, Q. Yan, J. Han, S. Qiu, Q. Li, X. Liang, and L. M. Peng, "Contact resistance effects in carbon nanotube thin film transistors," *Acta Physico-Chimica Sinica,* **32**, 4, pp. 1029–1035, 2016.
87. A. Javey, J. Guo, D. B. Farmer, Q. Wang, E. Yenilmez, R. G. Gordon, M. Lundstrom, and H. Dai, "Self-aligned ballistic molecular transistors and electrically parallel nanotube arrays," *Nano Letters,* **4**, 7, pp. 1319–1322, 2004.
88. Z. Zhang, X. Liang, S. Wang, K. Yao, Y. Hu, Y. Zhu, Q. Chen, W. Zhou, Y. Li, Y. Yao, J. Zhang, and L. M. Peng, "Doping-free fabrication of carbon nanotube based ballistic CMOS devices and circuits," *Nano Letters,* **7**, 12, pp. 3603–3607, 2007.
89. Z. Zhang, S. Wang, and L. M. Peng, "High-performance doping-free carbon-nanotube-based CMOS devices and integrated circuits," *Chinese Science Bulletin,* **57**, 2–3, pp. 135–148, 2012.
90. H. Kobayashi, K. Kishimoto, and Y. Nakato, "Reactions of hydrogen at the interface of palladium titanium-dioxide Schottky diodes as hydrogen sensors, studied by workfunction and electrical characteristic measurements," *Surface Science,* **306**, 3, pp. 393–405, 1994.
91. Y. Zhang, N. W. Franklin, R. Chen, and H. Dai, "Metal coating on suspended carbon nanotubes and its implication to metal-tube interaction," *Chemical Physics Letters,* **331**, 1, pp. 35–41, 2000.
92. Y. Zhang and H. Dai, "Formation of metal nanowires on suspended single-walled carbon nanotubes," *Applied Physics Letters,* **77**, 19, pp. 3015–3017, 2000.
93. A. N. Andriotis, M. Menon, and G. E. Froudakis, "Various bonding configurations of transition-metal atoms on carbon nanotubes: Their effect on contact resistance," *Applied Physics Letters,* **76**, 26, pp. 3890–3892, 2000.
94. A. Maiti and A. Ricca, "Metal-nanotube interactions: Binding energies and wetting properties," *Chemical Physics Letters,* **395**, 1–3, pp. 7–11, 2004.
95. V. Perebeinos and J. Tersoff, "Wetting transition for carbon nanotube arrays under metal contacts," *Physical Review Letters,* **114**, 8, p. 085501, 2015.
96. G. J. Brady, A. J. Way, N. S. Safron, H. T. Evensen, P. Gopalan, and M. S. Arnold, "Quasi-ballistic carbon nanotube array transistors with

current density exceeding Si and GaAs," *Science Advances,* **2**, 9, p. e1601240, 2016.

97. G. J. Brady, Y. Joo, M. Y. Wu, M. J. Shea, P. Gopalan, and M. S. Arnold, "Polyfluorene-sorted, carbon nanotube array field-effect transistors with increased current density and high on/off ratio," *ACS Nano,* **8**, 11, pp. 11614–11621, 2014.

98. A. D. Franklin, D. B. Farmer, and W. Haensch, "Defining and overcoming the contact resistance challenge in scaled carbon nanotube transistors," *ACS Nano,* **8**, 7, pp. 7333–7339, 2014.

99. J. Guo, A. Javey, H. Dai, and M. Lundstrom, "Performance analysis and design optimization of near ballistic carbon nanotube field-effect transistors," *IEEE International Electron Devices Meeting*, pp. 703–706, 2004.

100. B. W. Baugher, H. O. Churchill, Y. Yang, and P. Jarillo-Herrero, "Optoelectronic devices based on electrically tunable p-n diodes in a monolayer dichalcogenide," *Nature Nanotechnology,* **9**, 4, pp. 262–267, 2014.

101. Z. Arefinia and A. A. Orouji, "Quantum simulation study of a new carbon nanotube field-effect transistor with electrically induced source/drain extension," *IEEE Transactions on Device and Materials Reliability,* **9**, 2, pp. 237–243, 2009.

102. J. Guo and M. S. Lundstrom, "A computational study of thin-body, double-gate, Schottky barrier MOSFETs," *IEEE Transactions on Electron Devices,* **49**, 11, pp. 1897–1902, 2002.

103. J. Chen, C. Klinke, A. Afzali, and P. Avouris, "Self-aligned carbon nanotube transistors with charge transfer doping," *Applied Physics Letters,* **86**, 12, p. 123108, 2005.

104. Z. Zhang, S. Wang, L. Ding, X. Liang, T. Pei, J. Shen, H. Xu, Q. Chen, R. Cui, Y. Li, and L. M. Peng, "Self-aligned ballistic n-type single-walled carbon nanotube field-effect transistors with adjustable threshold voltage," *Nano Letters,* **8**, 11, pp. 3696–3701, 2008.

105. A. Javey, R. Tu, D. B. Farmer, J. Guo, R. G. Gordon, and H. Dai, "High performance n-type carbon nanotube field-effect transistors with chemically doped contacts," *Nano Letters,* **5**, 2, pp. 345–348, 2005.

106. M. Shiomi, Y. Mochizuki, Y. Imakita, T. Arie, S. Akita, and K. Takei, "Graphene and carbon nanotube heterojunction transistors with individual gate control," *ACS Nano,* **13**, 4, pp. 4771–4777, 2019.

107. Y. Yang, L. Ding, J. Han, Z. Zhang, and L. M. Peng, "High-performance complementary transistors and medium-scale integrated circuits

based on carbon nanotube thin films," *ACS Nano,* **11**, 4, pp. 4124–4132, 2017.

108. J. Zhang, C. Wang, Y. Fu, Y. He, and C. Zhou, "Air-stable conversion of separated carbon nanotube thin-film transistors from p-type to n-type using atomic layer deposition of high-kappa oxide and its application in CMOS logic circuits," *ACS Nano,* **5**, 4, pp. 3284–3292, 2011.
109. J. Yang, Q. L. Dong, Z. T. Jiang, and J. Zhang, "Electronic energy band structures of carbon nanotubes with spin-orbit coupling interaction," (in Chinese), *Acta Physica Sinica,* **60**, 7, p. 075202, 2011.
110. K. Gong, F. Du, Z. Xia, M. Durstock, and L. Dai, "Nitrogen-doped carbon nanotube arrays with high electrocatalytic activity for oxygen reduction," *Science,* **323**, 5915, pp. 760–764, 2009.
111. P. G. Collins, K. Bradley, M. Ishigami, and A. Zettl, "Extreme oxygen sensitivity of electronic properties of carbon nanotubes," *Science,* **287**, 5459, pp. 1801–1804, 2000.
112. H. Wang, H. Wang, Y. Chen, Y. Liu, J. Zhao, Q. Cai, and X. Wang, "Phosphorus-doped graphene and (8,0) carbon nanotube: Structural, electronic, magnetic properties, and chemical reactivity," *Applied Surface Science,* **273**, pp. 302–309, 2013.
113. A. Lekawa-Raus, L. Kurzepa, G. Kozlowski, S. C. Hopkins, M. Wozniak, D. Lukawski, B. A. Glowacki, and K. K. Koziol, "Influence of atmospheric water vapour on electrical performance of carbon nanotube fibres," *Carbon,* **87**, pp. 18–28, 2015.
114. X. Li, T. Tang, M. Li, and X. He, "Nitrogen-doped graphene films from simple photochemical doping for n-type field-effect transistors," *Applied Physics Letters,* **106**, 1, p. 013110, 2015.
115. M. Radosavljevic, M. Freitag, K. V. Thadani, and A. T. Johnson, "Nonvolatile molecular memory elements based on ambipolar nanotube field effect transistors," *Nano Letters,* **2**, 7, pp. 761–764, 2002.
116. Y. Ohno, Y. Kurokawa, S. Kishimoto, T. Mizutani, T. Shimada, M. Ishida, T. Okazaki, H. Shinohara, Y. Murakami, S. Maruyama, A. Sakai, and K. Hiraga, "Synthesis of carbon nanotube peapods directly on Si substrates," *Applied Physics Letters,* **86**, 2, p. 023109, 2005.
117. S. Ryu, L. Liu, S. Berciaud, Y. J. Yu, H. Liu, P. Kim, G. W. Flynn, and L. E. Brus, "Atmospheric oxygen binding and hole doping in deformed graphene on a SiO_2 substrate," *Nano Letters,* **10**, 12, pp. 4944–4951, 2010.

118. J. Guo, S. Datta, and M. Lundstrom, "A numerical study of scaling issues for Schottky-Barrier carbon nanotube transistors," *IEEE Transactions on Electron Devices,* **51**, 2, pp. 172–177, 2004.

119. J. Q. Li, W. S. Yue, Z. B. Guo, Y. Yang, X. B. Wang, A. A. Syed, and Y. F. Zhang, "Unique characteristics of vertical carbon nanotube field-effect transistors on silicon," *Nano-Micro Letters,* **6**, 3, pp. 287–292, 2014.

120. T. Park, K. Sim, J. Lee, and W. Yi, "Electron transfer properties of iodine-doped single-walled carbon nanotubes using field effect transistor," *Journal of Nanoscience and Nanotechnology,* **12**, 7, pp. 5812–5815, 2012.

121. C. Lau, T. Srimani, M. D. Bishop, G. Hills, and M. M. Shulaker, "Tunable n-type doping of carbon nanotubes through engineered atomic layer deposition HfOX films," *ACS Nano,* **12**, 11, pp. 10924–10931, 2018.

122. R. Saito, G. Dresselhaus, and M. S. Dresselhaus, "Tunneling conductance of connected carbon nanotubes," *Physical Review B,* **53**, 4, pp. 2044–2050, 1996.

123. C. Li, E. T. Thostenson, and T. W. Chou, "Dominant role of tunneling resistance in the electrical conductivity of carbon nanotube-based composites," *Applied Physics Letters,* **91**, 22, p. 223114, 2007.

124. M. C. Lemieux, S. Sok, M. E. Roberts, J. P. Opatkiewicz, D. Liu, S. N. Barman, N. Patil, S. Mitra, and Z. Bao, "Solution assembly of organized carbon nanotube networks for thin-film transistors," *ACS Nano,* **3**, 12, pp. 4089–4097, 2009.

125. N. Pimparkar, C. Kocabas, K. Seong Jun, J. A. Rogers, and M. A. Alam, "Limits of performance gain of aligned CNT over randomized network: Theoretical predictions and experimental validation," *IEEE Electron Device Letters,* **28**, 7, pp. 593–595, 2007.

126. Q. Cao, S. J. Han, G. S. Tulevski, Y. Zhu, D. D. Lu, and W. Haensch, "Arrays of single-walled carbon nanotubes with full surface coverage for high-performance electronics," *Nature Nanotechnology,* **8**, 3, pp. 180–186, 2013.

127. A. Ismach, L. Segev, E. Wachtel, and E. Joselevich, "Atomic-step-templated formation of single wall carbon nanotube patterns," *Angew Cheme International Edition England,* **43**, 45, pp. 6140–6143, 2004.

128. C. Kocabas, S. H. Hur, A. Gaur, M. A. Meitl, M. Shim, and J. A. Rogers, "Guided growth of large-scale, horizontally aligned arrays of single-walled carbon nanotubes and their use in thin-film transistors," *Small,* **1**, 11, pp. 1110–1116, 2005.

129. S. Han, X. Liu, and C. Zhou, "Template-free directional growth of single-walled carbon nanotubes on a- and r-plane sapphire," *Journal of the American Chemical Society,* **127**, 15, pp. 5294–5295, 2005.

130. X. Liu, S. Han, and C. Zhou, "Novel nanotube-on-insulator (NOI) approach toward single-walled carbon nanotube devices," *Nano Letters,* **6**, 1, pp. 34–39, 2006.

131. S. J. Kang, C. Kocabas, T. Ozel, M. Shim, N. Pimparkar, M. A. Alam, S. V. Rotkin, and J. A. Rogers, "High-performance electronics using dense, perfectly aligned arrays of single-walled carbon nanotubes," *Nature Nanotechnology,* **2**, 4, pp. 230–236, 2007.

132. S. W. Hong, T. Banks, and J. A. Rogers, "Improved density in aligned arrays of single-walled carbon nanotubes by sequential chemical vapor deposition on quartz," *Advanced Materials,* **22**, 16, pp. 1826–1830, 2010.

133. P. Stokes and S. I. Khondaker, "Local-gated single-walled carbon nanotube field effect transistors assembled by AC dielectrophoresis," *Nanotechnology,* **19**, 17, p. 175202, 2008.

134. F. N. Ishikawa, H. K. Chang, K. Ryu, P. C. Chen, A. Badmaev, L. Gomez De Arco, G. Shen, and C. Zhou, "Transparent electronics based on transfer printed aligned carbon nanotubes on rigid and flexible substrates," *ACS Nano,* **3**, 1, pp. 73–79, 2009.

135. C. L. Pint, Y. Q. Xu, S. Moghazy, T. Cherukuri, N. T. Alvarez, E. H. Haroz, S. Mahzooni, S. K. Doorn, J. Kono, M. Pasquali, and R. H. Hauge, "Dry contact transfer printing of aligned carbon nanotube patterns and characterization of their optical properties for diameter distribution and alignment," *ACS Nano,* **4**, 2, pp. 1131–1145, 2010.

136. C. L. Pint, Y. Q. Xu, M. Pasquali, and R. H. Hauge, "Formation of highly dense aligned ribbons and transparent films of single-walled carbon nanotubes directly from carpets," *ACS Nano,* **2**, 9, pp. 1871–1878, 2008.

137. M. M. Shulaker, H. Wei, N. Patil, J. Provine, H. Y. Chen, H. S. Wong, and S. Mitra, "Linear increases in carbon nanotube density through multiple transfer technique," *Nano Letters,* **11**, 5, pp. 1881–1886, 2011.

138. S. J. Kang, C. Kocabas, H. S. Kim, Q. Cao, M. A. Meitl, D. Y. Khang, and J. A. Rogers, "Printed multilayer superstructures of aligned single-walled carbon nanotubes for electronic, applications," *Nano Letters,* **7**, 11, pp. 3343–3348, 2007.

139. N. Patil, J. Deng, S. Mitra, and H. S. P. Wong, "Circuit-level performance benchmarking and scalability analysis of carbon nanotube transistor circuits," *IEEE Transactions on Nanotechnology,* **8**, 1, pp. 37–45, 2009.

140. J. H. Lee, W. S. Kang, G. H. Nam, S. W. Choi, and J. H. Kim, "Preparation of hierarchically aligned carbon nanotube films using the Langmuir–Blodgett technique," *Journal of Nanoscience and Nanotechnology,* **9**, 12, pp. 7080–7084, 2009.

141. X. Li, L. Zhang, X. Wang, I. Shimoyama, X. Sun, W. S. Seo, and H. Dai, "Langmuir–Blodgett assembly of densely aligned single-walled carbon nanotubes from bulk materials," *Journal of the American Chemical Society,* **129**, 16, pp. 4890–4891, 2007.

142. S. H. Jin, S. N. Dunham, J. Song, X. Xie, J. H. Kim, C. Lu, A. Islam, F. Du, J. Kim, J. Felts, Y. Li, F. Xiong, M. A. Wahab, M. Menon, E. Cho, K. L. Grosse, D. J. Lee, H. U. Chung, E. Pop, M. A. Alam, W. P. King, Y. Huang, and J. A. Rogers, "Using nanoscale thermocapillary flows to create arrays of purely semiconducting single-walled carbon nanotubes," *Nature Nanotechnology,* **8**, 5, pp. 347–355, 2013.

143. M. Park, S. Kim, H. Kwon, S. Hong, S. Im, and S. Y. Ju, "Selective dispersion of highly pure large-diameter semiconducting carbon nanotubes by a flavin for thin-film transistors," *ACS Applied Materials and Interfaces,* **8**, 35, pp. 23270–23280, 2016.

144. H. W. Lee, Y. Yoon, S. Park, J. H. Oh, S. Hong, L. S. Liyanage, H. Wang, S. Morishita, N. Patil, Y. J. Park, J. J. Park, A. Spakowitz, G. Galli, F. Gygi, P. H. Wong, J. B. Tok, J. M. Kim, and Z. Bao, "Selective dispersion of high purity semiconducting single-walled carbon nanotubes with regioregular poly(3-alkylthiophene)s," *Nature Communications,* **2**, 1, p. 541, 2011.

145. H. Wang, J. Mei, P. Liu, K. Schmidt, G. Jimenez-Oses, S. Osuna, L. Fang, C. J. Tassone, A. P. Zoombelt, A. N. Sokolov, K. N. Houk, M. F. Toney, and Z. Bao, "Scalable and selective dispersion of semiconducting arc-discharged carbon nanotubes by dithiafulvalene/thiophene copolymers for thin film transistors," *ACS Nano,* **7**, 3, pp. 2659–2668, 2013.

146. S. Park, G. Pitner, G. Giri, J. H. Koo, J. Park, K. Kim, H. Wang, R. Sinclair, H. S. Wong, and Z. Bao, "Large-area assembly of densely aligned single-walled carbon nanotubes using solution shearing and their application to field-effect transistors," *Advanced Materials*, **27**, 16, pp. 2656–2662, 2015.

147. K. S. Mistry, B. A. Larsen, and J. L. Blackburn, "High-yield dispersions of large-diameter semiconducting single-walled carbon nanotubes with tunable narrow chirality distributions," *ACS Nano,* **7**, 3, pp. 2231–2239, 2013.

148. X. He, W. Gao, L. Xie, B. Li, Q. Zhang, S. Lei, J. M. Robinson, E. H. Hároz, S. K. Doorn, W. Wang, R. Vajtai, P. Ajayan, W. W. Adams, R. H. Hauge,

and J. Kono, "Wafer-scale monodomain films of spontaneously aligned single-walled carbon nanotubes," *Nature Nanotechnology*, **11**, 7, pp. 633–638, 2016.

149. D. D. Tune, B. W. Stolz, M. Pfohl, and B. S. Flavel, "Dry shear aligning: A simple and versatile method to smooth and align the surfaces of carbon nanotube thin films," *Nanoscale*, **8**, 6, pp. 3232–3236, 2016.

150. T. R. Hebner, C. C. Wu, D. Marcy, M. H. Lu, and J. C. Sturm, "Ink-jet printing of doped polymers for organic light emitting devices," *Applied Physics Letters*, **72**, 5, pp. 519–521, 1998.

151. P. Chen, Y. Fu, R. Aminirad, C. Wang, J. Zhang, K. Wang, K. Galatsis, and C. Zhou, "Fully printed separated carbon nanotube thin film transistor circuits and its application in organic light emitting diode control," *Nano Letters*, **11**, 12, pp. 5301–5308, 2011.

152. V. D. B. Arjan, V. D. M. Markjan, T. Driessen, V. D. B. Marc, H. Reinten, H. Wijshoff, M. Versluis, and D. Lohse, "Velocity profile inside piezoacoustic inkjet droplets in flight: Comparison between experiment and numerical simulation," *Physical Review Applied*, **1**, 1, p. 014004, 2014.

153. T. M. Lee, J. H. Noh, C. H. Kim, J. Jo, and D. S. Kim, "Development of a gravure offset printing system for the printing electrodes of flat panel display," *Thin Solid Films*, **518**, 12, pp. 3355–3359, 2010.

154. T. Q. Trung and N.-E. Lee, "Recent Progress on Stretchable Electronic Devices with Intrinsically Stretchable Components," *Advanced Materials*, **29**, 3, p. 1603167, 2017.

155. K. Fukuda, Y. Yoshimura, T. Okamoto, Y. Takeda, D. Kumaki, Y. Katayama, and S. Tokito, "Reverse-offset printing optimized for scalable organic thin-film transistors with submicrometer channel lengths," *Advanced Electronic Materials*, **1**, 8, p. 1500145, 2015.

156. U. Zschieschang, J. W. Borchert, M. Geiger, F. Letzkus, J. N. Burghartz, and H. J. O. E. Klauk, "Stencil lithography for organic thin-film transistors with a channel length of 300 nm," *Organic Electronics*, **61**, pp. 65–69, 2018.

157. C. Wang, J. C. Chien, K. Takei, T. Takahashi, J. Nah, A. M. Niknejad, and A. Javey, "Extremely bendable, high-performance integrated circuits using semiconducting carbon nanotube networks for digital, analog, and radio-frequency applications," *Nano Letters*, **12**, 3, pp. 1527–1533, 2012.

158. X. Cao, Y. Cao, and C. Zhou, "Imperceptible and ultraflexible p-type transistors and macroelectronics based on carbon nanotubes," *ACS Nano*, **10**, 1, pp. 199–206, 2016.

159. H. Zhang, L. Xiang, Y. Yang, M. Xiao, J. Han, L. Ding, Z. Zhang, Y. Hu, and L.-M. Peng, "High-performance carbon nanotube complementary electronics and integrated sensor systems on ultrathin plastic foil," *ACS Nano,* **12**, 3, pp. 2773–2779, 2018.

160. J. B. Andrews, K. Mondal, T. V. Neumann, J. A. Cardenas, J. Wang, D. P. Parekh, Y. Lin, P. Ballentine, M. D. Dickey, and A. D. Franklin, "Patterned liquid metal contacts for printed carbon nanotube transistors," *ACS Nano,* **12**, 6, pp. 5482–5488, 2018.

161. Y. Chen, Y. Sun, Q. Zhu, B. Wang, X. Yan, S. Qiu, Q. Li, P. Hou, C. Liu, D. Sun, and H. Cheng, "High-throughput fabrication of flexible and transparent all-carbon nanotube electronics," *Advanced Science,* **5**, 5, p. 1700965, 2018.

162. C. Du, M. Zhang, Q. Huang, S. Zhang, and Y. Chai, "Ultralow-voltage all-carbon low-dimensional-material flexible transistors integrated by room-temperature photolithography incorporated filtration," *Nanoscale,* **11**, 32, pp. 15029–15036, 2019.

163. S. H. Tseng and N. H. Tai, "Fabrication of a transparent and flexible thin film transistor based on single-walled carbon nanotubes using the direct transfer method," *Applied Physics Letters,* **95**, 20, p. 204104, 2009.

164. Q. Cao and J. A. Rogers, "Ultrathin films of single-walled carbon nanotubes for electronics and sensors: A review of fundamental and applied aspects," *Advanced Materials,* **21**, 1, pp. 29–53, 2009.

165. T. Takenobu, T. Takahashi, T. Kanbara, K. Tsukagoshi, Y. Aoyagi, and Y. Iwasa, "High-performance transparent flexible transistors using carbon nanotube films," *Applied Physics Letters,* **88**, 3, p. 033511, 2006.

166. S. Wang, J. Xu, W. Wang, G. J. N. Wang, R. Rastak, F. Molinalopez, J. W. Chung, S. Niu, V. R. Feig, and J. J. N. Lopez, "Skin electronics from scalable fabrication of an intrinsically stretchable transistor array," *Nature,* **555**, 7694, pp. 83–88, 2018.

167. J. H. Cho, J. Lee, Y. Xia, B. Kim, Y. He, M. J. Renn, T. P. Lodge, and C. D. Frisbie, "Printable ion-gel gate dielectrics for low-voltage polymer thin-film transistors on plastic," *Nature Materials,* **7**, 11, pp. 900–906, 2008.

168. J. A. Rogers, S. Takao, and H. J. S. Yonggang, "Materials and mechanics for stretchable electronics," *Science,* **327**, 5973, pp. 1603–1607, 2010.

169. J. A. Rogers, Z. Bao, and T.-W. Lee, "Wearable bioelectronics: Opportunities for chemistry," *Accounts of Chemical Research,* **52**, 3, pp. 521–522, 2019.

170. P. Williams, S. Papadakis, M. Falvo, A. Patel, M. Sinclair, A. Seeger, A. Helser, R. Taylor, S. Washburn, and R. Superfine, "Controlled placement of an individual carbon nanotube onto a microelectromechanical structure," *Applied Physics Letters,* **80**, 14, pp. 2574–2576, 2002.

171. C. R. Chen, H. Qin, H. P. Cong, and S.-H. Yu, "A highly stretchable and real-time healable supercapacitor," *Advanced Materials,* **31**, 19, p. 1900573, 2019.

172. Z. Pei, Y. Liu, Q. Zhang, D. Zhao, J. Wang, Z. Yuan, W. Zhang, and S. Sang, "Highly sensitive, stretchable strain sensor based on Ag @ COOH-functionalized CNTs for stroke and pronunciation recognition," *Advanced Electronic Materials,* **5**, 7, p. 1900227, 2019.

173. K. Kim, Y. G. Park, B. G. Hyun, M. Choi, and J. U. Park, "Recent advances in transparent electronics with stretchable forms," *Advanced Materials,* **31**, 20, p. 1804690, 2019.

174. R. Schroeder, L. A. Majewski, and M. Grell, "High-performance organic transistors using solution-processed nanoparticle-filled high-k polymer gate insulators," *Advanced Materials,* **17**, 17, pp. 1535–1539, 2005.

175. S. H. Chae, W. J. Yu, J. J. Bae, D. L. Duong, D. Perello, H. Y. Jeong, Q. H. Ta, T. H. Ly, Q. A. Vu, M. Yun, X. Duan, and Y. H. Lee, "Transferred wrinkled Al_2O_3 for highly stretchable and transparent graphene-carbon nanotube transistors," *Nature Materials,* **12**, 5, pp. 403–409, 2013.

176. H. Zhang, J. Yu, X. Yang, G. Gao, S. Qin, J. Sun, M. Ding, C. Jia, Q. Sun, and Z. L. Wang, "Ion gel capacitively coupled tribotronic gating for multiparameter distance sensing," *ACS Nano,* **14**, 3, pp. 3461–3468, 2020.

177. J. Pu, Y. Yomogida, K.-K. Liu, L.-J. Li, Y. Iwasa, and T. Takenobu, "Highly flexible MoS2 thin-film transistors with ion gel dielectrics," *Nano Letters,* **12**, 8, pp. 4013–4017, 2012.

178. F. Xu, M. Y. Wu, N. S. Safron, S. S. Roy, R. M. Jacobberger, D. J. Bindl, J. H. Seo, T. H. Chang, Z. Ma, and M. S. Arnold, "Highly stretchable carbon nanotube transistors with ion gel gate dielectrics," *Nano Letters,* **14**, 2, pp. 682–686, 2014.

179. A. Chortos, C. Zhu, J. Y. Oh, X. Yan, I. Pochorovski, J. W. To, N. Liu, U. Kraft, B. Murmann, and Z. Bao, "Investigating limiting factors in stretchable all-carbon transistors for reliable stretchable electronics," *ACS Nano,* **11**, 8, pp. 7925–7937, 2017.

180. J. Liang, L. Li, D. Chen, T. Hajagos, Z. Ren, S. Y. Chou, W. Hu, and Q. Pei, "Intrinsically stretchable and transparent thin-film transistors based

on printable silver nanowires, carbon nanotubes and an elastomeric dielectric," *Nature Communications,* **6**, 1, p. 7647, 2015.

181. W. Wu, "Stretchable electronics: Functional materials, fabrication strategies and applications," *Science and Technology of Advanced Materials,* **20**, 1, pp. 187–224, 2019.

182. J. Y. Oh, S. Rondeau-Gagne, Y. C. Chiu, A. Chortos, F. Lissel, G. N. Wang, B. C. Schroeder, T. Kurosawa, J. Lopez, T. Katsumata, J. Xu, C. Zhu, X. Gu, W. G. Bae, Y. Kim, L. Jin, J. W. Chung, J. B. Tok, and Z. Bao, "Intrinsically stretchable and healable semiconducting polymer for organic transistors," *Nature,* **539**, 7629, pp. 411–415, 2016.

183. A. Chortos, G. I. Koleilat, R. Pfattner, D. Kong, P. Lin, R. Nur, T. Lei, H. Wang, N. Liu, Y. C. Lai, M.-G. Kim, J. W. Chung, S. Lee, and Z. Bao, "Mechanically durable and highly stretchable transistors employing carbon nanotube semiconductor and electrodes," *Advanced Materials,* **28**, 22, pp. 4441–4448, 2016.

184. M. F. De Volder, S. H. Tawfick, R. H. Baughman, and A. J. Hart, "Carbon nanotubes: Present and future commercial applications," *Science,* **339**, 6119, pp. 535–539, 2013.

185. H. Jinno, K. Fukuda, X. Xu, S. Park, Y. Suzuki, M. Koizumi, T. Yokota, I. Osaka, K. Takimiya, and T. Someya, "Stretchable and waterproof elastomer-coated organic photovoltaics for washable electronic textile applications," *Nature Energy,* **2**, 10, pp. 780–785, 2017.

186. G. Cai, J. Wang, K. Qian, J. Chen, S. Li, and P. S. Lee, "Extremely stretchable strain sensors based on conductive self-healing dynamic cross-links hydrogels for human-motion detection," *Advanced Science,* **4**, 2, p. 1600190, 2017.

187. S. Park, S. W. Heo, W. Lee, D. Inoue, Z. Jiang, K. Yu, H. Jinno, D. Hashizume, M. Sekino, and T. Yokota, "Self-powered ultra-flexible electronics via nano-grating-patterned organic photovoltaics," *Nature,* **561**, 7724, pp. 516–521, 2018.

188. M. Khatib, T. P. Huynh, Y. Deng, Y. D. Horev, W. Saliba, W. Wu, and H. Haick, "A freestanding stretchable and multifunctional transistor with intrinsic self-healing properties of all device components," *Small,* **15**, 2, p. e1803939, 2019.

189. C. Zhu, A. Chortos, Y. Wang, R. Pfattner, T. Lei, A. C. Hinckley, I. Pochorovski, X. Yan, J. W. F. To, J. Y. Oh, J. B. H. Tok, Z. Bao, and B. Murmann, "Stretchable temperature-sensing circuits with strain suppression based on carbon nanotube transistors," *Nature Electronics,* **1**, 3, pp. 183–190, 2018.

190. B. Chen, P. Zhang, L. Ding, J. Han, S. Qiu, Q. Li, Z. Zhang, and L. M. Peng, "Highly uniform carbon nanotube field-effect transistors and medium scale integrated circuits," *Nano Letters,* **16**, 8, pp. 5120–5128, 2016.

191. V. Derycke, R. Martel, J. Appenzeller, and P. Avouris, "Carbon nanotube inter- and intramolecular logic gates," *Nano Letters,* **1**, 9, pp. 453–456, 2001.

192. A. Javey, Q. Wang, W. Kim, and H. Dai, "Advancements in complementary carbon nanotube field-effect transistors," *IEEE International Electron Devices Meeting*, pp. 741–744, 2003.

193. X. Liang, S. Wang, X. Wei, L. Ding, Y. Zhu, Z. Zhang, Q. Chen, Y. Li, J. Zhang, and L. M. Peng, "Towards entire-carbon-nanotube circuits: The fabrication of single-walled-carbon-nanotube field-effect transistors with local multiwalled-carbon-nanotube interconnects," *Advanced Materials,* **21**, 13, pp. 1339–1343, 2009.

194. J. Tang, Q. Cao, D. B. Farmer, G. Tulevski, and S. J. Han, "High-performance carbon nanotube complementary logic with end-bonded contacts," *IEEE Transactions on Electron Devices,* **64**, 6, pp. 2744–2750, 2017.

195. Y. Zhang, T. Ichihashi, E. Landree, F. Nihey, and S. Iijima, "Heterostructures of single-walled carbon nanotubes and carbide nanorods," *Science,* **285**, 5434, pp. 1719–1722, 1999.

196. J. Tang, Q. Cao, G. Tulevski, K. A. Jenkins, L. Nela, D. B. Farmer, and S.-J. Han, "Flexible CMOS integrated circuits based on carbon nanotubes with sub-10 ns stage delays," *Nature Electronics,* **1**, 3, pp. 191–196, 2018.

197. A. Javey, Q. Wang, A. Ural, Y. Li, and H. Dai, "Carbon nanotube transistor arrays for multistage complementary logic and ring oscillators," *Nano Letters,* **2**, 9, pp. 929–932, 2002.

198. C. Wang, K. M. Ryu, A. Badmaev, N. Patil, A. Lin, S. Mitra, H. S. P. Wong, and C. Zhou, "Device study, chemical doping, and logic circuits based on transferred aligned single-walled carbon nanotubes," *Applied Physics Letters,* **93**, 3, p. 033101, 2008.

199. C. Wang, J. Zhang, K. Ryu, A. Badmaev, L. G. De Arco, and C. Zhou, "Wafer-scale fabrication of separated carbon nanotube thin-film transistors for display applications," *Nano Letters,* **9**, 12, pp. 4285–4291, 2009.

200. J. Zou, K. Zhang, J. Li, Y. Zhao, Y. Wang, S. K. Pillai, H. Volkan Demir, X. Sun, M. B. Chan-Park, and Q. Zhang, "Carbon nanotube driver circuit for 6 × 6 organic light emitting diode display," *Scientific Reports,* **5**, 1, p. 11755, 2015.

201. J. Zhang, Y. Fu, C. Wang, P. C. Chen, Z. Liu, W. Wei, C. Wu, M. E. Thompson, and C. Zhou, "Separated carbon nanotube macroelectronics for active matrix organic light-emitting diode displays," *Nano Letters,* **11**, 11, pp. 4852–4858, 2011.

202. X. Cao, H. Chen, X. Gu, B. Liu, W. Wang, Y. Cao, F. Wu, and C. Zhou, "Screen printing as a scalable and low-cost approach for rigid and flexible thin-film transistors using separated carbon nanotubes," *ACS Nano,* **8**, 12, pp. 12769–12776, 2014.

203. C. Wang, D. Hwang, Z. Yu, K. Takei, J. Park, T. Chen, B. Ma, and A. Javey, "User-interactive electronic skin for instantaneous pressure visualization," *Nature Materials,* **12**, 10, pp. 899–904, 2013.

204. L. Nela, J. Tang, Q. Cao, G. Tulevski, and S. J. Han, "Large-area high-performance flexible pressure sensor with carbon nanotube active matrix for electronic skin," *Nano Letters,* **18**, 3, pp. 2054–2059, 2018.

205. K. Balasubramanian and K. Kern, "25th anniversary article: Label-free electrical biodetection using carbon nanostructures," *Advanced Materials,* **26**, 8, pp. 1154–1175, 2014.

206. Z. Wang and Z. Dai, "Carbon nanomaterial-based electrochemical biosensors: An overview," *Nanoscale,* **7**, 15, pp. 6420–6431, 2015.

207. S. Liu and X. F. Guo, "Carbon nanomaterials field-effect-transistor-based biosensors," *NPG Asia Materials,* **4**, 8, pp. 1–10, 2012.

208. N. Yang, X. P. Chen, T. L. Ren, P. Zhang, and D. G. Yang, "Carbon nanotube based biosensors," *Sensors and Actuators B,* **207**, pp. 690–715, 2015.

209. F. Yuan, Y. Y. Deng, W. Y. Zhou, M. Zhang, and Z. G. Li, "A simple and novel method for the quantitative detection of 5-hydroxymethylcytosine using carbon nanotube field-effect transistors," *Nano Research,* **9**, 6, pp. 1701–1708, 2016.

210. K. Balasubramanian and M. Burghard, "Biosensors based on carbon nanotubes," *Analytical and Bioanalytical Chemistry,* **385**, 3, pp. 452–468, 2006.

211. P. He and L. Dai, "Aligned carbon nanotube-DNA electrochemical sensors," *Chemical Communications,* 3, pp. 348–349, 2004.

212. X. C. Dong, C. M. Lau, A. Lohani, S. G. Mhaisalkar, J. Kasim, Z. X. Shen, X. N. Ho, J. A. Rogers, and L. J. Li, "Electrical detection of femtomolar DNA via gold-nanoparticle enhancement in carbon-nanotube-network field-effect transistors," *Advanced Materials,* **20**, 12, pp. 2389–2393, 2008.

213. D. A. Heller, E. S. Jeng, T. K. Yeung, B. M. Martinez, A. E. Moll, J. B.

Gastala, and M. S. Strano, "Optical detection of DNA conformational polymorphism on single-walled carbon nanotubes," *Science,* **311**, 5760, pp. 508–511, 2006.

214. C. Staii, A. T. Johnson, Jr., M. Chen, and A. Gelperin, "DNA-decorated carbon nanotubes for chemical sensing," *Nano Letters,* **5**, 9, pp. 1774–1778, 2005.

215. H. Muguruma, Y. Shibayama, and Y. Matsui, "An amperometric biosensor based on a composite of single-walled carbon nanotubes, plasma-polymerized thin film, and an enzyme," *Biosensors and Bioelectronics,* **23**, 6, pp. 827–832, 2008.

216. J. C. Claussen, A. D. Franklin, A. Ul Haque, D. M. Porterfield, and T. S. Fisher, "Electrochemical biosensor of nanocube-augmented carbon nanotube networks," *ACS Nano,* **3**, 1, pp. 37–44, 2009.

217. W. Xue and T. Cui, "A thin-film transistor based acetylcholine sensor using self-assembled carbon nanotubes and SiO_2 nanoparticles," *Sensors & Actuators B,* **134**, 2, pp. 981–987, 2008.

218. Y. Zhang, Y. Pan, S. Su, L. Zhang, S. Li, and M. Shao, "A novel functionalized single-wall carbon nanotube modified electrode and its application in determination of dopamine and uric acid in the presence of high concentrations of ascorbic acid," *Electroanalysis,* **19**, 16, pp. 1695–1701, 2010.

Chapter 5

Graphene Transistors

Min Zhang, Shuo Zhang, Qiuyue Huang, Chao Xie, and Ni Li
School of Electronic and Computer Engineering, Peking University Shenzhen Graduate School, Shenzhen 518055, China
zhang.min@pku.edu.cn

5.1 Introduction

Silicon technology has dominated the development and innovations in electronics, from the microprocessor of the last century, to automation, computers, and smartphones today, as well as the artificial intelligence of tomorrow, with the scaling progress down to the nanometer regime. At the same time, when silicon transistor scaling no longer yields historical energy-efficiency benefits, research beyond silicon is expected. Among them, graphene and carbon nanotubes (CNTs) are two of the most promising technologies for future electronics. Two-dimensional (2D) graphene has excellent performance due to its unique structure. The term graphene, as a single carbon atom layer of a graphitic structure, was first introduced

Nanocarbon Electronics
Edited by Changjian Zhou, Min Zhang, and Cary Y. Yang

ISBN 978-981-4877-11-4 (Hardcover), 978-1-003-04308-9 (eBook)
www.jennystanford.com

in 1994 [1], 10 years before monolayer graphene was achieved [2]. However, the history of graphene can be traced back further. As shown in Fig. 5.1, the theoretical research of graphene was first explored by Wallace in 1947 [3]. During the following decades, it was shown theoretically that 2D crystals could not exist because the thermal fluctuations would destroy the long-range order. Even so, experimental research of producing graphene had never stopped. From the 1970s, a single layer of graphite was epitaxially grown on SiC substrate [4]. The single or few-layer graphene obtained had high crystallinity and mobility, but the strong interaction between the graphene and the substrate changed the electronic structure of the graphene, which rendered measurement of its properties difficult. Mechanical exfoliation method has been used to fabricate thin films of graphite since 1990 [5]. It was not until the work of Geim and Novoselov in 2004 that the first monolayer graphene was finally achieved [2]. The report of monolayer graphene was a key to opening the flood gate of research on graphene. Since then, the interest for graphene research has exploded and the number of papers about graphene has grown at a fast pace every year. Three years later, in 2007, the first top-gate graphene field-effect transistor (GFET) was proposed [6]. In this work, the carrier mobility of graphene was found to be comparable with that of silicon.

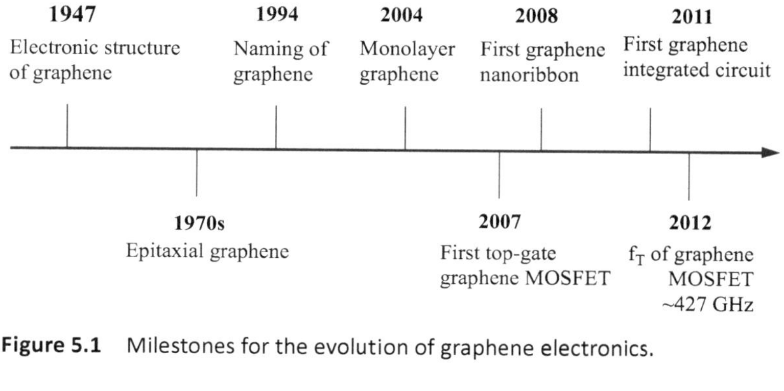

Figure 5.1 Milestones for the evolution of graphene electronics.

At the early stage, researchers focused on the excellent electrical properties of graphene, especially high mobility, and they expected to replace Si with graphene in the mainstream of the semiconductor electronics. However, carrier mobility is not the only parameter that determines FET performance. Graphene is called a gapless semiconductor (Fig. 5.2a), and devices made of large-area graphene cannot be switched off, resulting in a poor on/off current ratio (I_{on}/I_{off}). Then the gapless band structure of graphene has become an key obstacle to logic applications [7]. To solve this problem, bandgap engineering, which can modify the band structure and open a band-gap for graphene, has become a hot research topic.

In recent years, compared with the excessively optimistic expectations, more rational voices have been accepted gradually. That is, graphene would not replace silicon or the III–V compounds in the short term [8]. There are series of publications discussing the role of graphene in the field of electronics [9, 10]. Its properties of zero gap and super high mobility are more suitable for RF transistor applications. Besides, the other promising direction for graphene is to complement the conventional semiconductors instead of replacing them. Along this direction, intensive work has been conducted on graphene, involving transparent electrodes, flexible electronics, and printable electronics [11].

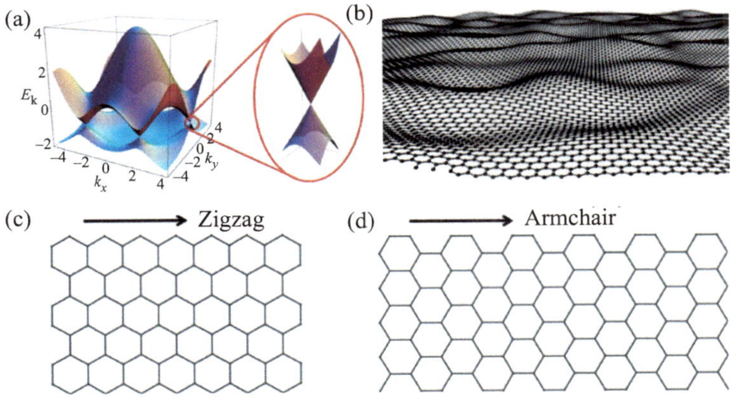

Figure 5.2 (a) Schematics of the graphene structure. Left: energy spectrum. Right: zoom in of the energy bands close to one of the Dirac points. (b) Electronic dispersion in the honeycomb lattice. (c) Zigzag edge. (d) Armchair edge. Figure (a) reprinted with permission from Ref. [9], Copyright 2009, American Physical Society; Fig. (b) reprinted with permission from Ref. [12], Copyright 2007, Springer Nature; and Figs. (c) and (d) reprinted with permission from Ref. [13], Copyright 2017, Springer Nature.

In this chapter, we review the crucial issues that limit the application of graphene to transistors. The state of the art of graphene transistors, especially key performance and advanced processes, will be reviewed. Novel structures of graphene transistors and applications of graphene in RF devices and circuits will also be discussed, followed by applications in flexible and transparent electronics.

5.2 Structure and Electronic Properties of Graphene

Graphene is defined as "a single-atom-thick sheet of hexagonally arranged, sp^2-bonded carbon atoms that is not an integral part of a carbon material, but is freely suspended or adhered on a foreign substrate" [14]. It can be considered the basic building block for all other sp^2-bonded carbon materials. For example, CNTs can be formed by rolling graphene in certain ways. However, many years ago, it was predicted that a long-range ordering 2D structure was impossible to be stable at any finite temperature, due to the displacements of atoms caused by thermal fluctuations in low-dimensional crystal lattices. The displacement was comparable to interatomic distances [15, 16]. The melting temperature of thin films rapidly decreased with the film thickness decreasing, and the film became instable at a thickness of dozens of atomic layers [17]. This conclusion was disproved by A. K. Geim and K. S. Novoselov in 2004 [2]. They achieved the first monolayer graphene by repeatedly sticking and tearing graphite with Scotch tape. Because there is only a single layer of graphite, the thickness of graphene is just around 0.335 nm [18], which is the thinnest known material. Indeed, graphene can be deemed as being in a metastable state. The strong interatomic bonds between the nearest carbon atoms ensure that the thermal fluctuations will not result in dislocation or other crystal defect even at an elevated temperature [15]. However, graphene was demonstrated not to be a perfect flat sheet (Fig. 5.2b). Meyer et al. observed broader diffraction peaks with increasing the tilt angle of suspended graphene on a microfabricated scaffold [12]. This phenomenon becomes much weaker in bilayer graphene, which has a much flatter surface with layers of graphene increasing.

Besides the surface state, edge state also plays an important role in the property of graphene. The edges of graphene can take two forms: armchair or zigzag, which are shown in Figs. 5.2c and d. Girit et al. investigated the dynamics of the edge reconstruction in real time and also demonstrated the stability of the armchair and zigzag arrangements. The long-term stability of zigzag edges was elucidated by a time-average analysis and is explained by a sample model [19]. The difference in edge states was demonstrated to determine the electronic properties of graphene quantum dots (GQDs) and graphene nanoribbons (GNRs). The 7–8 nm GQDs with zigzag-orientated edges were metallic, which was different from the other studies on semiconductor behavior. Besides, GNRs with more zigzag edges were demonstrated to have smaller energy gap [20]. Therefore, it is valuable to study the influence of the edge and how to control the edge. Cheng et al. reported that the edge state could be controlled by changing the reaction precursors and the growth or etching rates, since the etching rate of armchair edge was faster than that of zigzag edge [21].

Graphene's extraordinary electrical properties arise from its unique π-orbitals [9]. The valence shell of carbon contains four electrons, and three of them are bonded to the nearest neighbor carbon atoms. The fourth bond of the carbon atoms is dangling, shares electrons over the surface of the material, and forms the partially filled π-orbital. As Fig. 5.2a shows, the valence band and the conduction band of graphene intersect at six points, which is known as the Dirac point of the Brillouin zone. The unique energy-band structure of graphene makes it a zero-gap material. At the Dirac point, the density of states is zero and the linear dispersion relation results in an effective mass of zero. Therefore, electrons in graphene behave as massless Dirac fermions, giving rise to unprecedentedly high carrier mobility [22]. Graphene has a minimum carrier density of approximately 1×10^8 cm^{-2}, which allows a high on current of 3.32 mA/μm [23]. The carrier mobility measured for a high-quality single-layer graphene sheet reaches 200,000 cm^2 $V^{-1}s^{-1}$ [2]. Mobility of up to 2,000,000 cm^2 $V^{-1}s^{-1}$ was theorized to be achievable [24]. Graphene on SiO_2 was demonstrated to have a mobility over 20,000 cm^2 $V^{-1}s^{-1}$ [25]. Transistors on wafer scale using CVD graphene were also studied, and the mobility was over 3000 cm^2 $V^{-1}s^{-1}$ [26], much higher than that of Si-based device. Recently, CVD-grown graphene demonstrated high mobility over 350,000 $cm^2\ V^{-1}s^{-1}$, which approached that of pristine graphene made by mechanical exfoliating [27]. The high carrier mobility values of graphene suggest that the charge transport in graphene should be actually ballistic, in which carriers can travel along thousands of interatomic distances without scattering [2, 22, 28, 29].

Another notable property of graphene is ambipolar conduction, which has been demonstrated in several semiconductor materials, including graphene [22], CNT [30], and amorphous silicon [31]. However, graphene is the only one that possesses both high mobility and excellent scalability [32]. The carriers in graphene can be switched between holes and electrons by controlling the position of the Fermi level using the gate bias. This is impossible for conventional unipolar devices, whose majority carrier type is determined by the doping and cannot be changed once the devices are fabricated.

5.3 Issues of Graphene for FETs

Some critical issues impacting use of graphene in transistors must be addressed despite its appealing electronic properties. Ideally, the semiconductor in a transistor must have a sufficiently wide bandgap with high carrier mobilities, ability to form n- and p-types, low-resistance contacts not affecting carrier transport, few interface states at contacts with dielectrics, long-term thermal and chemical stability, and compatibility with Si CMOS process technology. Natural large-area graphene has a zero or very small bandgap and behaves like a semimetal, thus failing the first requirement. Therefore, research efforts are mainly concentrated on how to engineer graphene to function as a channel material. The key issues and potential solutions are discussed in this section.

5.3.1 Bandgap Engineering

As shown in Fig. 5.3, pristine graphene is gapless semiconductor. Fortunately, the band structure of graphene can be modified to open a bandgap. Researchers have reported several approaches, including confining graphene into pseudo-one-dimensional nanoribbons, applying a perpendicular electric field to bilayer graphene and adopting strain engineering, etc.

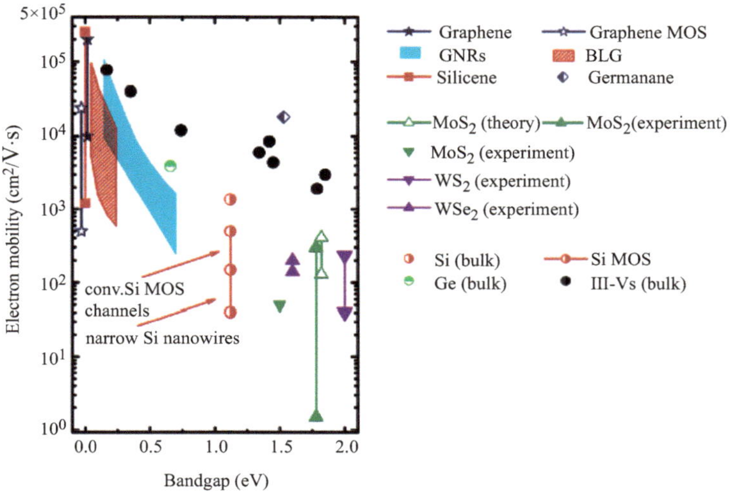

Figure 5.3 Room-temperature electron mobility versus bandgap for different materials. Reprinted from Ref. [33], Copyright 2015, with permission from the Royal Society of Chemistry.

5.3.1.1 Graphene nanoribbons

Graphene nanoribbon has been considered the most promising approach for opening a bandgap. By constraining large-area graphene into quasi-one-dimension, narrow GNRs can be formed. A great deal of theoretical work was done before experiments [34–36]. It is predicted that the size of bandgap is inversely proportional to the ribbon width of GNRs [37, 38]. GNRs with a width below 2 nm is excepted to obtain a bandgap over 1 eV [39]. Besides the ribbon width, the nanoribbon edge structure and roughness also have influence on the band structure [40, 41]. However, the origin of the bandgap of GNR is still controversial. Both lateral confinement [39] and Coulomb blockade [42] may be responsible for the formation of the bandgap.

The synthesis of graphene nanoribbon can fall into either top-down or bottom-up methods. At the early stage, top-down approaches were proposed, such as unzipping carbon nanotubes [43, 44] (Fig. 5.4a) and electron beam lithography [45, 46] (Fig. 5.4b). The first experiment of GNR was conducted in 2007 [47]. By electron beam lithography, graphene nanoribbons with variable width ranging from 50 nm to sub-10 nm were formed. However, the top-down fabrication approaches typically yield ribbons with width >10 nm, companied with rough edge and mixtures of different GNRs.

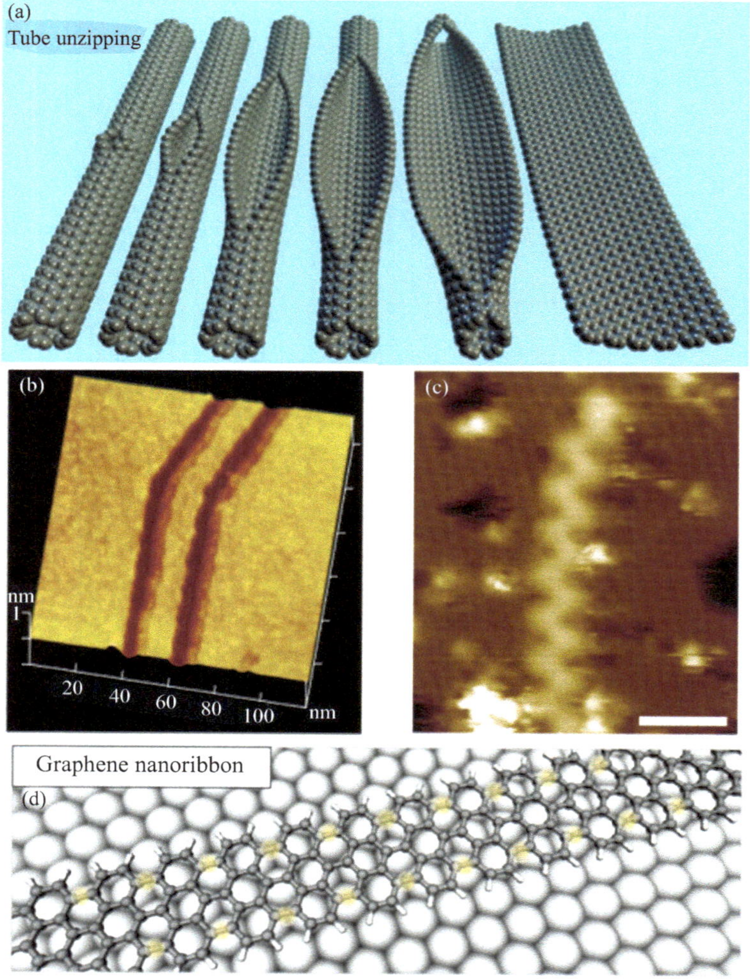

Figure 5.4 Top-down fabrication of GNRs: (a) CNT unzipping, (b) Electron beam lithograph. Bottom-up fabrication: (c) 1 nm wide GNR achieved by chemical synthesis, (d) 0.74 nm wide GNR achieved by using molecular precursors. Figures (a–d) reprinted from Ref. [43], Copyright 2009; Ref. [46], Copyright 2008; Ref. [49], Copyright 2014; Ref. [50], Copyright 2010; respectively, by permission from Springer Nature.

The major challenge is to achieve nanoribbons with 1–2 nm width and smooth edges. Unlike the top-down approaches, the bottom-up strategies achieve the construction of ribbons from small molecular precursors. By taking advantage of the precise control, the classic synthetic chemistry can achieve extreme narrow width for GNRs with better edge uniformity [48]. A bottom-up solution used pre-patterned SiC trenches to force graphene to bend between a (0001) face and a low-symmetry facet. A narrow graphene ribbon was produced with the localized strain, resulting in nanoribbons of 1 nm with a bandgap of 1.3 eV [49] (Fig. 5.4c). Another bottom-up method, which produced atomically precise graphene nanoribbons with different topology and width, was proposed. The method used surface-assisted coupling of molecular precursors into linear polyphenylenes and their subsequent cyclodehydrogenation. In this way, a GNR width as small as 0.74 nm was achieved and the GNR exhibited a bandgap of 1.6 eV [50] (Fig. 5.4d). Bottom-up methods consist of two main pathways. One is in solution using synthetic organic and polymer chemistry techniques, and the other is on substrates with modern surface science techniques.

Nanocrystallization is also a strategy to tune the properties and open bandgap for materials. Its principle is similar to graphene nanoribbons. Based on a liquid-phase precursor quenching method, nanocrystalline graphene films with average grain size of around 3.6~10.3 nm were synthesized in seconds [51]. The 3.6 nm grained film showed semiconducting behavior with a bandgap of ~50 meV, and the 10.3 nm grained film showed a bandgap of 10 meV. The inverse relationship between the bandgap and the average grain size indicates that the bandgap may originate from the quantum confinement and the crucial effect of the grain boundaries, as in the case of graphene nanoribbons.

5.3.1.2 Bilayer graphene

Although remarkable bandgaps have been obtained by graphene nanoribbons, the fabricating procedure is still too complicated, and also the process would result in a significant decrease in carrier mobility with the influence of edge scattering. Other alternatives were proposed, including applying a gate voltage perpendicular to the bilayer graphene to break the inversion symmetry of the two equivalent layers. In bilayer graphene, changing the applied gate voltages would in turn control the electron density, and the interlayer asymmetry of different potential energies. As Fig. 5.5 shows, when a perpendicular electric field is applied, the valence and conduction bands would separate from each other.

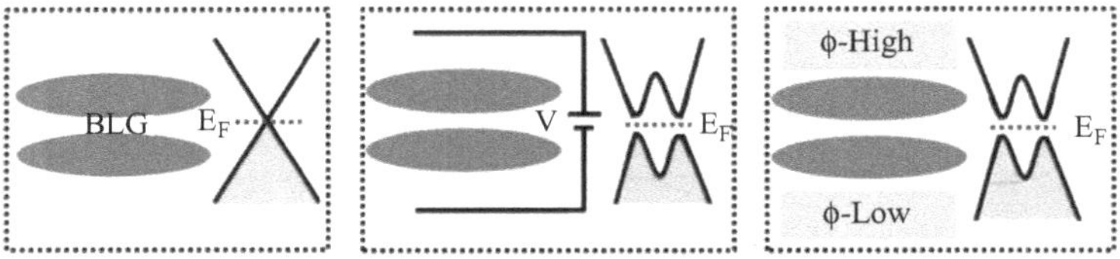

Figure 5.5 Bilayer graphene and its band structure under different conditions. Reprinted from Ref. [52], Copyright 2015, with permission from the Royal Society of Chemistry.

Recently, a widely tunable electronic bandgap in electrically gated bilayer graphene was reported. By using a dual-gate structure and infrared microspectroscopy, a gate-controlled, continuously tunable bandgap up to 250 meV was demonstrated [53]. The bandgap is controlled by changing the voltage that is introduced by a vertical external electric field. The schematic and mechanism of this design are shown in Fig. 5.6.

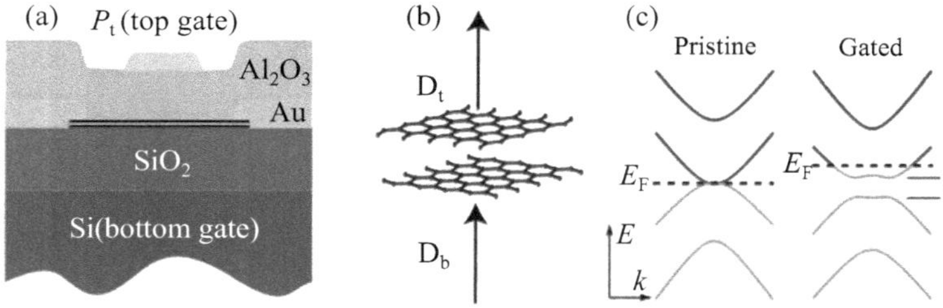

Figure 5.6 (a) Illustration of a cross-sectional side view of the gated device. (b) Sketch showing how the bilayer induces top (D_t) and bottom (D_b) electrical displacement fields. (c) Left, the electronic structure of a pristine bilayer has zero bandgap. Right, upon gating, the displacement fields induce a nonzero bandgap and a shift in the Fermi energy. Reprinted from Ref. [53], Copyright 2009, by permission from Springer Nature.

Heavy doping is another strategy to generate electric field, which avoids the external voltage and does not need the double-gate structure [54]. By chemical doping on both sides of a bilayer graphene, the inversion symmetry is broken, and the bandgap is created. Besides, opening an energy gap via graphene–substrate interactions also works. By sandwiching the bilayer graphene between two surface-functionalized boron nitride (BN) monolayers, spontaneous electric polarization can be produced, which forces electrons to transfer between two graphene layers and substrates. The transfer of electrons produced a significant chemical difference between two graphene layers and resulted in energy gaps ranging from 0.35 eV to 0.55 eV, calculated by the first principles [52]. Bilayer graphene could also be sandwiched between a pair of LiF_2 super-halogen and Li_3O super-alkali molecules, which enabled the energy band to open a gap of about 0.36 eV and 0.26 eV near the Fermi energy, making the system exhibit semiconducting properties. This was attributable to the strong dipole electric field between the LiF_2 super-halogen and Li_3O super-alkali molecules. It is also found that Li_3O and the adjacent layer of graphene provided an energy band at the bottom of the conduction band and that LiF_2 and the adjacent layer of graphene provided an energy band at the top of the valence band [55].

5.3.1.3 Strain engineering

Strain engineering has been widely used in semiconductor devices since the 1980s [56]. It is shown that strain can open bandgap in a metallic CNT and modify the bandgap in a semiconducting CNT with about 100 meV per 1% stretch [57, 58]. Therefore, it is reasonable to utilize strain to modify the electronic properties of graphene. Moreover, since the two carbon sublattices of graphene are not equivalent under uniaxial strain, it is possible to introduce a bandgap on graphene due to the break of sublattice symmetry [59–61]. Through the first-principle calculation of its electronic properties under deformation, the graphene bandgap can be modulated from zero to up to 0.9 eV [62]. As shown in Fig. 5.7, by observing the shift in the Raman spectroscopy (Raman 2D band and G band) for single-layer graphene under the uniaxial tensile strain, a bandgap opening of 300 meV for 1% strain can be calculated [63]. However, the applied strains, such as uniaxial, shear, isotropic planar strain, and local strain, have different influence on the graphene. Besides, the value of gap also depends strongly on the relative angle of applied strain [64, 65].

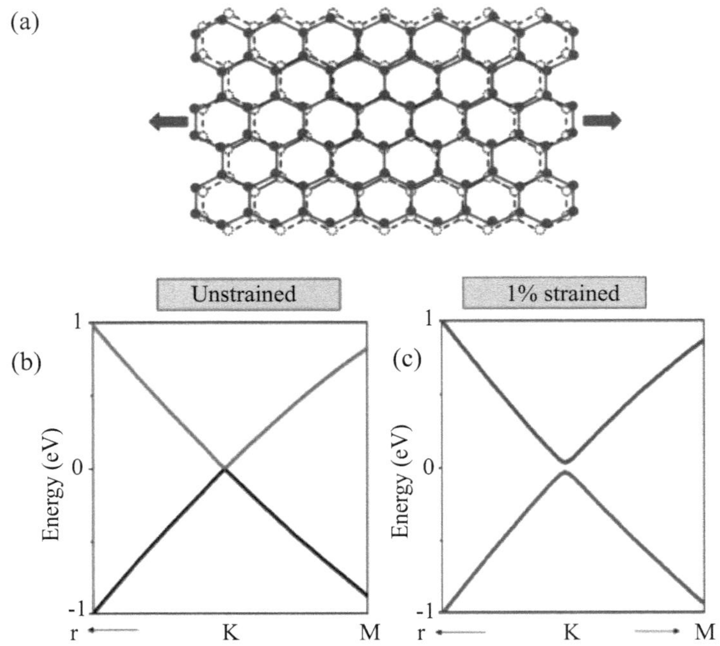

Figure 5.7 (a) Effect of uniaxial tensile stress on a graphene supercell. Calculated band structure of unstrained (b) and 1% tensile strained (c) graphene. Reprinted with permission from Ref. [63], Copyright 2008, American Chemical Society.

Besides the methods mentioned above, many other techniques have also been developed. As an example, substrate-induced bandgap opening on epitaxial graphene was investigated [59]. In the previous work, the graphene epitaxially grown on SiC substrate achieved a gap of 0.26 eV. The gap showed an inverse relationship with the thickness of sample. Graphene grown on hexagonal boron nitride (h-BN) and h-BN/Ni was reported with a bandgap up to 0.5 eV [66]. The origin of this gap can be attributed to the breaking of sublattice symmetry by the graphene–substrate interaction.

Substitutional doping is another promising route to open a bandgap. By doping nitrogen atoms into the graphene lattice, the gap originated from the substitutional doping was achieved, and this made the graphene exhibit n-type behavior [67]. In addition, molecular doping and chemical functionalization can also be used in bandgap engineering [68, 69].

5.3.2 Contact Resistance

Realizing low contact resistance between graphene channel and metal electrodes is crucial for improving the intrinsic transport speed [70] and I_{on}/I_{off} [71] of graphene logic devices. The contacts lead to large source resistance (R_S) and drain resistance (R_D), and limit the cutoff frequency (f_T) and maximum frequency (f_{max}) of transistors. The contact resistance of graphene transistors (~100–2000 Ω.μm [72–75]) is higher than that of Si MOSFETs (20–50 Ω.μm [76]). As shown in Fig. 5.8, there are three forms of contacts between metal and graphene. The contact resistance comes from three parts: (1) the sheet resistance of the overlapped graphene, (2) the difference of work functions between graphene and metals, and (3) the interface interactions. Compared with the excellent in-plane conductivity, graphene has extremely low out-of-plane conductivity. Carrier transporting from three-dimensional metal to two-dimensional graphene sheet is a problem, and most of the voltage is concentrated on the edges of the metal–graphene interface.

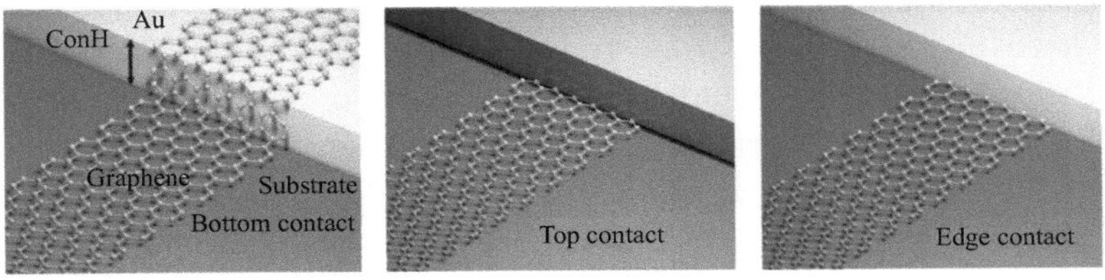

Figure 5.8 Schematic showing the concept of bottom, top, and edge contacts. "Top contact" refers to a situation where metal is evaporated or sputtered onto graphene. Reprinted with permission from Ref. [84], Copyright 2019, American Chemical Society.

Heavy doping is an effective method to realize metal–semiconductor Ohmic contact, but this technique is not chemically, mechanically, and electrically stable for graphene. Adopting a metal with a larger work function than that of graphene is expected to achieve the effect of doping. In such an ideal metal–graphene junction, graphene is doped with electrons if the work function of graphene is larger than that of metal ($W_G > W_M$), while the graphene would be doped with holes when W_G is smaller than W_M. Both the situations would lead to low contact resistance. However, the charge redistribution at the graphene–metal interface is not only the result of an electron transfer between the metal and the graphene. There is also a contribution from a metal–graphene chemical interaction. As a result, the true crossover point lies at $W_M - W_G = 0.9$ eV [77].

Values of contact resistance (R_C) vary from a few hundred to a few thousand Ωμm, depending on the metal selection, fabrication procedure, gate bias, and measurement conditions. Studies on the graphene–metal junction suggest that palladium (Pd) or gold (Au) contacts provide the best contact resistance. By the late 2010, the metal–graphene contact resistance of 110±20 Ωμm for Pd on exfoliated graphene was reported. The low-resistance stems from superior wetting of Pd to the graphene surface [72]. Now the lowest contact resistance is about 69 Ωμm, realized by adopting high-purity Pd as metal electrodes [78]. The super low resistance is attributed to the saturation of carrier injection efficiency between graphene and Pd, as the high-quality graphene has a sufficiently long carrier mean free path. Besides adopting appropriate metals, the contact resistance can also be reduced by removing the contamination on the surface [79]. Graphene–metal contact is highly sensitive to the fabrication technology and contamination. Organic residues are hard to be removed from graphene surface. Photoresist residue from lithography would be left and could not be cleaned up by oxygen plasma. Various methods have been proposed to clean the interfaces, such as controlled oxygen descum process [43]. In addition, patterning contact area [80, 81] and ultraviolet/ozone treatment [82, 83] were also demonstrated to be able to improve the contact resistance.

As an alternative, bottom-contact devices with a van der Waals interaction were proposed recently for use in various large-area GFET applications. By engineering the graphene–metal van der Waals contact, Liu. et al. demonstrated that ultralow contact resistance was achievable via a bottom-contact strategy incorporating a simple transfer process without any harsh thermal treatment (>150°C), and the contact resistance down to 65 Ωμm and most of values under 200 Ωμm [84].

5.3.3 Mobility Optimization

Graphene channel is only single-atom thick, and its zero-gap energy-band structure can provide enough carries for FET operation. It is characterized with an exceptionally high carrier mobility (μ) (>250,000 cm^2 V^{-1}s^{-1} at room temperature [85] and 6 × 10^6 cm^2 V^{-1}s^{-1} at 4 K [86]). However, graphene obtained by mechanical cleavage on top of an SiO_2 wafer exhibits μ ~10^4 cm^2 V^{-1}s^{-1} [87], far from the theoretical limit. The main influencing factors are the charged impurities on the substrate and the scattering. It is demonstrated that mobility of the suspended graphene, which was annealed by a high electrical current, can reach the value higher than 10^5 cm^2 V^{-1}s^{-1} [23, 88] and 10^6 cm^2 V^{-1}s^{-1} [89] at room and liquid-helium temperature, respectively. The reason is that the suspended graphene can reduce the charged impurities and other scattering centers from substrate. Apart from the suspended graphene, annealing can also remove contaminations. Besides contaminations, flexural phonons would become the leading scattering mechanism at temperature T >10 K, which can be eliminated by strain [89].

The structures of the transistor and gate dielectrics also affect mobility significantly. Structure roughness and impurities lead to an inhomogeneous distribution of charge [90]. Selection of appropriate substrates and gate dielectrics is very important to minimize the effect of electron–phonon scattering and polar optical phonons [91]. h-BN is such an ideal substrate material that has atomically smooth surface, large optical phonon modes, and a lattice constant comparable to the graphene. The graphene with an h-BN substrate achieves a mobility of 6 × 10^4 cm^2 V^{-1}s^{-1} [92]. Devices made of sandwiched graphene between two h-BN crystals exhibit ballistic transport and mobility more than 10^5 cm^2 V^{-1}s^{-1} at room temperature [85].

5.4 Performance of Graphene FETs

The first graphene FET was reported in 2004 with a back-gate structure [2], when the first monolayer graphene flake was observed. Three years later, the first graphene top-gate FET was reported [6] (Fig. 5.9). After that, enormous research works about graphene FETs have been proposed. The FETs are used for switching and amplification in circuits. Switching is the basis for digital logic, while amplification and gain are crucial for analog/RF applications. It is necessary for us to look at the figure of merit (FOM) for transistors before we continue discussing the status of graphene FETs.

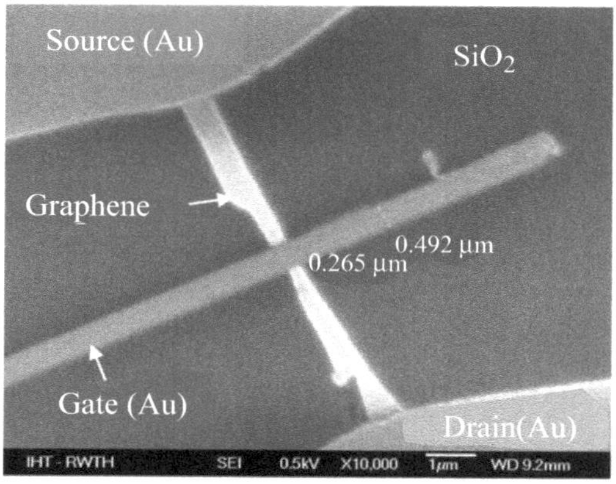

Figure 5.9 SEM image of a graphene transistor. Reprinted with permission from Ref. [6], Copyright 2007, IEEE.

5.4.1 Current Switching

The FOM related to switching is I_{on}/I_{off}, the on/off current ratio. For digital logic application, the I_{on}/I_{off} ratio of more than 10^6 is favored [93]. However, the I_{on}/I_{off} ratio values of MOSFETs with large-area gapless graphene channels are only 2–10, much smaller than the requirement and the values of the conventional Si devices (10^4–10^7). To increase the I_{on}/I_{off} to a suitable value, a wide bandgap is necessary.

Graphene nanoribbon with width below 10 nm was produced by a chemical method, opening a gap exceeding 300 meV [47]. The graphene FETs fabricated with this GNR showed a high I_{on}/I_{off} ratio of about 10^7 at room temperature. However, with the opening of the gap, the carrier mobility was significantly degraded to 200 cm^2 $V^{-1}s^{-1}$. Such a high I_{on}/I_{off} was also observed in a current annealed suspended GNR-FET, which derived from carbon nanotubes. The width of the GNR was approximately 20 nm, and the bandgap of the GNR was the order of 600 meV, with the I_{on}/I_{off} exceeding 10^4 [94]. The formation of such a large bandgap and I_{on}/I_{off} can be largely attributed to the creation of a narrow constriction in the suspended GNR. The performance of the above-mentioned GNR-FETs is shown in Table 5.1.

Table 5.1 GNR fabrication techniques and characteristics at room temperature

	Fabrication methods	Technique	Width (nm)	Bandgap (meV)	On/Off ratio	Mobility ($cm^2 V^{-1}s^{-1}$)	Ref
Top down	Conventional lithography	Electron beam lithography and etching	20–500	30	5 RT	—	[98]
		Mechanical exfoliation and lithography	10–100	20–300	—	—	[99]
		Oxygen plasma etch with chemically synthesized nanowires mask	8	170	160	—	[100]
		Mechanical exfoliation and reactive ion etching	45	110–340	—	5000	[101]
		Metal etch mask technique	30	—	—	800–6000	[102]
		Gas-phase etching chemistry	5	400	10000	—	[103]
	Local probe lithography	Scanning tunneling microscope lithography	2.5	500	—	—	[46]
		CVD and STM nanolithographic technique	3–12	200–300	—	—	[104]
	Chemical	Exfoliation and chemical treatment	<10	>300	10^7	100–200	[47]
	Unzipping CNTs	Gas-phase oxidation and mechanical sonication	10–20	10–15	—	1500	[105]
		Plasma etching of nanotubes partly embedded in a polymer film	10–20	—	10–100	—	[106]
	Particle cutting	Expitaxy and Fe nanotubes-assisted hydrogen etching	>1.1	120–1600		—	[107]

(*Continued*)

Table 5.1 (*Continued*)

	Fabrication methods	Technique	Width (nm)	Bandgap (meV)	On/Off ratio	Mobility ($cm^2 V^{-1}s^{-1}$)	Ref
Bottom up	Molecular precursor based	Using different specific molecular precurors	0.7–1.7	1200–3500		—	[108]
		Cyclodehydrogenation of semirigid polyphenylene precursors	—	1880	—	150–15000	[109]
		Small-molecule building block precursors	3–11	1400	—	—	[110]
		Polymerization of molecular precursors	1	1300	—	—	[49]
		Cyclodehydrogenation of precursor monomers	0.74	1600	—	—	[50]
		Topochemical polymerization and aromatization of a diacetylene precursor	1.36	1500	5	0.15	[111]
	CVD	Direct growth graphene nanoribbon from Ni nanobar	23	58.5	10000	40	[112]
		CVD growth of graphene nanoribbons on SiO_2 substrates	20	50	2	1000	[113]
		Growth on SiC trenches	1.4	>500	—		[114]
		CVD growth on Cu foil and paraffin as a transfer support layer	—	—	4.4	7438–14215	[115]
	Epitaxial graphene	Grown on single-crystal, semi-insulating SiC	10	140	10^6 at 4 K	800–1000	[116]

Similar to GNR, graphene nanomesh (GNM) is another approach to open a gap. The GNM with perfect square lattices of hexagonal nano-holes and different size, as well as neck widths, has great advantages over the pristine graphene. The I_{on}/I_{off} increases from 7 for pristine graphene FET to more than 7400 for GNM-FETs with a bandgap over 500 meV [95]. The highest on/off ratio up to now is achieved with the nanoperforated graphene as channel. The FET with nanopatterned graphene channel with 125 nm average hole diameter and 75 nm neck size exhibits an on/off ratio up to 10^8 at room temperature [96]. Apart from quantum confinement (GNRs and GNMs), many other techniques have been developed to improve I_{on}/I_{off}. A bandgap up to a few hundred meV can be created by the perpendicular electronic field in bilayer graphene. Current on/off ratio of around 2×10^3 at room temperature was achieved in dual-gate bilayer graphene FETs with an electrical bandgap of more than 130 meV [97]. This method is highly advantageous for large-scale integration compared to GNR-FETs. Surface doping on opposite sides of bilayer graphene can generate a gap (~350 meV) with only a gate instead of dual-gate structure. Experiments exhibit that current on/off ratio of this structure is up to 76.1, with a high mobility of ~3100 $cm^2\ V^{-1}s^{-1}$ [54].

Another FOM for FET switching is the subthreshold swing (SS), generally expressed in mV dec^{-1}. SS represents the inverse slope of the log I_{DS} versus V_{GS} curve. The smaller the SS, the better the transistor behaves as a switch. The SS of a graphene FET is expected to be smaller than 60 mV dec^{-1}, which is the value for conventional Si MOSFETs. The SS smaller than 60 mV dec^{-1} can be obtained in the tunneling devices, in which carrier transport through interband tunneling [117]. A computational analyzed small subthreshold swing of 20 meV dec^{-1} was achieved on a bilayer graphene tunnel TFT [118]. In addition, a modeled GNR tunnel FET was proposed and studied by three-dimensional atomistic simulations. The device showed a subthreshold swing of 14 meV dec^{-1} at room temperature [119].

5.4.2 Voltage Gain

Voltage gain $A_v > 1$ is required in analog voltage amplifiers and digital logic gates. However, most of the fabricated intrinsic graphene

FETs cannot provide a voltage gain larger than 1. The voltage gain depends on the ratio of transconductance g_m to output conductance g_{ds} of FET. Since a small g_{ds} can only be achieved when the transistor operates in saturation, the key of high-voltage gain is to obtain a good current saturation for graphene FETs. The primary reason for the weak saturation is the absence of a bandgap, which leads to a second linear region of the drain current [120]. For small values of V_{DS}, the drain current starts to saturate with the increase in V_{DS}. At the point, where the potential conditions at the drain end of the channel correspond to the Dirac point, the conduction type switches and the transistor enters a second linear region [8]. This behavior is the consequence of gapless channels and does not occur in the FETs with semiconductor materials.

Several approaches have been proposed to improve current saturation in graphene FETs such as patterning graphene into nanoribbons. However, this method would reduce the on current at the same time [47, 121], which in turn reduces the voltage gain. Therefore, it is necessary to find a way to open a bandgap, while the mobility of graphene does not decrease significantly. One of the effective methods to obtain $A_v > 1$ is increasing the transconductance by developing ultrathin high-κ gate dielectrics such as SiO_2 [6] and Al_2O_3 [122]. In order to achieve ultrathin and conformal dielectric films, which are pinhole free with minimal disorder or traps at the interface with graphene, atomic layer deposition (ALD) has been used for dielectric deposition. However, the chemical inertness and hydrophobicity of graphene resulted in nonuniform films [123, 124]. For optimal performance, surface treatment of graphene was proposed to optimize the ALD process, such as seeding monolayer with chemical functionalization oxidized metals [125], nitrogen dioxide [126], and solution-phase deposited organic seeding layers [127]. Recently, self-assembled nanodielectrics, as an alternative to ALD, were applied [128]. This method showed a high field-effect mobility, negligible hysteresis, low operating voltage, and desirable current saturation.

All transistors discussed so far in this section are graphene MOSFETs with a top-gate structure, because FETs with back gate suffer from very large parasitic capacitances. Recently, an alternative gate design was proposed, in which the gate lies on the side of channel [129, 130]. The side-gate graphene FET showed a high transconductance of 590 mS mm^{-1}, which is essential for high-voltage gain. Besides, voltage gain of graphene FETs can also be improved by using bilayer graphene, where a bandgap is introduced by a vertical electric field [131, 132]. Using h-BN as a substrate or encapsulating graphene is also a good idea to obtain better electrical performance of GFETs [133].

The first GFET with voltage gain $A_v > 1$ was demonstrated by adopting a thin and dense natural alumina passivation layer as the top-gate dielectric in a bilayer graphene FET [134]. The buffer-free alumina passivation dielectric is not only used for the purpose of high gating efficiency, but also to preserve the intrinsic behavior of graphene. In this device, a bandgap of 100 meV was created, and the voltage gain was improved to 6. However, in this device, $A_v > 1$ only appeared at cryogenic temperature. Compared with the room temperature thermal activation energy, the Schottky barrier is insufficient to effectively block thermal carrier injection at room temperature. The first $A_v > 1$ at room temperature was realized in a single misoriented bilayer graphene transistor with drop cast of solid polymer electrolyte as gate dielectric [135]. A complex six-finger-gate FET configuration achieved the voltage gain larger than unity for alternating current mode. Up to now, the highest A_v (~35) was achieved from a double-gated bilayer graphene FET with an Al_2O_3 acting as the top-gate dielectric [132].

5.5 Graphene FETs in CMOS Circuits

The short-channel effect in Si-based devices results in significant increase in power density and production cost [136]. The progress in digital logic relies on downscaling of size, controlling power dissipation, and improving performance. Thus, the ultrathin graphene channel with extremely high mobility is a powerful feature for digital logic. At the same time, graphene devices have the potential to be scaled to shorter channel without the penalties of reducing channel length [120]. The outstanding thermal property of graphene can help the devices to dissipate heat efficiently as the power density increases with the increasing number of transistors per chip.

To be used in CMOS logic circuits, graphene FETs with both n-channel and p-channel are needed. The first graphene integrated circuit was presented in 2011 and operated as a broadband RF mixer at frequencies up to 10 GHz [137] (Fig. 5.10a). Before that, logic gate with a single graphene transistor was demonstrated in 2009 [138]. The simple logic gate demonstrated that graphene can be used to realize functions like the conventional semiconductors. The first multi-GFETs circuit was realized on a complementary inverter, which is composed of two GFETs [139].

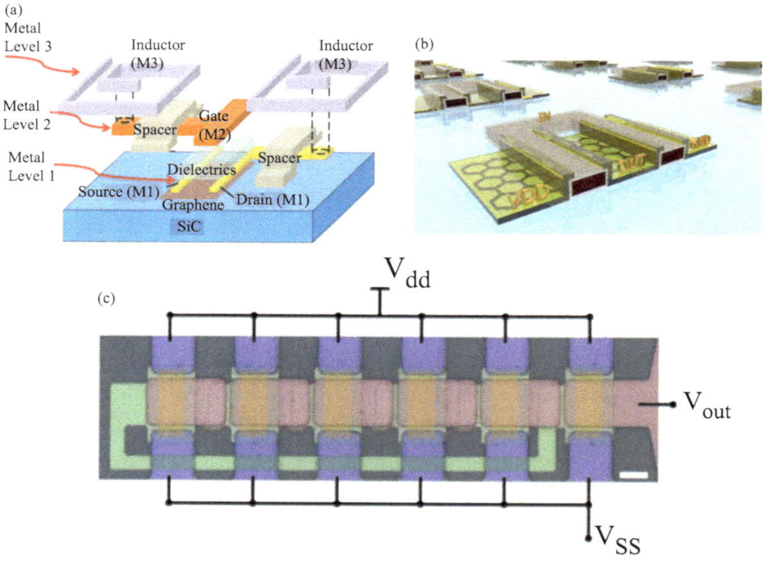

Figure 5.10 Schematic of graphene transistors circuits. (a) Graphene mixer circuit, (b) Inverters, (c) Ring oscillator. The scale bar is 10 μm. Figure (a) reprinted with permission from Ref. [137], Copyright 2011, American Association for the Advancement of Science; Fig. (b) reprinted with permission from Ref. [140], Copyright 2012, American Chemical Society; and Fig. (c) reprinted with permission from Ref. [136], Copyright 2013, Springer Nature.

The achievement is followed by two inverters connected in series with four graphene FETs [140] (Fig. 5.10b) and a high-frequency graphene ring oscillator with 12 graphene FETs [136] (Fig. 5.10c). In 2014, the ambient stable inverter was presented by integrating p- and n-type graphene transistors with simple polymer coating [141]. In the same year, the graphene-based room-temperature reversible logic gates were demonstrated [142]. Although super progresses have been achieved, the integrated circuit based on graphene devices is still in its infancy. The operation frequency of graphene-based circuits is comparable to those based on silicon, but still unable to compete with those based on III–V compounds.

5.6 Novel Graphene Transistor Structures

Although many efforts have been devoted to graphene electronics, a graphene transistor fabricated using a conventional device structure, with an acceptable on/off current ratio together with a large on-state current, remains unattainable. To address these problems, several novel structures have been proposed, such as bilayer pseudospin field-effect transistor (BiSFET) [143], vertical tunneling transistor [144], graphene-based transistors (GBT) [145], and graphene barrister (GB) [146].

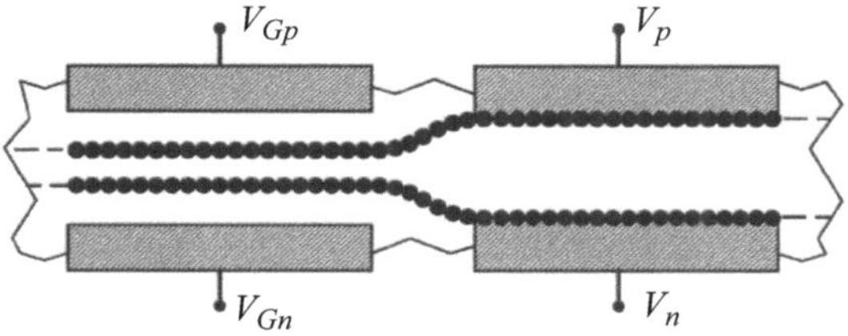

Figure 5.11 Schematic representation of a BiSFET (gate biases V_{Gp} and V_{Gn}, applied voltages V_p and V_n). Reprinted with permission from Ref. [143], Copyright 2010, IEEE.

The BiSFET was proposed by Banerjee et al. in 2009 [143] (Fig. 5.11). It consists of two graphene monolayers, one p-type and the other n-type, separated from each other by a thin tunnel dielectric. The graphene layers can be electrostatically coupled to gate through a thicker gate dielectric and form excitons. Under certain conditions, the electrons in one layer can pair with holes in the opposite layer, resulting in electron-hole pairs/excitons and forming a condensate. The condensation alters the quantum wave functions in the bilayer qualitatively and converts the isolated state into a coherent linear combination state of top and bottom layer components. The originally large tunneling resistance reduces with the layers effectively shorted by qualitative change, and a pronounced tunneling current can flow. With increasing potential difference between p-type and n-type graphene (V_{pn}), the tunneling current increases and reaches a maximum, while at a higher V_{pn}, the condensate is destroyed and the current decreases. The working voltage and power consumption of BiSFET logic gates are expected to be very low. According to a rough simulation, the consumption of BiSFET in an inverter circuit would be 8×10^{-21} J per cycle, 2–3 orders of magnitude less than Si technology. The vertical tunneling transistor is a viable alternative device, which has a rapid response and small size [144] (Fig. 5.12). Electrons tunneling through the several-nanometer-thick barrier can be extremely fast. The vertical tunneling graphene transistor is based on quantum tunneling from a graphene electrode through a thin insulating barrier. Both source and drain electrodes are made of graphene layers and separated by a thin (~1 nm) h-BN, which acts as the tunneling barrier. This graphene–BN–graphene layer is sandwiched between two other BN layers, which are used to provide higher quality graphene. Applying a voltage between the two graphene layers would cause a tunneling current through the BN tunnel barrier, and the voltage applied to the Si wafer could control the switch effect. The on/off ratios were around 50 from the experiments, which already exceeded those demonstrated for planar graphene FETs with a factor of 10.

Other novel geometries, such as GBT [145] (Fig. 5.13a) and GB [146] (Fig. 5.13b), have also been proposed. The hot-electron transistor is composed of a vertical emitter, base, and collector. The core of the device is a graphene layer serving as the transistor's base, sandwiched between the emitter-base insulator (EBI) and base-collector insulator (BCI). Depending on the base-emitter voltage, the transistor is either off or on. In the off state, carriers face a barrier with the existence of EBI, while in the on state, carriers tunnel through the EBI. Simulations show that GBTs have high on/off current ratios and high current gain, and the simulations of small-signal models indicate that it potentially allows THz operation. As shown in Fig. 5.13b, the graphene variable-barrier "barristor" is a three-terminal active device, which mimics a triode operation. The key of GB is an atomically sharp interface between graphene and hydrogenated silicon where a tunable Schottky barrier controls charge transport across a vertically stacked structure. By adjusting the gate voltage to control the graphene–silicon Schottky barrier, large modulation on the device current can be achieved, with on/off ratio up to 10^5.

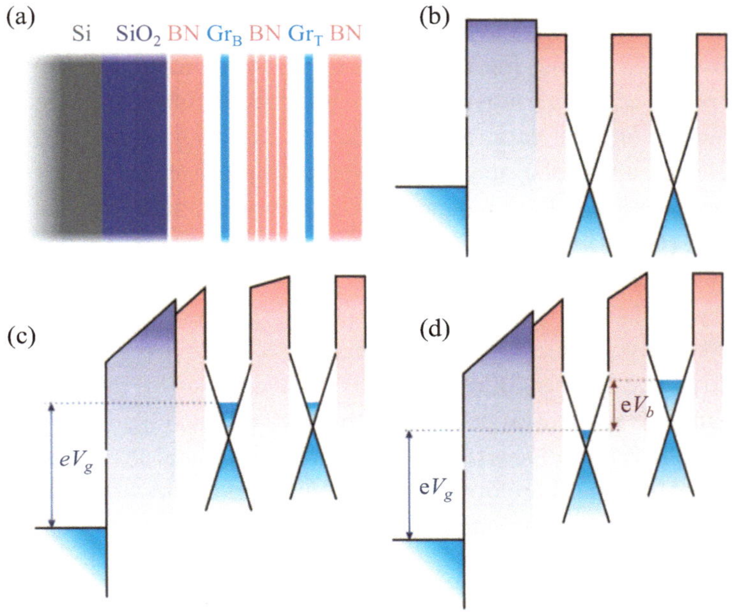

Figure 5.12 Graphene field-effect tunneling transistor. (a) Schematic structure of experimental devices. (b) The corresponding band structure with no gate voltage applied. (c) The same band structure for a finite gate voltage V_G and zero bias V_B. (d) Both V_G and V_B are finite. Reprinted with permission from Ref. [144], Copyright 2012, American Association for the Advancement of Science.

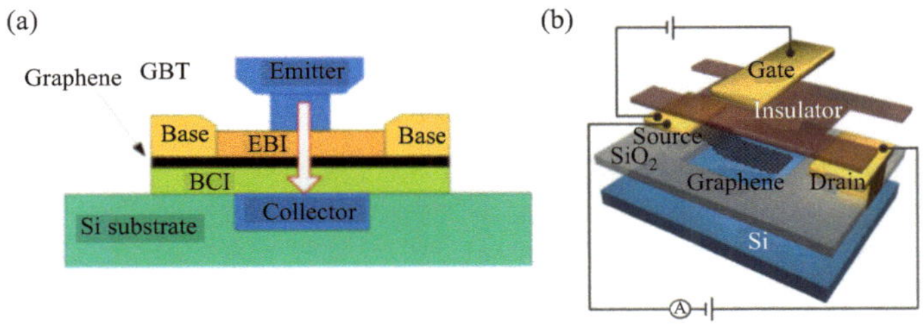

Figure 5.13 Schematic cross sections of (a) GBT and (b) GB . Figure (a) reprinted with permission from Ref. [145], Copyright 2012, IEEE, and Fig. (b) reprinted from Ref. [146], Copyright 2012, American Association for the Advancement of Science.

In addition to the structures described above, more novel structures have been proposed in recent years. The electrically configurable GFET with a graded-potential gate was presented in 2015 [147]. It used a graded-potential gate to continuously shift the Dirac points along the channel of GFET, thus produced a pseudo-bandgap within a controllable energy range. By tuning the electrical gradient of the gate, a GFET device can transform reversibly between ambipolar and n- and p-type unipolar characteristics. As shown in Fig. 5.14a, the special structure of graded-potential gate GFET (GPG–GFET) is its two terminal electrodes (T1 and T2) placed above the source and drain regions, which can cause a linearly varying electric field perpendicular to the channel plane by applying different voltages. As shown in Fig. 5.14b, there are three regions with different Fermi-level positions. In region I or III, the graphene channel is electrostatically doped to p- or n-type, respectively. And in region 2, it produces the off state with the minimal channel carrier concentration. Therefore, this GPG–GFET can electrically configure the device in operation to different desired function.

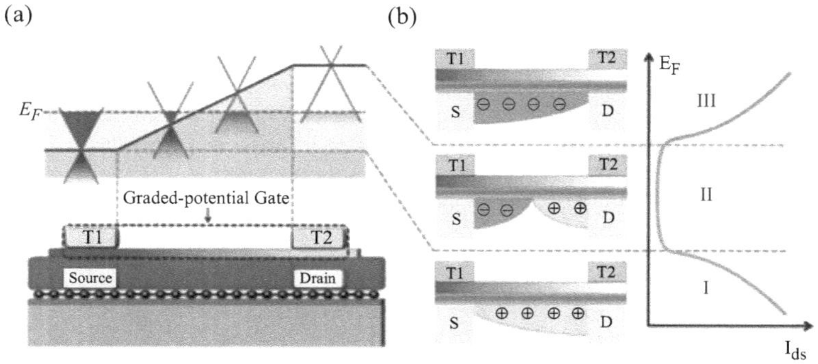

Figure 5.14 Graded-potential gate GFET. (a) Schematic device structure and the band diagram of a GPG–GFET. (b) Schematic carrier concentration in the GPG–GFET channel and simulated transfer characteristic of GPG–GFET. Reprinted with permission from Ref. [147], Copyright 2015, American Chemical Society.

In 2016, the ferroelectric-driven GFET based on vertical tunneling heterostructure was presented [148] (Fig. 5.15). Different from the traditional tunneling FETs, this vertical graphene heterostructure FET (VGHFET) used an ultrathin ferroelectric film to replace the conventional tunnel barrier of insulator or semiconductor. The key operation principle of the VGHFET is related to the changes in the density of states (DOS) and Fermi energy (E_F) level of the graphene, as well as the effective barrier height of the tunnel barrier adjacent to the bottom graphene layer (Gr_B) when varying the gate voltage (V_g) between the Si substrate and Gr_B. When $V_g < 0$, the barrier height will increase from Φ_0 to Φ_{1+}, which represents the off state of the FET. When increasing V_d in excess of the coercive voltage (V_c) of the ferroelectric (FE), the polarization (P) switching will occur, which may cause a negligible shift in the potential barrier height compared to the overall barrier height since the original barrier height (Φ_{1+}) is large enough. Therefore, the current in this off state (I_{off}) will only change slightly after P switching. On the contrary, a positive V_g will cause a reduction in the barrier height from Φ_0 to Φ_{2+}, which represents the on state of the FET. Also, when increasing V_d in excess of V_c, the polarization switching will occur and an extra barrier change $\Delta\Phi_B$ will appear, which can further decrease the overall barrier height. Since the original barrier height of Φ_{2+} before P switching is quite low, such an additional reduction of $\Delta\Phi_B$ will have an appreciable effect on the overall barrier height, leading to a great enhancement in I_{on} and causing a remarkable I_{on}/I_{off} ratio (7×10^7).

Considering the important works about the novel structures of graphene transistor in recent years, most of them are either with a vertical structure [148–157] or hybrid planar and vertical structures with a Schottky barrier [158]. Among them, heterostructures [148–156], especially those based on graphene and other two-dimensional crystals [152–155], were always used to improve the GFET performance. At the same time, graphene-based flexible transistors have also become a hot research field [157, 159].

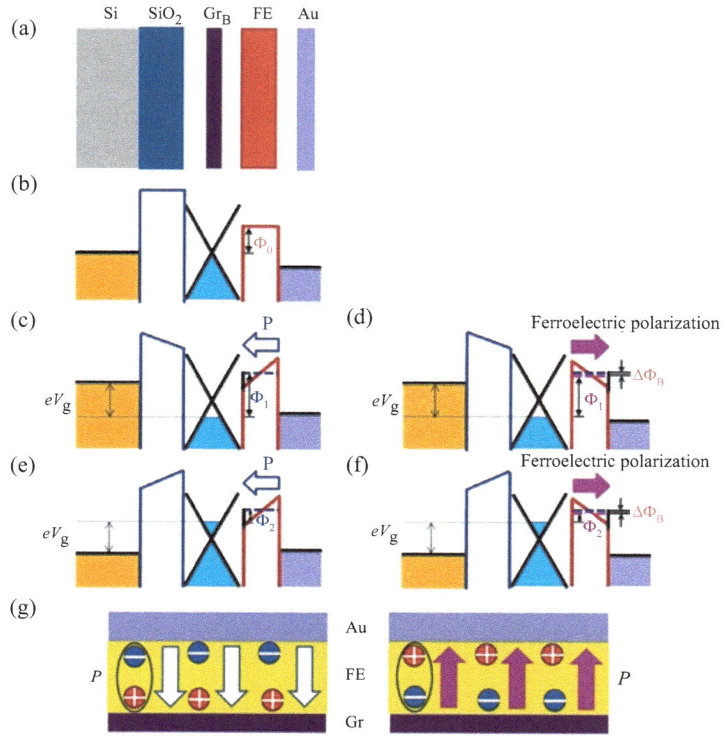

Figure 5.15 Schematic of energy band and operation principle of VGHFET with the tunnel barrier of ferroelectric (FE) and bottom graphene layer (Gr_B). (a) Schematic structure of device. (b) Energy-band diagram when $V_g = 0$ and $V_d = 0$. (c) Energy-band diagram when $V_g < 0$ and $V_d < V_c$ (coercive voltage of FE). (d) Energy-band diagram when $V_g < 0$ and $V_d > V_c$. (e) Energy-band diagram when $V_g > 0$ and $V_d < V_c$. (f) Energy-band diagram when $V_g > 0$ and $V_d > V_c$. (g) Schematic illustration of domain behavior upon P switching. Reprinted from Ref. [148], Copyright 2016, with permission from WILEY-VCH Verlag GmbH & Co. KgaA, Weinheim.

5.7 Radio-Frequency Graphene Electronics

Graphene FETs suffer from low on/off ratio, which limits their applications in logic circuits, but this is not fatal in analog/RF FETs. On the other hand, the high mobility, high carrier saturation velocity, and large critical current density make graphene a promising material for high-frequency FETs. Further, the limit of f_T beyond 1 THz has been predicted by simulations, although it has not yet been demonstrated experimentally [160]. Two parameters of FETs are crucial in RF field, current-gain cutoff frequency, and maximum frequency. The cutoff frequency is the limiting frequency, at which a signal can be carried through, while the maximum operating frequency is the highest radio frequency (RF) that can still be used for transmission.

The first graphene FET with a cutoff frequency f_T in the gigahertz range was reported in late 2008 by Meric et al. [161]. In this work, the first RF graphene FET device was fabricated with f_T performance exceeding 14 GHz and a 500 nm length channel. In 2009, graphene FETs with 350 nm gate length and a cutoff frequency of 50 GHz were presented [162], and then 100 GHz with a 240 nm epitaxial-graphene channel was reported in 2010 by the same team [166]. In the same year, a 144 nm gate graphene FET with f_T = 100–300 GHz was proposed [163]. Finally, in 2012, a graphene FET with a 40 nm gate length shows a maximum f_T of 350 GHz [164], and a graphene FET of 67 nm gate length shows an f_T of 427 GHz by fabricating self-aligned graphene transistors on glass with the transferred gate stacks [165].

In terms of f_T, graphene FETs perform competitively. They outperform the Si MOSFETs with similar gate length and compete well with InP HEMTs as well as GaAs mHEMTs down to gate lengths of about 60 nm (Fig. 5.16). However, regarding f_{max}, the picture looks less promising for graphene FETs. While f_{max} of either InP HEMTs or GaAs mHEMTs exceeds 1 THz and f_{max} of Si MOSFETs is 420 GHz, the best graphene RF FET shows an f_{max} of about 105 GHz [166]. The main reasons for this poor f_{max} performance are the weak and unsatisfying saturation of the drain current, and the large parasitic effects in these devices (contact and/or gate resistance). The reason for the weak saturation is the absence of bandgap. So to improve f_{max} of graphene transistors, it is urgently needed to reduce the parasitic effect and introduce a bandgap. A numerical simulation for bilayer graphene-based FETs exhibited that bilayer graphene FET showed better current saturation due to the electrostatically induced bandgap of up to 220 meV, which predicted an f_{max} in the THz regime [160].

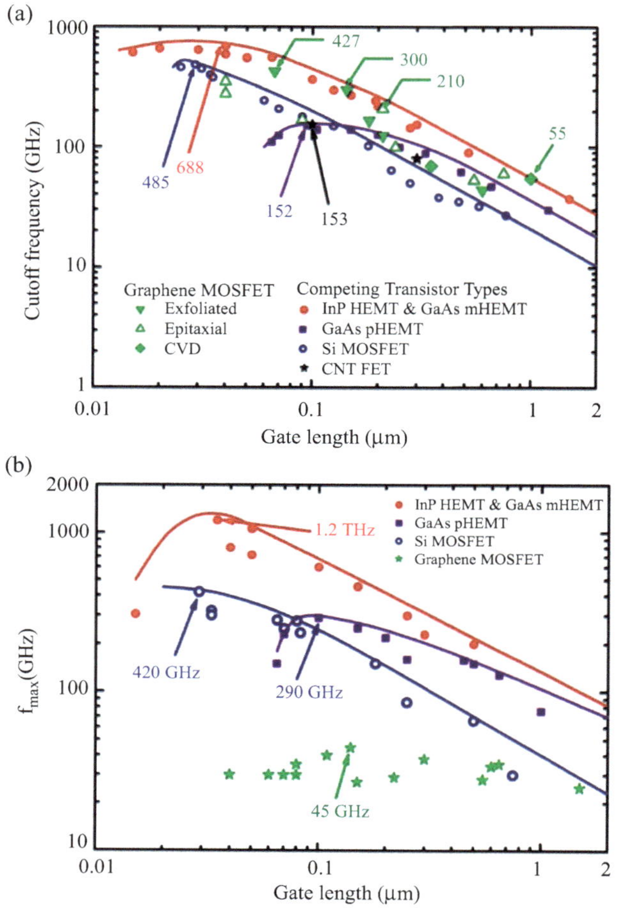

Figure 5.16 (a) Cutoff frequency and (b) maximum frequency versus gate length of oscillation in different types of RF MOSFETs (gate length). Reprinted with permission from Ref. [8], Copyright 2013, IEEE.

Although graphene shows poorer f_{max} performance compared to the Si and III–V semiconductors, flexible GFETs have demonstrated significantly higher flexibility since graphene is intrinsically flexible, which makes it suitable for flexible RF electronics. Recently, flexible RF electronics have attracted great interest [167–173]. At the same time, some RF devices have less stringent requirements for frequency. Graphene can play a role in these fields with its high mobility, carrier saturation velocity, and critical current density.

Graphene-based heterojunction transistors have also been proposed to achieve high RF performance. In a vertical silicon–graphene–germanium transistor, a Schottky emitter constructed by single-crystal silicon and single-layer graphene was achieved. Such Schottky emitter showed a current of 692 A cm^{-2} and a capacitance of 41 nF cm^{-2}. Thus, the alpha cutoff frequency of the transistor was expected to increase from about 1 MHz to above 1 GHz by using the Schottky emitter [174].

In addition, another promising direction based on the ambipolar nature of gapless graphene was proposed. Ambipolar devices have been used in the RF mixers and frequency multiplies, which rely on the nonlinear response. By using ambipolar devices, the number of transistors required would be reduced significantly and the zero bandgap is not a barrier anymore. Besides the defects of intrinsic graphene, another challenge for RF applications is the integration of graphene-based circuits. So far, inverters [139, 140], voltage amplifiers [175], mixers [176], frequency doubler [177], and oscillators [136, 178] have been realized. They have covered the major active building blocks of electronics. However, integrated circuits with complexity have not been realized, because they still have to face the aforementioned critical problems and also require a mature fabrication process.

5.8 Flexible Graphene Electronics

Flexible electronics are providing solutions for many new applications in human–machine interface, healthcare, energy system, E-skin, and new display. While a clear market exists, the challenges for flexible electronics still exist. The traditional Si devices can be flexible by thinning down below 25 μm, but the short lifetime and poor reliability due to the fragility of Si limit their further development [179, 180]. Organic semiconductors can provide flexible and transparent devices in large-area production, but they suffer from poor transport property, their mobility is typically below 10 cm^2 $V^{-1}s^{-1}$, and they cannot bear high-temperature environment [181, 182]. Currently, the dominant material of transparent conductive film is indium tin oxide (ITO), a mixture of In_2O_3 and SnO_2. However, ITO is brittle and may crack when it is bend. Besides, ITO is sensitive to acid and environment, and also it is hard to be patterned. Many alternatives have been proposed, such as metal nanowire mesh [183], metallic nanowires [184], CNTs [185], and graphene. As an ideal candidate for flexible electronics, graphene has unique electrical, mechanical, and optical properties. The single layer of graphene with a thickness of 0.335 nm endows it with a constant transparency of about 97.7% [186]. Besides, graphene is mechanically flexible, which comes from its strong carbon–carbon bond. The single-layer graphene without defect was measured with a Young's modulus of 1 TPa and intrinsic fracture strength of 130 GPa.

5.8.1 Preparation of Flexible Graphene Thin Film

Different strategies can be used to prepare graphene films. They can be mainly divided into two groups: solution-based method and CVD-based method. Generally, in the solution-based methods, graphene oxide (GO) solutions are transferred to the target substrates and finally reduced to graphene. In 2008, Eda et al. first used the vacuum filtration method to obtain a uniform and controllable deposition of 1–5 nm graphene thin film, from the reduction of GO. The lowest sheet resistance value obtained was ~43 kΩ/□, much higher than that of ITO [187]. Improved methods were proposed to prepare stable colloidal suspension graphene in aqueous medium by surface modification of graphene with 4-benzenediazonium sulfonate. The film fabricated by spin-coating technology exhibited a resistivity of 1.1 kΩ/□ with the transparency of 89% [188]. Chemical reduction method was reported to adopt sodium–ammonia solution as the reducing agent. The solvated electrons in sodium–ammonia solution can effectively facilitate the de-oxygenation and the restoration. With the help of low-temperature solution processing, high-quality graphene film with low sheet resistance (~350 Ω/□) and high transmittance (~80%) was fabricated [189]. Apart from vacuum filtration and spin-coating methods, the Langmuir–Blodgett process was demonstrated to have the advantage on forming monolayer GO. Transparent conductive films that were reduced from the ultra-large GO sheets exhibited a sheet resistance of 1.1 kΩ/□ at 91% transparency [190]. Liquid-phase exfoliation (LPE) methods were proposed to produce graphene flakes in mass production by dispersion and exfoliation of graphite in organic solvents [191].

Highly conductive and transparent graphene films were also fabricated from direct exfoliation of graphite using a nonionic semiconducting polymer. After the film was treated with nitric acid and thionyl chloride, the film showed a low sheet resistance of 300 Ω/□ with 74% transmittance at 550 nm [192]. Voiry et al. fabricated pristine graphene by reducing graphene oxide using microwaves pulses [193]. This method can form high-quality graphene simply and rapidly. Compared with highly ordered pyrolytic graphite (HOPG), dispersed graphene, thermally reduced graphene oxide (rGO), the microwave-reduced graphene oxide (MW-rGO) have less disorder or impurities, as Fig. 5.17 shows.

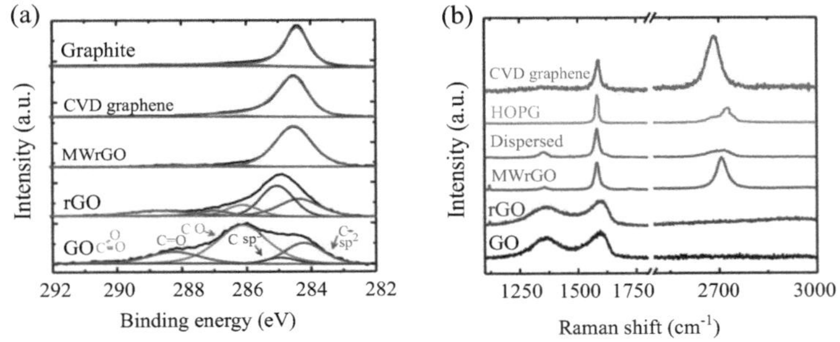

Figure 5.17 (a) X-ray photoelectron spectra. (b) Raman spectra of microwave-reduced graphene oxide (MW-rGO) and other graphene-based samples. Reprinted from Ref. [193], Copyright 2016, American Association for the Advancement of Science.

Solution-based methods have the advantages of availability and are suitable for mass production. But the preparation of GO solution would inevitably introduce lots of defects and oxygen-containing functional groups when they are treated with strong oxidizer or acid. That makes the conductivity of the solution-based graphene poor.

Compared with rGO, which is reduced from GO solution, graphene grown by CVD has better crystallinity and conductivity. Graphene on the metallic substrate needs to be transferred to flexible substrates, since the substrates for CVD-grown graphene are normally metals, such as copper and nickel, which have poor flexibility and transparency. However, the transfer process would cause graphene broken to some extent. Therefore, reliable transfer of large-area graphene to the flexible substrate is a challenging step after the CVD-based graphene growth. In some cases, the graphene is transferred by etching off the metal substrates. For example, in the first reported CVD-based method by Kim et al., they transferred the graphene by etching the nickel layers in $FeCl_3$ aqueous solution. The transferred graphene films showed sheet resistance of ~280 Ω/□, with ~80% optical transparency [194]. Wang et al. demonstrated a unique clean-lifting transfer (CLT) technology using a controllable electrostatic attraction force to transfer graphene films onto the target substrates. The graphene electrode, which consisted of an HNO_3-doped four-layered graphene film, had a transparency of ~90% and an average sheet resistance of 50 Ω/□. The CLT technique does not use any organic support or adhesives but still needs sacrifice of the substrate material [195]. In addition, these methods usually cannot avoid contamination from solutions, wrinkling, and breakage of graphene. For large-scale applications, Bae et al. reported a roll-to-roll method to transfer graphene onto PET substrate and created a 30-inch flexible and transparent film, as shown in Fig. 5.18. Its sheet resistance was as low as 125 Ω/□, and the transmittance reached 97.4%. For the four-layer graphene film fabricated by layer-by-layer stacking, the transmittance reached 90% and the resistance was only 30 Ω/□ [196]. Furthermore, a 100 m long graphene transparent conductive film was fabricated by low pressure thermal CVD synthesis and direct roll-to-roll transfer process. The fabricated film showed a sheet resistance as low as 150 Ω/□ [197]. However, in the approaches mentioned above, graphene is inevitably contaminated by the organic adhesive from the thermal release tape during the transfer process, which could degrade the electrical properties of graphene.

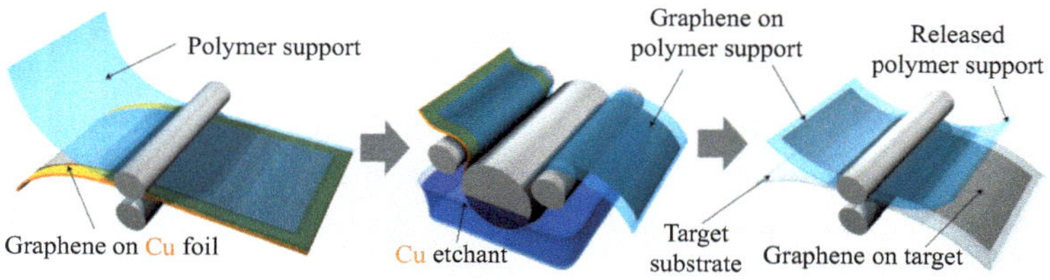

Figure 5.18 Schematic of the roll-based production of graphene films. Reprinted from Ref. [196], Copyright 2010, by permission from Springer Nature.

Generally, multilayer graphene would enhance the conduction performance, while decrease the transparency. Application of graphene needs a tradeoff between the conductivity and transparency. In the case of high transparency, chemical doping is an effectivity strategy to improve the electrical performance. Amal Kasry et al. prepared p-doped graphene transparent conductive films with nitric acid. The doping decreased the sheet resistance by a factor of 3, yielding films comprising eight stacked layers with a sheet resistance of 90 Ω/□ at a transmittance of 80% [198]. $FeCl_3$ as a dopant to reduce the sheet resistance of few-layer graphene was studied. After doping, graphene shows p-type conducting behavior and its conductance is enhanced, resulting in a sheet resistance of 40 Ω/□ for three-layer graphene and 20 Ω/□ for four-layer graphene, with transmittance ≥90% for both cases [199]. However, in most cases, the stability of the dopant is unknown. Innovative doping techniques need to be investigated. Hybridizations of graphene and other conductive material, which combine their advantages, have caught researchers' attention. A hybrid CNT–graphene film was fabricated by graphene oxide deposition on top of CNT films. After annealing and chemically doping by gold chloride, the film demonstrated high performance with sheet resistance of 73 Ω/□ at 90% transmittance [200]. Lower sheet resistance was achieved in graphene/metallic grids, with 3 Ω/□ at ~80% transmittance, and ~20 Ω/□ at 90% transmittance [201]. Hybrid nanostructure based on 2D graphene and networks of one-dimensional (1D) metal nanowires showed low sheet resistance (33 Ω/□) with high transmittance (94% in visible range) [202].

To avoid the disadvantages resulted from transferring, transfer-free CVD methods have been investigated to directly grow graphene on semiconductors or dielectrics [203]. Three kinds of strategies have been studied, including metal catalyst-free direct growth [204], metal-assisted growth [205], and plasma-enhanced growth [206]. The metal catalyst-free direct growth method can separately control the nucleation and growth of graphene, which has high applicability for different substrates. But the growth time is often too long to be applied in industrial production. The metal-assisted growth method compromises the merits of metal catalysis and direct growth method. Chen et al. used gaseous Cu molecules evaporating from the surface of the tube to transform an amorphous carbon film into a crystalline graphene layer, which enabled a one-step process to fabricate an all-carbon device [205]. However, these methods are either time consuming or metal contaminated. The plasma-enhanced growth method can rapidly prepare high-quality graphene at lower temperature, as shown in Fig. 5.19, but its controllability is unsatisfactory.

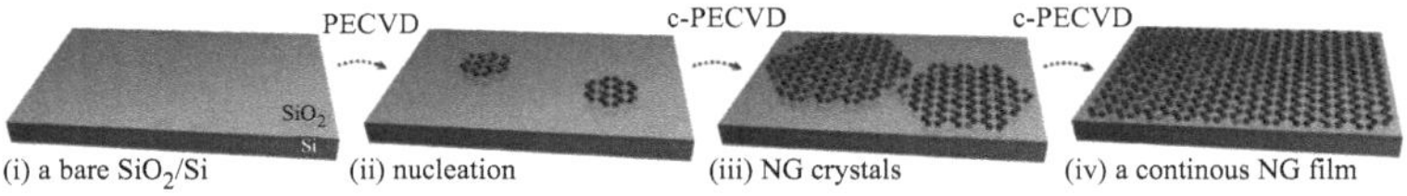

Figure 5.19 Schematic of the plasma-enhanced nitrogen-doped graphene (NG) growth process. Reprinted with permission from Ref. [206], Copyright 2015, American Chemical Society.

Although great progress has been achieved in the past few years, still some challenges remain to be overcome. The existing methods mentioned above should be further refined for industrial production. Stable transfer and doping, which can avoid contamination, need to be studied. Integration of films and other components is another challenging topic to be explored in depth.

5.8.2 Flexible Graphene Transistor

The high flexibility, high electrical conductivity, and high optical transparency make graphene an excellent electrode, which can outperform ITO (~20 Ω/□ with transmittance of ~90%). Furthermore, benefited by special structure design, the graphene electrode can obtain good stretchability with high conductance. As shown in Fig. 5.20, Liu et al. fabricated stretchable all-carbon transistors with trilayer graphene/graphene scrolls as electrodes, which can maintain 60% of its original current output at 120% strain [207].

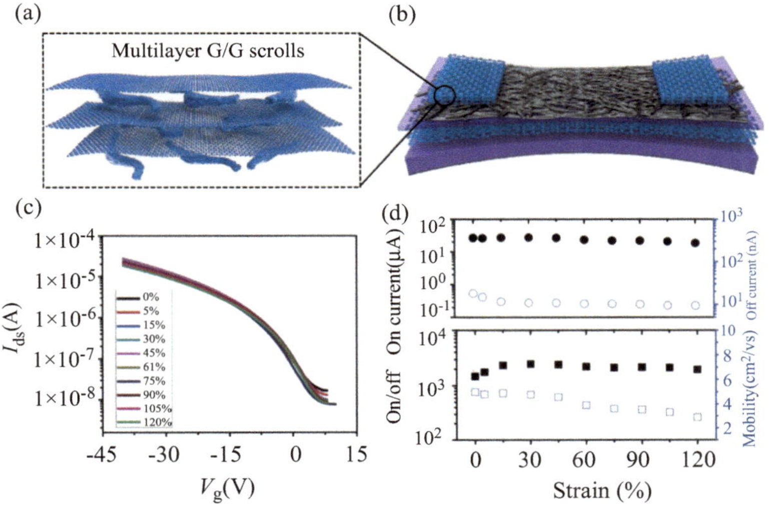

Figure 5.20 (a) Schematic of multilayer graphene/graphene scrolls structure. (b) Schematic of the stretchable all-carbon transistor with trilayer graphene/graphene scrolls as electrodes. (c–d) Transfer curves, on and off current, on/off ratio, and mobility of the all-carbon transistor at different strains. Adapted from Ref. [207], Copyright 2017, the authors.

Graphene serving as active channel layer shows high mobility and ambipolar transport characteristics. Oh et al. fabricated a flexible ambipolar transistor with high mobility and on/off ratio by combining the advantages of graphene channel and typical organic semiconductor [208].

An all-graphene thin-film transistor consisting of graphene electrodes, a semiconducting reduced graphene oxide channel, and dielectric composed of graphene oxide was demonstrated by Soni et al. [209]. Exposed under different intensities of UV light, graphene oxide reduction can be tuned. Figure 5.21 illustrates the different reductions realized by using masks with different gray-scale values during exposure.

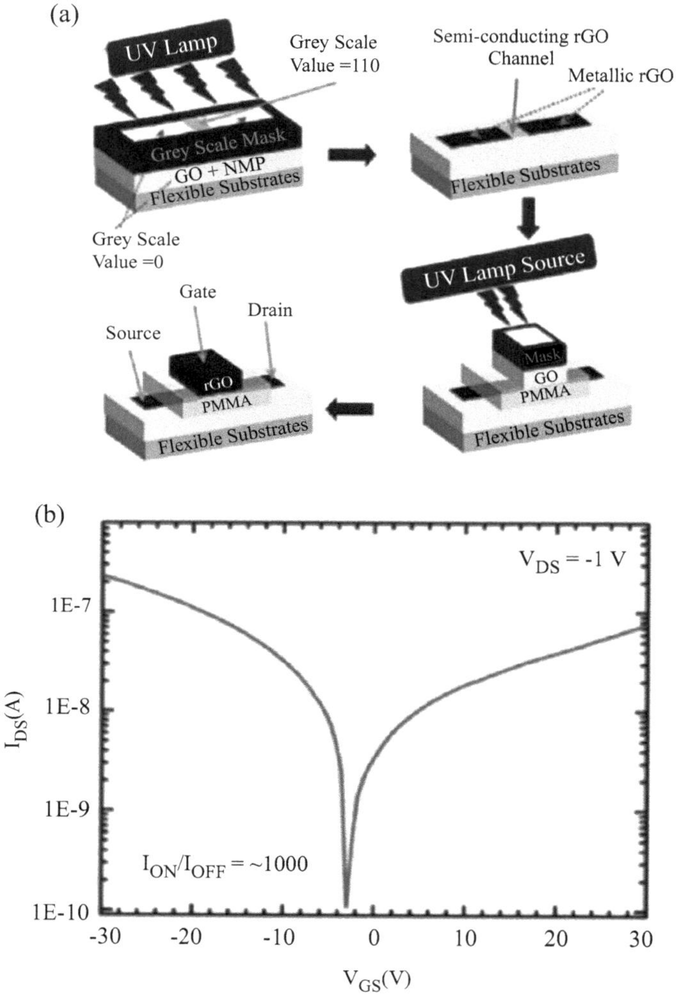

Figure 5.21 (a) Process flow of the all-graphene thin-film transistor. (b) Transfer characteristic at $V_{DS} = -1$ V. Reprinted from Ref. [209], Copyright 2018, with permission from Elsevier Ltd.

Considering the intriguing properties of graphene, it can serve as both metallic electrode and active channel under different conditions. By patterning the monolithic graphene layer to act as active channel and source/drain electrodes, Lee et al. fabricated an all-graphene-based flexible thin-film transistor, which can be bent at 4.13 mm bending radius [210]. To avoid the direct leakage current from gate electrode to source/drain electrodes, Kim et al. proposed the coplanar-gate graphene transistors with a simple configuration of using a monolayer graphene serving as both source/drain electrodes (the region not covered with ion gel dielectric) and active channel (the region covered with ion gel dielectric) at the same time, which is shown in Fig. 5.22 [211]. Benefiting from the long-range polarizability of ion gel dielectrics, the conductance of graphene channel can be tuned by the offset gate electrode. This structure not only simplified the process, but also improved the contact performance between channel and source/drain electrodes. The fabricated flexible graphene transistor exhibited high mobility of hole and electron, centered on 892 and 628 $cm^2V^{-1}s^{-1}$, respectively. After bending for 5000 times at the radius of about 3 mm, only a 20% change yielded for the mobilities. Besides, almost no appreciable change of the electric performance was observed after storing the transistors under ambient conditions with temperature of 27°C and humidity of 20% for 60 days. However, because the channel is charged by gate voltage, the off-state current is pretty large, which limits application in the real scene. In brief, limited by the bandgap and the other problems mentioned above, flexible graphene thin-film transistors developed slowly in recent years.

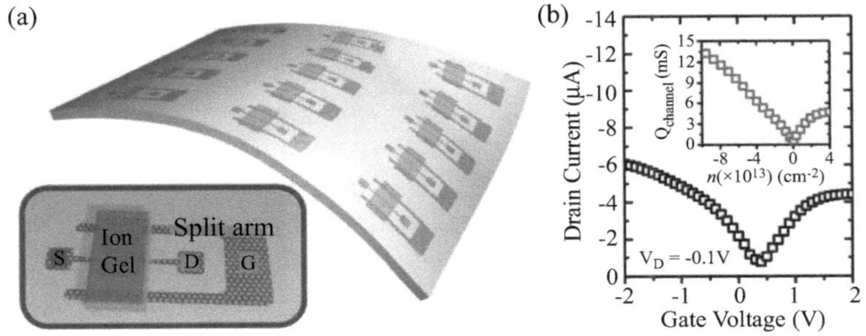

Figure 5.22 (a) Schematic of the flexible coplanar-gate graphene transistor structure. (b) Transfer characteristic of a coplanar-gate graphene transistor and the inset with the corresponding channel conductivity ($\sigma_{channel}$) as a function of the carrier density (n). Reprinted with permission from Ref. [211], Copyright 2012, American Chemical Society.

Further, flexible and transparent conductive graphene films are very promising in various fields, including solar cells, organic light-emitting diodes (OLEDs), touch panels, displays, and sensors. It is rewarding to see that the applications of graphene in commercial products have become reality. Hopefully, more and more complex products using graphene can be realized in the future.

5.9 Summary

Graphene transistors have progressed significantly in its short history since 2004. The performance of graphene FETs in digital logic and RF amplifiers has been impressive. A bandgap up to 1.6 eV has been achieved for GNRs [50]. An I_{on}/I_{off} ratio of ~10^7 [47] and voltage gain of ~35 [132] were demonstrated. $I_{on}/I_{off} > 10^4$ and $A_v > 10$ are the two essential criteria for graphene FETs to work as building blocks for logic and amplifier circuits, respectively. Unfortunately, as the mobility decreases significantly with the increase in bandgap, the high values of I_{on}/I_{off} and A_v cannot be achieved at the same time by a single transistor. FETs with large-area graphene channels still cannot be switched off, limiting their logic applications. When compared with conventional Si and III–V compounds for RF device application, the ambipolar conduction of graphene creates new opportunities, which can realize RF circuit functions with fewer components.

Although notable progress has been achieved, challenges still remain for graphene to be used at the industrial scale. First, the material properties, including bandgap, carrier mobility, and sheet size, require further intensive study. Second, the structure of graphene transistors is crucial for its performance. Novel structures for graphene devices should be designed and investigated for applications in next-generation electronics. Further, research on integration of graphene-based circuits is still in its infancy. On the other hand, graphene-based transparent electrodes and flexible electronics have been developed even faster and are closer to applications.

On the other hand, research on synergistic combination of 2D materials with silicon chips has demonstrated a heterogeneous platform with massively enhanced potential. Graphene and related 2D materials offer prospects of unprecedented advances in device performance in the atomic scale, while silicon technology provides a mature and sophisticated platform. Integration can be achieved via three-dimensional monolithic construction of multifunctional high-rise 2D silicon chips, enabling enhanced performance by exploiting the vertical direction and the functional diversification of the silicon platform for computational and non-computational applications [212]. The latest ITRS roadmap strongly recommends intensified research on graphene and proposes a research and development schedule for carbon-based nanoelectronics [213]. The competition to reach the goal is ongoing, and the prospects for graphene devices remain promising.

References

1. H. P. Boehm, R. Setton, and E. Stumpp, "Nomenclature and terminology of graphite intercalation compounds," *Pure and Applied Chemistry,* **66**, 9, pp. 1893–1901, 1994.
2. K. S. Novoselov, A. K. Geim, S. V. Morozov, D. Jiang, Y. Zhang, S. V. Dubonos, I. V. Grigorieva, and A. A. Firsov, "Electric field effect in atomically thin carbon films," *Science,* **306**, 5696, pp. 666–669, 2004.
3. P. R. Wallace, "The band theory of graphite," *Physical Review,* **71**, 9, pp. 622–634, 1947.
4. O. Chuhei and N. Ayato, "Ultra-thin epitaxial films of graphite and hexagonal boron nitride on solid surfaces," *Journal of Physics: Condensed Matter,* **9**, 1, p. 1, 1997.
5. I. Forbeaux, J. M. Themlin, and J. M. Debever, "Heteroepitaxial graphite on 6H-SiC(0001): Interface formation through conduction-band electronic structure," *Physical Review B,* **58**, 24, pp. 16396–16406, 1998.
6. M. C. Lemme, T. J. Echtermeyer, M. Baus, and H. Kurz, "A graphene field-effect device," *IEEE Electron Device Letters,* **28**, 4, pp. 282–284, 2007.
7. V. Ryzhii, M. Ryzhii, A. Satou, and T. Otsuji, "Current-voltage characteristics of a graphene-nanoribbon field-effect transistor," *Journal of Applied Physics,* **103**, 9, p. 094510, 2008.
8. F. Schwierz, "Graphene transistors: Status, prospects, and problems," *Proceedings of the IEEE,* **101**, 7, pp. 1567–1584, 2013.
9. A. H. Castro Neto, F. Guinea, N. M. R. Peres, K. S. Novoselov, and A. K. Geim, "The electronic properties of graphene," *Reviews of Modern Physics,* **81**, 1, pp. 109–162, 2009.

10. N. Savage, "Materials science: Super carbon," *Nature,* **483**, 7389, pp. S30–S31, 2012.

11. X. Li, Y. Zhu, W. Cai, M. Borysiak, B. Han, D. Chen, R. D. Piner, L. Colombo, and R. S. Ruoff, "Transfer of large-area graphene films for high-performance transparent conductive electrodes," *Nano Letters,* **9**, 12, pp. 4359–4363, 2009.

12. J. C. Meyer, A. K. Geim, M. I. Katsnelson, K. S. Novoselov, T. J. Booth, and S. Roth, "The structure of suspended graphene sheets," *Nature,* **446**, 7131, pp. 60–63, 2007.

13. T. Zhang, S. Wu, R. Yang, and G. Zhang, "Graphene: Nanostructure engineering and applications," *Frontiers of Physics,* **12**, 1, p. 127206, 2017.

14. A. Bianco, H.-M. Cheng, T. Enoki, Y. Gogotsi, R. H. Hurt, N. Koratkar, T. Kyotani, M. Monthioux, C. R. Park, J. M. D. Tascon, and J. Zhang, "All in the graphene family: A recommended nomenclature for two-dimensional carbon materials," *Carbon,* **65**, pp. 1–6, 2013.

15. N. D. Mermin, "Crystalline order in two dimensions," *Physical Review,* **176**, 1, pp. 250–254, 1968.

16. N. D. Mermin and H. Wagner, "Absence of ferromagnetism or antiferromagnetism in one- or two-dimensional isotropic Heisenberg models," *Physical Review Letters,* **17**, 22, pp. 1133–1136, 1966.

17. J. A. Venables, G. D. T. Spiller, and M. Hanbucken, "Nucleation and growth of thin films," *Reports on Progress in Physics,* **47**, 4, pp. 399–459, 1984.

18. J. S. Bunch, S. S. Verbridge, J. S. Alden, A. M. van der Zande, J. M. Parpia, H. G. Craighead, and P. L. McEuen, "Impermeable atomic membranes from graphene sheets," *Nano Letters,* **8**, 8, pp. 2458–2462, 2008.

19. Ç. Ö. Girit, J. C. Meyer, R. Erni, M. D. Rossell, C. Kisielowski, L. Yang, C.-H. Park, M. F. Crommie, M. L. Cohen, S. G. Louie, and A. Zettl, "Graphene at the edge: Stability and dynamics," *Science,* **323**, 5922, pp. 1705–1708, 2009.

20. K. A. Ritter and J. W. Lyding, "The influence of edge structure on the electronic properties of graphene quantum dots and nanoribbons," *Nature Materials,* **8**, 3, pp. 235–242, 2009.

21. T. Ma, W. Ren, X. Zhang, Z. Liu, Y. Gao, L.-C. Yin, X.-L. Ma, F. Ding, and H.-M. Cheng, "Edge-controlled growth and kinetics of single-crystal graphene domains by chemical vapor deposition," *Proceedings of the National Academy of Sciences,* **110**, 51, pp. 20386–20391, 2013.

22. K. S. Novoselov, A. K. Geim, S. V. Morozov, D. Jiang, M. I. Katsnelson, I. V. Grigorieva, S. V. Dubonos, and A. A. Firsov, "Two-dimensional gas of massless Dirac fermions in graphene," *Nature,* **438**, 7065, pp. 197–200, 2005.

23. K. I. Bolotin, K. J. Sikes, Z. Jiang, M. Klima, G. Fudenberg, J. Hone, P. Kim, and H. L. Stormer, "Ultrahigh electron mobility in suspended graphene," *Solid State Communications,* **146**, 9–10, pp. 351–355, 2008.

24. E. H. Hwang, S. Adam, and S. D. Sarma, "Carrier transport in two-dimensional graphene layers," *Physical Review Letters,* **98**, 18, p. 186806, 2007.

25. V. E. Dorgan, M. H. Bae, and E. Pop, "Mobility and saturation velocity in graphene on SiO_2," *Applied Physics Letters,* **97**, 8, p. 082112, 2010.

26. S. Rahimi, L. Tao, S. F. Chowdhury, S. Park, A. Jouvray, S. Buttress, N. Rupesinghe, K. Teo, and D. Akinwande, "300 mm wafer-scalable high-performance polycrystalline chemical vapor deposited graphene transistors," *ACS Nano,* **8**, 10, pp. 10471–10479, 2014.

27. L. Banszerus, M. Schmitz, S. Engels, J. Dauber, M. Oellers, F. Haupt, K. Watanabe, T. Taniguchi, B. Beschoten, and C. Stampfer, "Ultrahigh-mobility graphene devices from chemical vapor deposition on reusable copper," *Science Advances,* **1**, 6, p. e1500222, 2015.

28. K. S. Novoselov, D. Jiang, F. Schedin, T. J. Booth, V. V. Khotkevich, S. V. Morozov, and A. K. Geim, "Two-dimensional atomic crystals," *Proceedings of the National Academy of Sciences of the United States of America,* **102**, 30, pp. 10451–10453, 2005.

29. Y. Zhang, Y.-W. Tan, H. L. Stormer, and P. Kim, "Experimental observation of the quantum Hall effect and Berry's phase in graphene," *Nature,* **438**, 7065, pp. 201–204, 2005.

30. R. Martel, V. Derycke, C. Lavoie, J. Appenzeller, K. K. Chan, J. Tersoff, and P. Avouris, "Ambipolar electrical transport in semiconducting single-wall carbon nanotubes," *Physical Review Letters,* **87**, 25, p. 256805, 2001.

31. G. W. Neudeck, H. F. Bare, and K. Y. Chung, "Modeling of ambipolar a-Si:H thin-film transistors," *IEEE Transactions on Electron Devices,* **34**, 2, pp. 344–350, 1987.

32. H. Wang, A. Hsu, K. K. Kim, J. Kong, and T. Palacios, "Graphene electronics for RF applications," *IEEE MTT-S International Microwave Symposium,* pp. 1–4, 2011.

33. F. Schwierz, J. Pezoldt, and R. Granzner, "Two-dimensional materials and their prospects in transistor electronics," *Nanoscale,* **7**, 18, pp. 8261–8283, 2015.

34. L. Brey and H. A. Fertig, "Electronic states of graphene nanoribbons studied with the Dirac equation," *Physical Review B,* **73**, 23, p. 235411, 2006.

35. N. M. R. Peres, A. H. Castro Neto, and F. Guinea, "Conductance quantization in mesoscopic graphene," *Physical Review B,* **73**, 19, p. 195411, 2006.

36. K. Wakabayashi, M. Fujita, H. Ajiki, and M. Sigrist, "Electronic and magnetic properties of nanographite ribbons," *Physical Review B,* **59**, 12, pp. 8271–8282, 1999.

37. M. Y. Han, J. C. Brant, and P. Kim, "Electron transport in disordered graphene nanoribbons," *Physical Review Letters,* **104**, 5, p. 056801, 2010.

38. P. Kim, M. Y. Han, A. F. Young, I. Meric, and K. L. Shepard, "Graphene nanoribbon devices and quantum heterojunction devices," *IEEE International Electron Devices Meeting*, pp. 220–223, 2009.

39. L. Yang, C. H. Park, Y. W. Son, M. L. Cohen, and S. G. Louie, "Quasiparticle energies and band gaps in graphene nanoribbons," *Physical Review Letters,* **99**, 18, p. 186801, 2007.

40. K. Nakada, M. Fujita, G. Dresselhaus, and M. S. Dresselhaus, "Edge state in graphene ribbons: Nanometer size effect and edge shape dependence," *Physical Review B,* **54**, 24, pp. 17954–17961, 1996.

41. Y. Miyamoto, K. Nakada, and M. Fujita, "First-principles study of edge states of H-terminated graphitic ribbons," *Physical Review B,* **59**, 15, pp. 9858–9861, 1999.

42. F. Sols, F. Guinea, and A. H. Neto, "Coulomb blockade in graphene nanoribbons," *Physical Review Letters,* **99**, 16, p. 166803, 2007.

43. D. V. Kosynkin, A. L. Higginbotham, A. Sinitskii, J. R. Lomeda, A. Dimiev, B. K. Price, and J. M. Tour, "Longitudinal unzipping of carbon nanotubes to form graphene nanoribbons," *Nature,* **458**, 7240, pp. 872–876, 2009.

44. Y. S. Li, J. L. Liao, S. Y. Wang, and W. H. Chiang, "Intercalation-assisted longitudinal unzipping of carbon nanotubes for green and scalable synthesis of graphene nanoribbons," *Scientific Reports,* **6**, 1, p. 22755, 2016.

45. B. Sommer, J. Sonntag, A. Ganczarczyk, D. Braam, G. Prinz, A. Lorke, and M. Geller, "Electron-beam induced nano-etching of suspended graphene," *Scientific Reports,* **5**, 1, p. 7781, 2015.

46. L. Tapaszto, G. Dobrik, P. Lambin, and L. P. Biro, "Tailoring the atomic structure of graphene nanoribbons by scanning tunnelling microscope lithography," *Nature Nanotechnology,* **3**, 7, pp. 397–401, 2008.

47. X. Li, X. Wang, L. Zhang, S. Lee, and H. Dai, "Chemically derived, ultrasmooth graphene nanoribbon semiconductors," *Science,* **319**, 5867, pp. 1229–1232, 2008.

48. L. Chen, Y. Hernandez, X. Feng, and K. Mullen, "From nanographene and graphene nanoribbons to graphene sheets: Chemical synthesis," *Angew Chemie International Edition England,* **51**, 31, pp. 7640–7654, 2012.

49. T. H. Vo, M. Shekhirev, D. A. Kunkel, M. D. Morton, E. Berglund, L. Kong, P. M. Wilson, P. A. Dowben, A. Enders, and A. Sinitskii, "Large-scale solution synthesis of narrow graphene nanoribbons," *Nature Communications,* **5**, 1, p. 3189, 2014.

50. J. Cai, P. Ruffieux, R. Jaafar, M. Bieri, T. Braun, S. Blankenburg, M. Muoth, A. P. Seitsonen, M. Saleh, X. Feng, K. Mullen, and R. Fasel, "Atomically precise bottom-up fabrication of graphene nanoribbons," *Nature,* **466**, 7305, pp. 470–473, 2010.

51. T. Zhao, C. Xu, W. Ma, Z. Liu, T. Zhou, Z. Liu, S. Feng, M. Zhu, N. Kang, D. M. Sun, H. M. Cheng, and W. Ren, "Ultrafast growth of nanocrystalline graphene films by quenching and grain-size-dependent strength and bandgap opening," *Nature Communications,* **10**, 1, p. 4854, 2019.

52. J. Hao, C. Huang, H. Wu, Y. Qiu, Q. Gao, Z. Hu, E. Kan, and L. Zhang, "A promising way to open an energy gap in bilayer graphene," *Nanoscale,* **7**, 40, pp. 17096–17101, 2015.

53. Y. Zhang, T. T. Tang, C. Girit, Z. Hao, M. C. Martin, A. Zettl, M. F. Crommie, Y. R. Shen, and F. Wang, "Direct observation of a widely tunable bandgap in bilayer graphene," *Nature,* **459**, 7248, pp. 820–823, 2009.

54. S. Y. Lee, D. L. Duong, Q. A. Vu, Y. Jin, P. Kim, and Y. H. Lee, "Chemically modulated band gap in bilayer graphene memory transistors with high on/off ratio," *ACS Nano,* **9**, 9, pp. 9034–9042, 2015.

55. Y. Ma, D. Li, X. Feng, H. Zhang, and C. Liang, "Electronic structure and transport properties of bilayer graphene adsorbed by LiF_2 super-halogen clusters and Li_3O super-alkali clusters," *Journal of Physics D,* **52**, 24, p. 245302, 2019.

56. C. Jacoboni, C. Canali, G. Ottaviani, and A. Alberigi Quaranta, "A review of some charge transport properties of silicon," *Solid-State Electronics,* **20**, 2, pp. 77–89, 1977.

57. E. D. Minot, Y. Yaish, V. Sazonova, J. Y. Park, M. Brink, and P. L. McEuen, "Tuning carbon nanotube band gaps with strain," *Physical Review Letters,* **90**, 15, p. 156401, 2003.

58. T. W. Tombler, C. Zhou, L. Alexseyev, J. Kong, H. Dai, L. Liu, C. S. Jayanthi, M. Tang, and S.-Y. Wu, "Reversible electromechanical characteristics of carbon nanotubes under local-probe manipulation," *Nature,* **405**, 6788, pp. 769–772, 2000.

59. S. Y. Zhou, G. H. Gweon, A. V. Fedorov, P. N. First, W. A. de Heer, D. H. Lee, F. Guinea, A. H. Castro Neto, and A. Lanzara, "Substrate-induced bandgap opening in epitaxial graphene," *Nature Materials,* **6**, 10, pp. 770–775, 2007.

60. R. M. Ribeiro, N. M. R. Peres, J. Coutinho, and P. R. Briddon, "Inducing energy gaps in monolayer and bilayer graphene: Local density approximation calculations," *Physical Review B,* **78**, 7, p. 075442, 2008.

61. G. Giovannetti, P. A. Khomyakov, G. Brocks, P. J. Kelly, and J. van den Brink, "Substrate-induced band gap in graphene on hexagonal boron nitride: Ab initio density functional calculations," *Physical Review B,* **76**, 7, p. 073103, 2007.

62. G. Cocco, E. Cadelano, and L. Colombo, "Gap opening in graphene by shear strain," *Physical Review B,* **81**, 24, p. 241412, 2010.

63. Z. H. Ni, T. Yu, Y. H. Lu, Y. Y. Wang, Y. P. Feng, and Z. X. Shen, "Uniaxial strain on graphene: Raman spectroscopy study and band-gap opening," *ACS Nano,* **2**, 11, pp. 2301–2305, 2008.

64. G. Gui, D. Morgan, J. Booske, J. Zhong, and Z. Ma, "Local strain effect on the band gap engineering of graphene by a first-principles study," *Applied Physics Letters,* **106**, 5, p. 053113, 2015.

65. M. A. Bissett, M. Tsuji, and H. Ago, "Strain engineering the properties of graphene and other two-dimensional crystals," *Physical Chemistry Chemical Physics,* **16**, 23, pp. 11124–11138, 2014.

66. T. Kawasaki, T. Ichimura, H. Kishimoto, A. A. Akbar, T. Ogawa, and C. Oshima, "Double atomic layers of graphene/monolayer h-BN on Ni(111) studied by scanning tunneling microscopy and scanning tunneling spectroscopy," *Surface Review and Letters,* **09**, 03–04, pp. 1459–1464, 2002.

67. D. Wei, Y. Liu, Y. Wang, H. Zhang, L. Huang, and G. Yu, "Synthesis of N-doped graphene by chemical vapor deposition and its electrical properties," *Nano Letters,* **9**, 5, pp. 1752–1758, 2009.

68. T. O. Wehling, K. S. Novoselov, S. V. Morozov, E. E. Vdovin, M. I. Katsnelson, A. K. Geim, and A. I. Lichtenstein, "Molecular doping of graphene," *Nano Letters,* **8**, 1, pp. 173–177, 2008.

69. D. C. Elias, R. R. Nair, T. M. G. Mohiuddin, S. V. Morozov, P. Blake, M. P. Halsall, A. C. Ferrari, D. W. Boukhvalov, M. I. Katsnelson, A. K. Geim,

and K. S. Novoselov, "Control of graphene's properties by reversible hydrogenation: Evidence for graphane," *Science,* **323**, 5914, pp. 610–613, 2009.

70. J. S. Moon, M. Antcliffe, H. C. Seo, D. Curtis, S. Lin, A. Schmitz, I. Milosavljevic, A. A. Kiselev, R. S. Ross, D. K. Gaskill, P. M. Campbell, R. C. Fitch, K. M. Lee, and P. Asbeck, "Ultra-low resistance ohmic contacts in graphene field effect transistors," *Applied Physics Letters,* **100**, 20, p. 203512, 2012.

71. A. Hsu, H. Wang, K. K. Kim, J. Kong, and T. Palacios, "Impact of graphene interface quality on contact resistance and RF device performance," *IEEE Electron Device Letters,* **32**, 8, pp. 1008–1010, 2011.

72. F. Xia, V. Perebeinos, Y. M. Lin, Y. Wu, and P. Avouris, "The origins and limits of metal-graphene junction resistance," *Nature Nanotechnology,* **6**, 3, pp. 179–184, 2011.

73. J. A. Robinson, M. LaBella, M. Zhu, M. Hollander, R. Kasarda, Z. Hughes, K. Trumbull, R. Cavalero, and D. Snyder, "Contacting graphene," *Applied Physics Letters,* **98**, 5, p. 053103, 2011.

74. H. Allen, W. Han, K. Ki Kang, K. Jing, and P. Tomás, "High frequency performance of graphene transistors grown by chemical vapor deposition for mixed signal applications," *Japanese Journal of Applied Physics,* **50**, 7R, p. 070114, 2011.

75. A. Venugopal, L. Colombo, and E. M. Vogel, "Contact resistance in few and multilayer graphene devices," *Applied Physics Letters,* **96**, 1, p. 013512, 2010.

76. K. Seong-Dong, P. Cheol-Min, and J. C. S. Woo, "Advanced model and analysis of series resistance for CMOS scaling into nanometer regime-Part 1: Theoretical derivation," *IEEE Transactions on Electron Devices,* **49**, 3, pp. 457–466, 2002.

77. G. Giovannetti, P. A. Khomyakov, G. Brocks, V. M. Karpan, J. van den Brink, and P. J. Kelly, "Doping graphene with metal contacts," *Physical Review Letters,* **101**, 2, p. 026803, 2008.

78. H. Zhong, Z. Zhang, B. Chen, H. Xu, D. Yu, L. Huang, and L. Peng, "Realization of low contact resistance close to theoretical limit in graphene transistors," *Nano Research,* **8**, 5, pp. 1669–1679, 2015.

79. J. Lee, Y. Kim, H.-J. Shin, C. Lee, D. Lee, C.-Y. Moon, J. Lim, and S. Chan Jun, "Clean transfer of graphene and its effect on contact resistance," *Applied Physics Letters,* **103**, 10, p. 103104, 2013.

80. J. T. Smith, A. D. Franklin, D. B. Farmer, and C. D. Dimitrakopoulos, "Reducing contact resistance in graphene devices through contact area patterning," *ACS Nano,* **7**, 4, pp. 3661–3667, 2013.

81. L. Anzi, A. Mansouri, P. Pedrinazzi, E. Guerriero, M. Fiocco, A. Pesquera, A. Centeno, A. Zurutuza, A. Behnam, E. A. Carrion, E. Pop, and R. Sordan, "Ultra-low contact resistance in graphene devices at the Dirac point," *2D Materials,* **5**, 2, p. 025014, 2018.

82. C. Wei Chen, F. Ren, G.-C. Chi, S.-C. Hung, Y. P. Huang, J. Kim, I. I. Kravchenko, and S. J. Pearton, "UV ozone treatment for improving contact resistance on graphene," *Journal of Vacuum Science & Technology B,* **30**, 6, p. 060604, 2012.

83. W. Li, Y. Liang, D. Yu, L. Peng, K. P. Pernstich, T. Shen, A. R. Hight Walker, G. Cheng, C. A. Hacker, C. A. Richter, Q. Li, D. J. Gundlach, and X. Liang, "Ultraviolet/ozone treatment to reduce metal-graphene contact resistance," *Applied Physics Letters,* **102**, 18, p. 183110, 2013.

84. F. Liu, W. T. Navaraj, N. Yogeswaran, D. H. Gregory, and R. Dahiya, "Van der Waals contact engineering of graphene field-effect transistors for large-area flexible electronics," *ACS Nano,* **13**, 3, pp. 3257–3268, 2019.

85. A. S. Mayorov, R. V. Gorbachev, S. V. Morozov, L. Britnell, R. Jalil, L. A. Ponomarenko, P. Blake, K. S. Novoselov, K. Watanabe, T. Taniguchi, and A. K. Geim, "Micrometer-scale ballistic transport in encapsulated graphene at room temperature," *Nano Letters,* **11**, 6, pp. 2396–2399, 2011.

86. J. Baringhaus, M. Ruan, F. Edler, A. Tejeda, M. Sicot, I. Taleb, A.-P. Li, Z. Jiang, E. H. Conrad, C. Berger, C. Tegenkamp, and W. A. de Heer, "Exceptional ballistic transport in epitaxial graphene nanoribbons," *Nature,* **506**, 7488, pp. 349–354, 2014.

87. A. K. Geim and K. S. Novoselov, "The rise of graphene," *Nature Materials,* **6**, 3, pp. 183–191, 2007.

88. X. Du, I. Skachko, A. Barker, and E. Y. Andrei, "Approaching ballistic transport in suspended graphene," *Nature Nanotechnology,* **3**, 8, pp. 491–495, 2008.

89. E. V. Castro, H. Ochoa, M. I. Katsnelson, R. V. Gorbachev, D. C. Elias, K. S. Novoselov, A. K. Geim, and F. Guinea, "Limits on charge carrier mobility in suspended graphene due to flexural phonons," *Physical Review Letters,* **105**, 26, p. 266601, 2010.

90. Y. W. Tan, Y. Zhang, K. Bolotin, Y. Zhao, S. Adam, E. H. Hwang, S. Das Sarma, H. L. Stormer, and P. Kim, "Measurement of scattering rate and minimum conductivity in graphene," *Physical Review Letters,* **99**, 24, p. 246803, 2007.

91. C. Berger, Z. Song, X. Li, X. Wu, N. Brown, C. Naud, D. Mayou, T. Li, J. Hass, A. N. Marchenkov, E. H. Conrad, P. N. First, and W. A. de Heer, "Electronic

confinement and coherence in patterned epitaxial graphene," *Science,* **312**, 5777, pp. 1191–1196, 2006.

92. C. R. Dean, A. F. Young, I. Meric, C. Lee, L. Wang, S. Sorgenfrei, K. Watanabe, T. Taniguchi, P. Kim, K. L. Shepard, and J. Hone, "Boron nitride substrates for high-quality graphene electronics," *Nature Nanotechnology,* **5**, 10, pp. 722–726, 2010.

93. A. C. Ferrari, F. Bonaccorso, V. Fal'ko, K. S. Novoselov, S. Roche, P. Boggild, S. Borini, F. H. L. Koppens, V. Palermo, N. Pugno, J. A. Garrido, R. Sordan, A. Bianco, L. Ballerini, M. Prato, E. Lidorikis, J. Kivioja, C. Marinelli, T. Ryhanen, A. Morpurgo, J. N. Coleman, V. Nicolosi, L. Colombo, A. Fert, M. Garcia-Hernandez, A. Bachtold, G. F. Schneider, F. Guinea, C. Dekker, M. Barbone, Z. P. Sun, C. Galiotis, A. N. Grigorenko, G. Konstantatos, A. Kis, M. Katsnelson, L. Vandersypen, A. Loiseau, V. Morandi, D. Neumaier, E. Treossi, V. Pellegrini, M. Polini, A. Tredicucci, G. M. Williams, B. H. Hong, J. H. Ahn, J. M. Kim, H. Zirath, B. J. van Wees, H. van der Zant, L. Occhipinti, A. Di Matteo, I. A. Kinloch, T. Seyller, E. Quesnel, X. L. Feng, K. Teo, N. Rupesinghe, P. Hakonen, S. R. T. Neil, Q. Tannock, T. Lofwander, and J. Kinaret, "Science and technology roadmap for graphene, related two-dimensional crystals, and hybrid systems," *Nanoscale,* **7**, 11, pp. 4598–4810, 2015.

94. M. W. Lin, C. Ling, Y. Zhang, H. J. Yoon, M. M. Cheng, L. A. Agapito, N. Kioussis, N. Widjaja, and Z. Zhou, "Room-temperature high on/off ratio in suspended graphene nanoribbon field-effect transistors," *Nanotechnology,* **22**, 26, p. 265201, 2011.

95. S. Berrada, V. Hung Nguyen, D. Querlioz, J. Saint-Martin, A. Alarcón, C. Chassat, A. Bournel, and P. Dollfus, "Graphene nanomesh transistor with high on/off ratio and good saturation behavior," *Applied Physics Letters,* **103**, 18, p. 183509, 2013.

96. M. Dragoman, A. Dinescu, and D. Dragoman, "Room temperature nanostructured graphene transistor with high on/off ratio," *Nanotechnology,* **28**, 1, p. 015201, 2017.

97. F. Xia, D. B. Farmer, Y. M. Lin, and P. Avouris, "Graphene field-effect transistors with high on/off current ratio and large transport band gap at room temperature," *Nano Letters,* **10**, 2, pp. 715–718, 2010.

98. Z. Chen, Y.-M. Lin, M. J. Rooks, and P. Avouris, "Graphene nano-ribbon electronics," *Physica E,* **40**, 2, pp. 228–232, 2007.

99. M. Y. Han, B. Ozyilmaz, Y. Zhang, and P. Kim, "Energy band-gap engineering of graphene nanoribbons," *Physical Review Letters,* **98**, 20, p. 206805, 2007.

100. J. Bai, X. Duan, and Y. Huang, "Rational fabrication of graphene nanoribbons using a nanowire etch mask," *Nano Letters,* **9**, 5, pp. 2083–2087, 2009.

101. C. Stampfer, J. Guttinger, S. Hellmuller, F. Molitor, K. Ensslin, and T. Ihn, "Energy gaps in etched graphene nanoribbons," *Physical Review Letters,* **102**, 5, p. 056403, 2009.

102. P. Gallagher, K. Todd, and D. Goldhaber-Gordon, "Disorder-induced gap behavior in graphene nanoribbons," *Physical Review B,* **81**, 11, p. 115409, 2010.

103. X. Wang and H. Dai, "Etching and narrowing of graphene from the edges," *Nature Chemistry,* **2**, 8, pp. 661–665, 2010.

104. G. Z. Magda, X. Jin, I. Hagymasi, P. Vancso, Z. Osvath, P. Nemes-Incze, C. Hwang, L. P. Biro, and L. Tapaszto, "Room-temperature magnetic order on zigzag edges of narrow graphene nanoribbons," *Nature,* **514**, 7524, pp. 608–611, 2014.

105. L. Jiao, X. Wang, G. Diankov, H. Wang, and H. Dai, "Facile synthesis of high-quality graphene nanoribbons," *Nature Nanotechnology,* **5**, 5, pp. 321–325, 2010.

106. L. Jiao, L. Zhang, X. Wang, G. Diankov, and H. Dai, "Narrow graphene nanoribbons from carbon nanotubes," *Nature,* **458**, 7240, pp. 877–880, 2009.

107. Y. Y. Li, M. X. Chen, M. Weinert, and L. Li, "Direct experimental determination of onset of electron-electron interactions in gap opening of zigzag graphene nanoribbons," *Nature Communications,* **5**, 1, p. 4311, 2014.

108. S. Linden, D. Zhong, A. Timmer, N. Aghdassi, J. H. Franke, H. Zhang, X. Feng, K. Mullen, H. Fuchs, L. Chi, and H. Zacharias, "Electronic structure of spatially aligned graphene nanoribbons on Au(788)," *Physical Review Letters,* **108**, 21, p. 216801, 2012.

109. A. Narita, X. Feng, Y. Hernandez, S. A. Jensen, M. Bonn, H. Yang, I. A. Verzhbitskiy, C. Casiraghi, M. R. Hansen, A. H. Koch, G. Fytas, O. Ivasenko, B. Li, K. S. Mali, T. Balandina, S. Mahesh, S. De Feyter, and K. Mullen, "Synthesis of structurally well-defined and liquid-phase-processable graphene nanoribbons," *Nature Chemistry,* **6**, 2, pp. 126–132, 2014.

110. Y.-C. Chen, D. G. de Oteyza, Z. Pedramrazi, C. Chen, F. R. Fischer, and M. F. Crommie, "Tuning the band gap of graphene nanoribbons synthesized from molecular precursors," *ACS Nano,* **7**, 7, pp. 6123–6128, 2013.

111. R. S. Jordan, Y. Wang, R. D. McCurdy, M. T. Yeung, K. L. Marsh, S. I. Khan, R. B. Kaner, and Y. Rubin, "Synthesis of graphene nanoribbons via

the topochemical polymerization and subsequent aromatization of a diacetylene precursor," *Chem,* **1**, 1, pp. 78–90, 2016.

112. T. Kato and R. Hatakeyama, "Site- and alignment-controlled growth of graphene nanoribbons from nickel nanobars," *Nature Nanotechnology,* **7**, 10, pp. 651–656, 2012.

113. I. Martin-Fernandez, D. Wang, and Y. Zhang, "Direct growth of graphene nanoribbons for large-scale device fabrication," *Nano Letters,* **12**, 12, pp. 6175–6179, 2012.

114. R. M. Jacobberger, B. Kiraly, M. Fortin-Deschenes, P. L. Levesque, K. M. McElhinny, G. J. Brady, R. Rojas Delgado, S. Singha Roy, A. Mannix, M. G. Lagally, P. G. Evans, P. Desjardins, R. Martel, M. C. Hersam, N. P. Guisinger, and M. S. Arnold, "Direct oriented growth of armchair graphene nanoribbons on germanium," *Nature Communications,* **6**, 1, p. 8006, 2015.

115. W. S. Leong, H. Wang, J. Yeo, F. J. Martin-Martinez, A. Zubair, P. C. Shen, Y. Mao, T. Palacios, M. J. Buehler, J. Y. Hong, and J. Kong, "Paraffin-enabled graphene transfer," *Nature Communications,* **10**, 1, p. 867, 2019.

116. W. S. Hwang, P. Zhao, K. Tahy, L. O. Nyakiti, V. D. Wheeler, R. L. Myers-Ward, C. R. Eddy, D. K. Gaskill, J. A. Robinson, W. Haensch, H. Xing, A. Seabaugh, and D. Jena, "Graphene nanoribbon field-effect transistors on wafer-scale epitaxial graphene on SiC substrates," *APL Materials,* **3**, 1, p. 011101, 2015.

117. A. C. Seabaugh and Q. Zhang, "Low-voltage tunnel transistors for beyond CMOS logic," *Proceedings of the IEEE,* **98**, 12, pp. 2095–2110, 2010.

118. P. Zhao, J. Chauhan, and J. Guo, "Computational study of tunneling transistor based on graphene nanoribbon," *Nano Letters,* **9**, 2, pp. 684–688, 2009.

119. G. Fiori and G. Iannaccone, "Ultralow-voltage bilayer graphene tunnel FET," *IEEE Electron Device Letters,* **30**, 10, pp. 1096–1098, 2009.

120. F. Schwierz, "Graphene transistors," *Nature Nanotechnology,* **5**, 7, pp. 487–496, 2010.

121. E. U. Stützel, M. Burghard, K. Kern, F. Traversi, F. Nichele, and R. Sordan, "A graphene nanoribbon memory cell," *Small,* **6**, 24, pp. 2822–2825, 2010.

122. L. Liao, J. Bai, Y. Qu, Y. Huang, and X. Duan, "Single-layer graphene on Al_2O_3/Si substrate: Better contrast and higher performance of graphene transistors," *Nanotechnology,* **21**, 1, p. 015705, 2010.

123. X. Wang, S. M. Tabakman, and H. Dai, "Atomic layer deposition of metal oxides on pristine and functionalized graphene," *Journal of the American Chemical Society,* **130**, 26, pp. 8152–8153, 2008.

124. Y. Xuan, Y. Q. Wu, T. Shen, M. Qi, M. A. Capano, J. A. Cooper, and P. D. Ye, "Atomic-layer-deposited nanostructures for graphene-based nanoelectronics," *Applied Physics Letters,* **92**, 1, p. 013101, 2008.

125. A. Pirkle, R. M. Wallace, and L. Colombo, "In situ studies of Al_2O_3 and HfO_2 dielectrics on graphite," *Applied Physics Letters,* **95**, 13, p. 133106, 2009.

126. Y.-M. Lin, K. A. Jenkins, A. Valdes-Garcia, J. P. Small, D. B. Farmer, and P. Avouris, "Operation of graphene transistors at gigahertz frequencies," *Nano Letters,* **9**, 1, pp. 422–426, 2009.

127. Y.-M. Lin, C. Dimitrakopoulos, K. A. Jenkins, D. B. Farmer, H.-Y. Chiu, A. Grill, and P. Avouris, "100-GHz transistors from wafer-scale epitaxial graphene," *Science,* **327**, 5966, pp. 662–662, 2010.

128. V. K. Sangwan, D. Jariwala, K. Everaerts, J. J. McMorrow, J. He, M. Grayson, L. J. Lauhon, T. J. Marks, and M. C. Hersam, "Wafer-scale solution-derived molecular gate dielectrics for low-voltage graphene electronics," *Applied Physics Letters,* **104**, 8, p. 083503, 2014.

129. B. Hähnlein, B. Händel, J. Pezoldt, H. Töpfer, R. Granzner, F. Schwierz, and A. Ph., "Side-gate graphene field-effect transistors with high transconductance," *Applied Physics Letters,* **101**, 9, p. 093504, 2012.

130. A. Di Bartolomeo, F. Giubileo, L. Iemmo, F. Romeo, S. Russo, S. Unal, M. Passacantando, V. Grossi, and A. M. Cucolo, "Leakage and field emission in side-gate graphene field effect transistors," *Applied Physics Letters,* **109**, 2, p. 023510, 2016.

131. S.-L. Li, H. Miyazaki, H. Hiura, C. Liu, and K. Tsukagoshi, "Enhanced logic performance with semiconducting bilayer graphene channels," *ACS Nano,* **5**, 1, pp. 500–506, 2011.

132. B. N. Szafranek, G. Fiori, D. Schall, D. Neumaier, and H. Kurz, "Current saturation and voltage gain in bilayer graphene field effect transistors," *Nano Letters,* **12**, 3, pp. 1324–1328, 2012.

133. S. M. Kim, A. Hsu, M. H. Park, S. H. Chae, S. J. Yun, J. S. Lee, D. H. Cho, W. Fang, C. Lee, T. Palacios, M. Dresselhaus, K. K. Kim, Y. H. Lee, and J. Kong, "Synthesis of large-area multilayer hexagonal boron nitride for high material performance," *Nature Communications,* **6**, 1, p. 8662, 2015.

134. S. L. Li, H. Miyazaki, A. Kumatani, A. Kanda, and K. Tsukagoshi, "Low operating bias and matched input-output characteristics in graphene logic inverters," *Nano Letters,* **10**, 7, pp. 2357–2362, 2010.

135. A. Sagar, K. Balasubramanian, M. Burghard, K. Kern, and R. Sordan, "Polymer-electrolyte gated graphene transistors for analog and digital phase detection," *Applied Physics Letters,* **99**, 4, p. 043307, 2011.

136. D. Schall, M. Otto, D. Neumaier, and H. Kurz, "Integrated ring oscillators based on high-performance graphene inverters," *Scientific Reports,* **3**, p. 2592, 2013.

137. Y.-M. Lin, A. Valdes-Garcia, S.-J. Han, D. B. Farmer, I. Meric, Y. Sun, Y. Wu, C. Dimitrakopoulos, A. Grill, P. Avouris, and K. A. Jenkins, "Wafer-scale graphene integrated circuit," *Science,* **332**, 6035, pp. 1294–1297, 2011.

138. R. Sordan, F. Traversi, and V. Russo, "Logic gates with a single graphene transistor," *Applied Physics Letters,* **94**, 7, p. 073305, 2009.

139. F. Traversi, V. Russo, and R. Sordan, "Integrated complementary graphene inverter," *Applied Physics Letters,* **94**, p. 223312, 2009.

140. L. G. Rizzi, M. Bianchi, A. Behnam, E. Carrion, E. Guerriero, L. Polloni, E. Pop, and R. Sordan, "Cascading wafer-scale integrated graphene complementary inverters under ambient conditions," *Nano Letters,* **12**, 8, pp. 3948–3953, 2012.

141. J. M. Yun, S. Park, Y. H. Hwang, E.-S. Lee, U. Maiti, H. Moon, B.-H. Kim, B.-S. Bae, Y.-H. Kim, and S. O. Kim, "Complementary p- and n-type polymer doping for ambient stable graphene inverter," *ACS Nano,* **8**, 1, pp. 650–656, 2014.

142. D. Dragoman and M. Dragoman, "Enhanced architectures for room-temperature reversible logic gates in graphene," *Applied Physics Letters,* **105**, 11, p. 113109, 2014.

143. S. K. Banerjee, L. F. Register, E. Tutuc, D. Reddy, and A. H. MacDonald, "Bilayer pseudospin field-effect transistor (BiSFET): A proposed new logic device," *IEEE Electron Device Letters,* **30**, 2, pp. 158–160, 2010.

144. L. Britnell, R. V. Gorbachev, R. Jalil, B. D. Belle, F. Schedin, A. Mishchenko, T. Georgiou, M. I. Katsnelson, L. Eaves, S. V. Morozov, N. M. R. Peres, J. Leist, A. K. Geim, K. S. Novoselov, and L. A. Ponomarenko, "Field-effect tunneling transistor based on vertical graphene heterostructures," *Science,* **335**, 6071, pp. 947–950, 2012.

145. W. Mehr, J. Dabrowski, J. C. Scheytt, G. Lippert, Y.-H. Xie, M. C. Lemme, M. Ostling, and G. Lupina, "Vertical Graphene Base Transistor," *IEEE Electron Device Letters,* **33**, 5, pp. 691–693, 2012.

146. H. Yang, J. Heo, S. Park, H. J. Song, D. H. Seo, K.-E. Byun, P. Kim, I. Yoo, H.-J. Chung, and K. Kim, "Graphene barristor, a triode device with a gate-controlled Schottky barrier," *Science,* **336**, 6085, pp. 1140–1143, 2012.

147. X. Wang, X. Jiang, T. Wang, J. Shi, M. Liu, Q. Zeng, Z. Cheng, and X. Qiu, "Electrically configurable graphene field-effect transistors with a graded-potential gate," *Nano Letters,* **15**, 5, pp. 3212–3216, 2015.

148. S. Yuan, Z. Yang, C. Xie, F. Yan, J. Dai, S. P. Lau, H. L. W. Chan, and J. Hao, "Ferroelectric-driven performance enhancement of graphene field-effect transistors based on vertical tunneling heterostructures," *Advanced Materials,* **28**, 45, pp. 10048–10054, 2016.

149. Y. Liu, J. Sheng, H. Wu, Q. He, H.-C. Cheng, M. I. Shakir, Y. Huang, and X. Duan, "High-current-density vertical-tunneling transistors from graphene/highly doped silicon heterostructures," *Advanced Materials,* **28**, 21, pp. 4120–4125, 2016.

150. J. Kang, D. Jariwala, C. R. Ryder, S. A. Wells, Y. Choi, E. Hwang, J. H. Cho, T. J. Marks, and M. C. Hersam, "Probing out-of-plane charge transport in black phosphorus with graphene-contacted vertical field-effect transistors," *Nano Letters,* **16**, 4, pp. 2580–2585, 2016.

151. H. Hlaing, C.-H. Kim, F. Carta, C.-Y. Nam, R. A. Barton, N. Petrone, J. Hone, and I. Kymissis, "Low-voltage organic electronics based on a gate-tunable injection barrier in vertical graphene-organic semiconductor heterostructures," *Nano Letters,* **15**, 1, pp. 69–74, 2015.

152. T. Georgiou, R. Jalil, B. D. Belle, L. Britnell, R. V. Gorbachev, S. V. Morozov, Y.-J. Kim, A. Gholinia, S. J. Haigh, O. Makarovsky, L. Eaves, L. A. Ponomarenko, A. K. Geim, K. S. Novoselov, and A. Mishchenko, "Vertical field-effect transistor based on graphene-WS2 heterostructures for flexible and transparent electronics," *Nature Nanotechnology,* **8**, 2, pp. 100–103, 2013.

153. W. J. Yu, Z. Li, H. Zhou, Y. Chen, Y. Wang, Y. Huang, and X. Duan, "Vertically stacked multi-heterostructures of layered materials for logic transistors and complementary inverters," *Nature Materials,* **12**, 3, pp. 246–252, 2013.

154. Y. Choi, J. Kang, D. Jariwala, M. S. Kang, T. J. Marks, M. C. Hersam, and J. H. Cho, "Low-voltage complementary electronics from ion-gel-gated vertical van der Waals heterostructures," *Advanced Materials,* **28**, 19, pp. 3742–3748, 2016.

155. J. Shim, H. S. Kim, Y. S. Shim, D.-H. Kang, H.-Y. Park, J. Lee, J. Jeon, S. J. Jung, Y. J. Song, W.-S. Jung, J. Lee, S. Park, J. Kim, S. Lee, Y.-H. Kim, and J.-H. Park, "Extremely large gate modulation in vertical graphene/WSe2 heterojunction barristor based on a novel transport mechanism," *Advanced Materials,* **28**, 26, pp. 5293–5299, 2016.

156. C. Zeng, E. B. Song, M. Wang, S. Lee, C. M. Torres, Jr., J. Tang, B. H. Weiller, and K. L. Wang, "Vertical graphene-base hot-electron transistor," *Nano Letters,* **13**, 6, pp. 2370–2375, 2013.

157. Y. Liu, H. Zhou, R. Cheng, W. Yu, Y. Huang, and X. Duan, "Highly flexible electronics from scalable vertical thin film transistors," *Nano Letters,* **14**, 3, pp. 1413–1418, 2014.

158. J. Kunc, Y. Hu, J. Palmer, Z. Guo, J. Hankinson, S. H. Gamal, C. Berger, and W. A. de Heer, "Planar edge Schottky barrier-tunneling transistors using epitaxial graphene/SiC junctions," *Nano Letters,* **14**, 9, pp. 5170–5175, 2014.

159. B. M. Blaschke, M. Lottner, S. Drieschner, A. B. Calia, K. Stoiber, L. Rousseau, G. Lissourges, and J. A. Garrido, "Flexible graphene transistors for recording cell action potentials," *2D Materials,* **3**, 2, p. 025007, 2016.

160. G. Fiori and G. Iannaccone, "Insights on radio frequency bilayer graphene FETs," *IEEE International Electron Devices Meeting,* pp. 17.3.1–17.3.4, 2012.

161. I. Meric, N. Baklitskaya, P. Kim, and K. L. Shepard, "RF performance of top-gated, zero-bandgap graphene field-effect transistors," *IEEE International Electron Devices Meeting*, pp. 513–516, 2008.

162. Y. M. Lin, K. Jenkins, D. Farmer, A. Valdes-Garcia, P. Avouris, C. Y. Sung, H. Y. Chiu, and B. Ek, "Development of graphene FETs for high frequency electronics," *IEEE International Electron Devices Meeting*, pp. 216–219, 2009.

163. L. Liao, Y. C. Lin, M. Bao, R. Cheng, J. Bai, Y. Liu, Y. Qu, K. L. Wang, Y. Huang, and X. Duan, "High-speed graphene transistors with a self-aligned nanowire gate," *Nature,* **467**, 7313, pp. 305–308, 2010.

164. Y. Wu, K. A. Jenkins, A. Valdes-Garcia, D. B. Farmer, Y. Zhu, A. A. Bol, C. Dimitrakopoulos, W. Zhu, F. Xia, P. Avouris, and Y. M. Lin, "State-of-the-art graphene high-frequency electronics," *Nano Letters,* **12**, 6, pp. 3062–3067, 2012.

165. R. Cheng, J. Bai, L. Liao, H. Zhou, Y. Chen, L. Liu, Y. C. Lin, S. Jiang, Y. Huang, and X. Duan, "High-frequency self-aligned graphene transistors with transferred gate stacks," *Proceedings of the National Academy of Sciences of the United States of America,* **109**, 29, pp. 11588–11592, 2012.

166. Z. Feng, C. Yu, J. Li, Q. Liu, Z. He, X. Song, J. Wang, and S. J. C. Cai, "An ultra clean self-aligned process for high maximum oscillation frequency graphene transistors," *Carbon*, **75**, 1, pp. 249–254, 2014.

167. C.-H. Yeh, Y.-W. Lain, Y.-C. Chiu, C.-H. Liao, D. R. Moyano, S. S. H. Hsu, and P.-W. Chiu, "Gigahertz flexible graphene transistors for microwave integrated circuits," *ACS Nano,* **8**, 8, pp. 7663–7670, 2014.

168. N. Petrone, I. Meric, T. Chari, K. L. Shepard, and J. Hone, "Graphene field-effect transistors for radio-frequency flexible electronics," *IEEE Journal of the Electron Devices Society,* **3**, 1, pp. 44–48, 2015.

169. Y. Liang, X. Liang, Z. Zhang, W. Li, X. Huo, and L. Peng, "High mobility flexible graphene field-effect transistors and ambipolar radio-frequency circuits," *Nanoscale,* **7**, 25, pp. 10954–10962, 2015.

170. X. Huang, T. Leng, M. Zhu, X. Zhang, J. Chen, K. Chang, M. Aqeeli, A. K. Geim, K. S. Novoselov, and Z. Hu, "Highly flexible and conductive printed graphene for wireless wearable communications applications," *Scientific Reports,* **5**, 1, p. 18298, 2015.

171. S. Park, S. H. Shin, M. N. Yogeesh, A. L. Lee, S. Rahimi, and D. Akinwande, "Extremely high-frequency flexible graphene thin-film transistors," *IEEE Electron Device Letters,* **37**, 4, pp. 512–515, 2016.

172. W. Wei, E. Pallecchi, S. Haque, S. Borini, V. Avramovic, A. Centeno, Z. Amaia, and H. Happy, "Mechanically robust 39 GHz cut-off frequency graphene field effect transistors on flexible substrates," *Nanoscale,* **8**, 29, pp. 14097–14103, 2016.

173. Y. H. Jung, H. Zhang, S. J. Cho, and Z. Ma, "Flexible and stretchable microwave microelectronic devices and circuits," *IEEE Transactions on Electron Devices,* **64**, 5, pp. 1881–1893, 2017.

174. C. Liu, W. Ma, M. Chen, W. Ren, and D. Sun, "A vertical silicon-graphene-germanium transistor," *Nature Communications,* **10**, 1, p. 4873, 2019.

175. E. Guerriero, L. Polloni, L. G. Rizzi, M. Bianchi, G. Mondello, and R. Sordan, "Graphene audio voltage amplifier," *Small,* **8**, 3, pp. 357–361, 2012.

176. O. Habibpour, J. Vukusic, and J. Stake, "A 30-GHz integrated subharmonic mixer based on a multichannel graphene FET," *IEEE Transactions on Microwave Theory and Techniques,* **61**, 2, pp. 841–847, 2013.

177. M. E. Ramón, K. N. Parrish, S. F. Chowdhury, C. W. Magnuson, H. C. P. Movva, R. S. Ruoff, S. K. Banerjee, and D. Akinwande, "Three-gigahertz graphene frequency doubler on quartz operating beyond the transit frequency," *IEEE Transactions on Nanotechnology,* **11**, 5, pp. 877–883, 2012.

178. E. Guerriero, L. Polloni, M. Bianchi, A. Behnam, E. Carrion, L. G. Rizzi, E. Pop, and R. Sordan, "Gigahertz integrated graphene ring oscillators," *ACS Nano,* **7**, 6, pp. 5588–5594, 2013.

179. J. Yoon, A. J. Baca, S.-I. Park, P. Elvikis, J. B. Geddes, L. Li, R. H. Kim, J. Xiao, S. Wang, T.-H. Kim, M. J. Motala, B. Y. Ahn, E. B. Duoss, J. A. Lewis, R. G. Nuzzo, P. M. Ferreira, Y. Huang, A. Rockett, and J. A. Rogers, "Ultrathin silicon solar microcells for semitransparent, mechanically flexible and microconcentrator module designs," *Nature Materials,* **7**, 11, pp. 907–915, 2008.

180. Y. Zhai, L. Mathew, R. Rao, D. Xu, and S. K. Banerjee, "High-performance flexible thin-film transistors exfoliated from bulk wafer," *Nano Letters,* **12**, 11, pp. 5609–5615, 2012.

181. T. Sekitani, T. Yokota, U. Zschieschang, H. Klauk, S. Bauer, K. Takeuchi, M. Takamiya, T. Sakurai, and T. Someya, "Organic nonvolatile memory transistors for flexible sensor arrays," *Science,* **326**, 5959, pp. 1516–1519, 2009.

182. S. R. Forrest, "The path to ubiquitous and low-cost organic electronic appliances on plastic," *Nature,* **428**, 6986, pp. 911–918, 2004.

183. J.-Y. Lee, S. T. Connor, Y. Cui, and P. Peumans, "Solution-processed metal nanowire mesh transparent electrodes," *Nano Letters,* **8**, 2, pp. 689–692, 2008.

184. S. De, T. M. Higgins, P. E. Lyons, E. M. Doherty, P. N. Nirmalraj, W. J. Blau, J. J. Boland, and J. N. Coleman, "Silver nanowire networks as flexible, transparent, conducting films: Extremely high DC to optical conductivity ratios," *ACS Nano,* **3**, 7, pp. 1767–1774, 2009.

185. Y. Zhu and F. Xu, "Buckling of aligned carbon nanotubes as stretchable conductors: A new manufacturing strategy," *Advanced Materials,* **24**, 8, pp. 1073–1077, 2012.

186. R. R. Nair, P. Blake, A. N. Grigorenko, K. S. Novoselov, T. J. Booth, T. Stauber, N. M. R. Peres, and A. K. Geim, "Fine structure constant defines visual transparency of graphene," *Science,* **320**, 5881, pp. 1308–1308, 2008.

187. G. Eda, G. Fanchini, and M. Chhowalla, "Large-area ultrathin films of reduced graphene oxide as a transparent and flexible electronic material," *Nature Nanotechnology,* **3**, 5, pp. 270–274, 2008.

188. R. Karthick, M. Brindha, M. Selvaraj, and S. Ramu, "Stable colloidal dispersion of functionalized reduced graphene oxide in aqueous medium for transparent conductive film," *Journal of Colloid and Interface Science,* **406**, 1, pp. 69–74, 2013.

189. H. Feng, R. Cheng, X. Zhao, X. Duan, and J. Li, "A low-temperature method to produce highly reduced graphene oxide," *Nature Communications,* **4**, 1, p. 1539, 2013.

190. X. Y. Lin, J. J. Jia, N. Yousefi, X. Shen, and J. K. Kim, "Excellent optoelectrical properties of graphene oxide thin films deposited on a flexible substrate by Langmuir–Blodgett assembly," *Journal of Materials Chemistry C,* **1**, 41, pp. 6869–6877, 2013.

191. Y. Hernandez, V. Nicolosi, M. Lotya, F. M. Blighe, Z. Sun, S. De, I. T. McGovern, B. Holland, M. Byrne, Y. K. Gun'Ko, J. J. Boland, P. Niraj, G. Duesberg, S. Krishnamurthy, R. Goodhue, J. Hutchison, V. Scardaci, A. C. Ferrari, and J. N. Coleman, "High-yield production of graphene by liquid-phase exfoliation of graphite," *Nature Nanotechnology,* **3**, 9, pp. 563–568, 2008.

192. M. S. Kang, K. T. Kim, J. U. Lee, and W. H. Jo, "Direct exfoliation of graphite using a non-ionic polymer surfactant for fabrication of transparent and conductive graphene films," *Journal of Materials Chemistry C,* **1**, 9, pp. 1870–1875, 2013.

193. D. Voiry, J. Yang, J. Kupferberg, R. Fullon, C. Lee, H. Y. Jeong, H. S. Shin, and M. Chhowalla, "High-quality graphene via microwave reduction of solution-exfoliated graphene oxide," *Science,* **353**, 6306, pp. 1413–1416, 2016.

194. K. S. Kim, Y. Zhao, H. Jang, S. Y. Lee, J. M. Kim, K. S. Kim, J.-H. Ahn, P. Kim, J.-Y. Choi, and B. H. Hong, "Large-scale pattern growth of graphene films for stretchable transparent electrodes," *Nature,* **457**, 7230, pp. 706–710, 2009.

195. D. Y. Wang, I. S. Huang, P. H. Ho, S. S. Li, Y. C. Yeh, D. W. Wang, W. L. Chen, Y. Y. Lee, Y. M. Chang, C. C. Chen, C. T. Liang, and C. W. Chen, "Clean-lifting transfer of large-area residual-free graphene films," *Advanced Materials,* **25**, 32, pp. 4521–4526, 2013.

196. S. Bae, H. Kim, Y. Lee, X. Xu, J.-S. Park, Y. Zheng, J. Balakrishnan, T. Lei, H. Ri Kim, Y. I. Song, Y.-J. Kim, K. S. Kim, B. Ozyilmaz, J.-H. Ahn, B. H. Hong, and S. Iijima, "Roll-to-roll production of 30-inch graphene films for transparent electrodes," *Nature Nanotechnology,* **5**, 8, pp. 574–578, 2010.

197. T. Kobayashi, M. Bando, N. Kimura, K. Shimizu, K. Kadono, N. Umezu, K. Miyahara, S. Hayazaki, S. Nagai, Y. Mizuguchi, Y. Murakami, and D. Hobara, "Production of a 100-m-long high-quality graphene transparent conductive film by roll-to-roll chemical vapor deposition and transfer process," *Applied Physics Letters,* **102**, 2, p. 023112, 2013.

198. A. Kasry, M. A. Kuroda, G. J. Martyna, G. S. Tulevski, and A. A. Bol, "Chemical doping of large-area stacked graphene films for use as transparent, conducting electrodes," *ACS Nano,* **4**, 7, pp. 3839–3844, 2010.

199. W. Liu, J. Kang, and K. Banerjee, "Characterization of $FeCl_3$ intercalation doped CVD few-layer graphene," *IEEE Electron Device Letters,* **37**, 9, pp. 1246–1249, 2016.

200. A. L. Gorkina, A. P. Tsapenko, E. P. Gilshteyn, T. S. Koltsova, T. V. Larionova, A. Talyzin, A. S. Anisimov, I. V. Anoshkin, E. I. Kauppinen, O. V. Tolochko, and A. G. Nasibulin, "Transparent and conductive hybrid graphene/carbon nanotube films," *Carbon,* **100**, 1, pp. 501–507, 2016.

201. Y. Zhu, Z. Sun, Z. Yan, Z. Jin, and J. M. Tour, "Rational design of hybrid graphene films for high-performance transparent electrodes," *ACS Nano,* **5**, 8, pp. 6472–6479, 2011.

202. M.-S. Lee, K. Lee, S.-Y. Kim, H. Lee, J. Park, K.-H. Choi, H.-K. Kim, D.-G. Kim, D.-Y. Lee, S. Nam, and J.-U. Park, "High-performance, transparent, and stretchable electrodes using graphene–metal nanowire hybrid structures," *Nano Letters,* **13**, 6, pp. 2814–2821, 2013.

203. H. Wang and G. Yu, "Direct CVD graphene growth on semiconductors and dielectrics for transfer-free device fabrication," *Advanced Materials,* **28**, 25, pp. 4956–4975, 2016.

204. J. Chen, Y. Guo, L. Jiang, Z. Xu, L. Huang, Y. Xue, D. Geng, B. Wu, W. Hu, G. Yu, and Y. Liu, "Near-equilibrium chemical vapor deposition of high-quality single-crystal graphene directly on various dielectric substrates," *Advanced Materials,* **26**, 9, pp. 1348–1353, 2014.

205. Y. Z. Chen, H. Medina, H. C. Lin, H. W. Tsai, T. Y. Su, and Y. L. Chueh, "Large-scale and patternable graphene: Direct transformation of amorphous carbon film into graphene/graphite on insulators via Cu mediation engineering and its application to all-carbon based devices," *Nanoscale,* **7**, 5, pp. 1678–1687, 2015.

206. D. Wei, L. Peng, M. Li, H. Mao, T. Niu, C. Han, W. Chen, and A. T. S. Wee, "Low temperature critical growth of high quality nitrogen doped graphene on dielectrics by plasma-enhanced chemical vapor deposition," *ACS Nano,* **9**, 1, pp. 164–171, 2015.

207. N. Liu, A. Chortos, T. Lei, L. Jin, T. R. Kim, W.-G. Bae, C. Zhu, S. Wang, R. Pfattner, X. Chen, R. Sinclair, and Z. Bao, "Ultratransparent and stretchable graphene electrodes," *Science Advances,* **3**, 9, p. e1700159, 2017.

208. G. Oh, J. S. Kim, J. H. Jeon, E. Won, J. W. Son, D. H. Lee, C. K. Kim, J. Jang, T. Lee, and B. H. Park, "Graphene/pentacene barristor with ion-gel gate dielectric: Flexible ambipolar transistor with high mobility and on/off ratio," *ACS Nano,* **9**, 7, pp. 7515–7522, 2015.

209. M. Soni, P. Kumar, J. Pandey, S. K. Sharma, and A. Soni, "Scalable and site-specific functionalization of reduced graphene oxide for circuit elements and flexible electronics," *Carbon,* **128**, 1, pp. 172–178, 2018.

210. S. K. Lee, H. Y. Jang, S. Jang, E. Choi, B. H. Hong, J. Lee, S. Park, and J. H. Ahn, "All graphene-based thin film transistors on flexible plastic substrates," *Nano Letters,* **12**, 7, pp. 3472–3476, 2012.

211. B. J. Kim, S.-K. Lee, M. S. Kang, J.-H. Ahn, and J. H. Cho, "Coplanar-gate transparent graphene transistors and inverters on plastic," *ACS Nano,* **6**, 10, pp. 8646–8651, 2012.

212. D. Akinwande, C. Huyghebaert, C. H. Wang, M. I. Serna, S. Goossens, L. J. Li, H. P. Wong, and F. H. L. Koppens, "Graphene and two-dimensional materials for silicon technology," *Nature,* **573**, 7775, pp. 507–518, 2019.

213. *ITRS—The International Technology Roadmap for Semiconductors.* Available: http://www.itrs2.net.

Chapter 6

Nanocarbons for Flexible Sensing Applications

Morgan J. Anderson,[a] Shobhit Kareer,[b]
Seyed Esmaeil Mahdavi Ardakani,[b] Seamus D. Thomson,[c]
Jessica E. Koehne,[d] and Jeongwon Park[b,e]

[a]*Department of Chemistry, Kansas State University, Manhattan, KS 66502, USA*
[b]*School of Electrical Engineering and Computer Science, University of Ottawa, Ontario K1N 6N5, Canada*
[c]*NASA Postdoctoral Program Fellow, NASA Ames Research Center, Moffett Field, CA 94035, USA*
[d]*NASA Ames Research Center, Center for Nanotechnology, Moffett Field, CA 94035, USA*
[e]*Department of Electrical and Biomedical Engineering, University of Nevada, Reno NV 89557, USA*
jessica.e.koehne@nasa.gov, jepark@unr.edu

6.1 Introduction and Overview for Flexible Nanocarbon Electronics

Research in the field of flexible and wearable electronics has attracted a great deal of attention from across the world. Flexible sensors and

Nanocarbon Electronics
Edited by Changjian Zhou, Min Zhang, and Cary Y. Yang

ISBN 978-981-4877-11-4 (Hardcover), 978-1-003-04308-9 (eBook)
www.jennystanford.com

electronics offer a wide array of applications ranging from wearable sensors for continuous health monitoring to on-demand fabrication of complex circuits. The ability to rapidly fabricate by printing methodologies is an attractive feature in this field, as it drastically reduces cost and allows for a high level of device tunability. Integration of nanocarbon materials such as graphene, reduced graphene oxide, and carbon nanotubes (CNTs) has further expanded the capabilities of flexible electronics by enabling direct manipulation of material properties for a variety of functions. Overall, these features render a wide range of applications and expanding global market that is estimated to grow to $77.3 billion by 2029 [1].

In this chapter, we will discuss the fabrication and applications of flexible nanocarbon devices. This includes discussion on the preparation, manipulation, and fabrication of these materials into sensors and electronics, as well as their integration into fully functional devices. Flexible electronic components and periphery—focusing on flexible nanocarbon transistors—will also be examined. Additionally, we will discuss applications of flexible nanocarbon electronics to sensor applications as well as functional, wearable sensors. Finally, we will examine the future directions of flexible nanocarbon devices and their potential impacts.

6.2 Flexible Device Fabrication

Flexible nanocarbon electronics are typically fabricated by two techniques: (1) traditional lithographic microfabrication processes and (2) direct printing using nanocarbon-based inks.

Lithographic fabrication techniques provide high spatial resolution but require highly specialized equipment and thorough, precise characterization. In general, electronic components are composed of several metallic, insulating, and semiconducting layers embedded in a soft polymer matrix, along with any additional nanocarbon materials. These devices often utilize graphene or CNTs grown by chemical vapor deposition (CVD). Lithographic patterning offers high spatial resolution but requires expensive materials and long processing times. Despite these limitations, the infrastructure for lithographic processing is ubiquitous and readily scalable.

Printed flexible electronics utilize simple, well-established technologies such as screen printing and inkjet printing. Here nanocarbon materials are typically composed of reduced graphene oxide or CNTs that are suspended in organic or aqueous solvents to form a nanocarbon ink. The nanocarbon ink is subsequently patterned directly onto the flexible substrate. Common substrates include cellulose paper, polyimide (PI) films, and polydimethylsiloxane (PDMS). Screen printed devices are made by spreading the ink directly onto the substrate through a prefabricated template that establishes the print design. Inkjet printing utilizes commercial printers and allowing patterns to be rapidly altered with computer software from print to print, allowing for rapid prototyping. Compared to lithography, these printing techniques tend to be rapid and inexpensive but have lower spatial resolution.

An important consideration for all flexible electronics is the integration and interfacing of multiple components to make the final device. Device integration has been approached by interfacing components onto flexible substrates. This is accomplished using conventional electronic components, creating all flexible components, or a combination of both conventional and flexible components.

In this section, we present a general overview of the considerations for selection of materials, fabrication strategies, and integration approaches for the various types of nanocarbon-based flexible electronics and sensors.

6.2.1 Flexible Substrate Selection

Substrate choice is often dictated by the application and deployment requirements of the device. However, several universal properties such as low cost, high flexibility, chemical stability, and resistance to thermal deformation are advantageous. Polymeric materials such as PI and polyethylene terephthalate (PET) are known to have these respective properties and have long been used in printed electronics [2]. Recently, paper substrates have been used to fabricate low-cost, flexible printed circuit boards [3] and bioanalytical sensors [4]. Additional advantages of paper substrates are that they can be folded into three-dimensional structures and used to deliver fluid by capillary action [5]. Flexible bioanalytical sensors have also been deployed as tattoo style sensors that can be worn directly on the skin [6]. These sensors have the advantage that they have conformal contact with the measurement surface.

6.2.2 Fabrication Techniques

6.2.2.1 Lithography

Lithographic processing of flexible nanocarbon devices involves three main steps: (1) CVD of the nanocarbon material from a suitable catalyst, (2) transfer of the nanocarbon to the desired flexible substrate and removal of the catalyst when necessary, and (3) fabrication of the additional supporting features of the device. Some of the most commonly used substrates for flexible electronics (i.e., PI, polyethylene naphthalate (PEN), PET, etc.) are incompatible with the CVD process because of the elevated temperatures that are required. This limitation dictates the fabrication process for flexible electronics. Typically, nanocarbons are secondarily transferred to a flexible substrate after CVD. Any additional patterning or deposition can be performed on the flexible substrate before or after nanocarbon transfer.

6.2.2.2 Nanocarbon ink preparation

The vast majority of flexible nanocarbon sensors and electronic devices are produced using inks containing suspended nanocarbons. These nanocarbon inks generally contain CNTs, exfoliated graphene, or reduced graphene oxide. However, these materials are generally difficult to disperse in liquids due to their strong van der Waals forces, which cause agglomeration [7]. Ultrasonication has been a successful technique used to break up large aggregates into smaller aggregates and disperse nanocarbon materials [7, 8]. To break up the smaller aggregates and to improve overall dispersion stability, further efforts have been made to determine methods to overcome the high van der Waals forces. One of the earliest methods to minimize these forces in CNTs was the modification of the outer sidewall with carboxylic acids [9]. This chemical modification is commonly achieved using mixtures of strong oxidizing acids, such as sulfuric acid, nitric acid, and hydrogen peroxide, to create structural defects and increase oxygen content at those defects. However, this method results in CNTs with decreased conductivity [10]. Another method to minimize material aggregation is to employ solvents that minimize the surface tension between the solvent and nanomaterial, typically nonpolar organic solvents [11–14]. However, chemical incompatibilities with flexible substrate materials and deposition equipment such as printers often restrict the use of organic solvents. Therefore, development of water-based ink formulations is an area of much research. Due to the hydrophobicity of carbon nanomaterials and the hydrophilic nature of aqueous solvents, additions of dispersants, including surfactants and wrapping polymers, are needed to make aggregate-free suspensions. These dispersants act to physically separate nanomaterials by individually coating each nanoparticle. Commonly used surfactant dispersants are sodium dodecyl sulfate, poly(styrene sulfonate), sodium dodecylbenzene sulfonate, Triton X-100, and Pluronic polymers [9, 12, 15–17]. Additionally, pyrenes have been demonstrated to disperse both CNTs and graphene through π–π stacking interactions [18]. Further considerations for electronic applications include the role dispersants play on the connectivity of the carbon nanomaterial to itself and ultimately on the sheet resistance of the deposited ink. As a result, conducting polymers such as poly(3,4-ethylenedioxythiophene)-poly(styrenesulfonate), PEDOT:PSS, have been employed as dispersants [19]. Finally, ink viscosity, surface tension, adhesion, and transparency must be evaluated and tuned for each deposition technique, substrate choice, and end application [17, 20].

6.2.2.3 Nanocarbon printing

Several printing techniques have been used to fabricate flexible, nanocarbon-based sensors and electronic devices. Two of the most straightforward printing techniques are stencil printing and screen printing. In stencil printing, patterns are created by pressing thick paste inks through a stencil using a squeegee and onto a flexible substrate. Mechanically induced bending or buckling of the stencil is known to occur during printing and can reduce printing reliability and ultimately feature density. Screen printing does not have this problem because the ink is applied to a mesh screen, which is held at a fixed distance above the stencil; therefore, the squeegee is not in contact with the stencil [21]. Previously, screen printed graphene lines on a PI substrate showed high electrical conductivity (~1.86×10^4 S m^{-1}) and good mechanical properties [22]. Graphene electrodes fabricated by screen printing were demonstrated as electrochemical sensors for a variety of biologically relevant analytes [23]. Graphene oxide electrodes were screen printed onto textiles for influenza detection [24]. Lastly, traditional screen printed carbon electrodes were modified with CNTs to improve electrochemical performance based on the catalytic activity of CNTs [25, 26].

Inkjet printing is another promising fabrication technique for flexible electronics and sensors. Typically, inks are administered to flexible substrates by either thermal or piezoelectric drop-on-demand inkjet printers. In addition to the rheological properties of the ink, the print head features are essential to ensuring high-quality prints. Nozzle diameter, applied jet voltage, and distance of print head to the substrate are all tunable parameters that affect the droplet size and thus printed feature size and resolution [27, 28]. Temperature, relative humidity, and solvent choice all affect the stability of the ink and its likelihood of drying and clogging the nozzle [28]. Inkjet printing of carbon nanomaterials has been demonstrated using a variety of commercial printers: Cannon [29], HP [30], and Epson [31, 32], and research-grade printers, Picojet-1000W [33–36] and Dimatix [29, 37–40]. Research-grade printers such as Dimatix allow the user to view the formation and release of ink droplets from the print head and, therefore, afford greater print and ink diagnostics [27, 28].

6.2.2.4 Nanocarbon textiles

Flexible nanocarbon electronics and sensors have the advantage over rigid electronics and sensors in that they can readily integrated into textiles without compromising textile fluidity. This can be accomplished using the following methods: (1) nanocarbon coating of traditional threads, (2) spinning nanocarbon threads, or (3) direct printing nanocarbons onto textile. Nanocarbon inks have been used to coat the outside of a traditional thread to make them conductive. Graphene inks have been used to coat elastomer, polypropylene, and copper threads to create strain gauges, capacitive touch sensors, and capacitors [41–43]. Inks are typically applied by dipping threads into a nanocarbon ink or by CVD growth of the nanomaterial on the outside of the thread, if the thread material is compatible with the temperatures required for CVD. CNT and graphene nanocarbons have been spun directly into threads and woven to make electronic fabrics. Typically, CVD-grown nanocarbons are pulled directly off their growth substrate, much in the same way of drawing a thread [44]. Improvements in fiber strength are achieved by twisting the fibers during the pulling process [45]. Finally, direct printing onto textile fabric is a simple approach that does not require electronic fiber weaving for device fabrication and textile generation. As an example, heterostructures of graphene and hexagonal boron nitride were previously inkjet printed onto textile and demonstrated as a field-effect transistor (FET) that maintained performance after multiple washing cycles [46].

6.2.3 Device Integration

6.2.3.1 Flexible hybrid devices

Many of the advancements in the field of flexible electronics have been made by integrating printed electronics with traditional complementary metal–oxide-semiconductor (CMOS)-based components onto flexible substrates. This allows for a highly tunable fabrication of flexible devices that merge low-cost printing with the performance of CMOS. These devices are commonly referred to as flexible hybrid electronics (FHE) as they are composed of both soft, flexible components and hard, rigid components on flexible substrates. The overall flexible architecture of FHEs lends itself to applications ranging from structural health monitoring for construction and aviation industries to human health monitoring devices that can be worn on the skin.

In their 2018 review, Tong et al. defined hybrid electronics as being composed of the following modules: (1) sensory, (2) energy, (3) processing/computation/memory, and (4) communication [47]. In FHE, rigid components are replaced by flexible components when appropriate and placed on flexible substrates. Traditional modules such as sensors and antennas for communication are commonly printed on the surface of flexible substrates. Energy harvesting devices are sometimes printed or can be assembled onto the flexible substrate, such as external rigid batteries. Computation modules are often external and rigid integrated circuits taking advantage of CMOS performance. An example of a flexible hybrid electronic device is shown in Fig. 6.1. This example highlights the use of both printed, flexible components with rigid hardware and integrated circuitry [48].

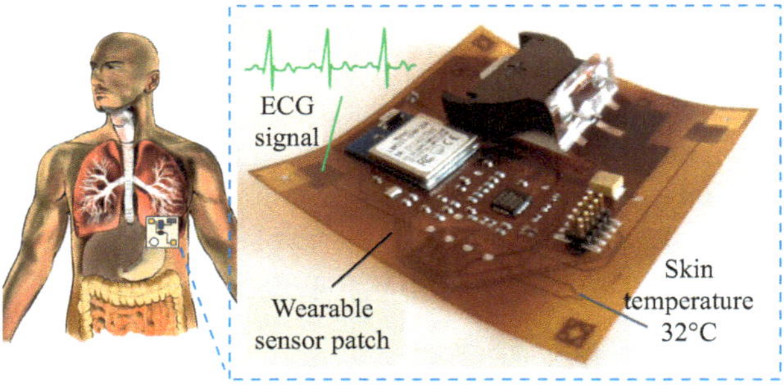

Figure 6.1 Proposed overview and system design of wearable sensor patch enabled by FHE. (a) Illustration of the wearable sensor patch being worn on the skin. (b) Photograph of the component side of the patch, including a battery holder, Bluetooth system-on-a-chip, analog front end, and test points. Reprinted from Ref. [48], Copyright 2016, with permission from John Wiley and Sons.

6.2.3.2 Fully flexible devices

To achieve integrated devices that are completely flexible, rigid components must be eliminated and replaced by flexible components. This means that the sensors must be wirelessly interfaced to external devices. Several groups have achieved this by printing the entire device onto flexible substrates [49–51]. These devices are commonly passive sensors composed of a sensing module and a communication module. Yang et al. fabricated a fully flexible, passive gas sensor using a printed radio frequency identification (RFID) antenna with a bridged single-walled carbon nanotube (SWCNT) composite, which acts as a gas sensor. When gases, such as NO_2, interact with the nanacarbons, the individual molecules of the gases adsorb to the material. This results in backscattering and a corresponding change in the current transferred to the RFID antenna [49]. In addition to chemical sensing, fully flexible, passive biosensing devices have also been demonstrated [52, 53]. Recently, a contact lens was patterned using graphene-silver nanowire hybrids and graphene to make an antenna, FET, and strain and glucose sensors [52]. Additionally, gold and graphene structures were patterned on silk to create a flexible sensor that is capable of detecting bacteria on tooth enamel [53].

To create more sophisticated sensors and to allow for active sensing, the addition of computational modules is needed. Typically, printed integrated circuits have performance disadvantages compared to traditional CMOS. Developments in ultrathin silicon have filled this gap. Recent advances in ultrathin silicon integrated circuits enable the use of flexible integrated circuits in flexible hybrid electronic devices [54]. This approach uses the undisputed benefits of CMOS integrated circuits but thinned down to be flexible. For silicon to be flexible, it must be 50 μm or thinner [10]. Wafer thinning is typically performed by wafer grinding, dicing-by-thinning, or by processing a silicon-on-insulator wafer [54–57]. An example of flexible silicon fabricated from a silicon-on-insulator starting wafer is shown in Fig. 6.2. Here, a thin layer of Si is deposited onto a polymer support. Devices are then patterned onto the exposed silicon. Once fabrication is complete, the polymer support is etched away, leaving only the patterned, thin silicon. This approach allows for typical fabrication protocols with minimal post-processing. Integration of these flexible silicon integrated circuits is generally performed by pick and place followed by embedding or laminating on a flexible support such as PI [58].

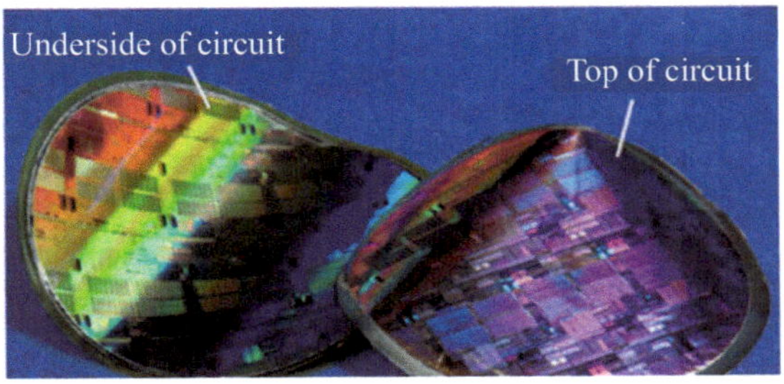

Figure 6.2 Flexible, ultra-thin, and high-performance CMOS circuit. Reprinted from Ref. [54], Copyright 2018, by permission from Springer Nature.

One example of such a device is a wireless contact lens sensor designed to measure intraocular pressure (shown in Fig. 6.3). Leonardi et al. created a microfabricated, embedded strain gauge with gold loop antenna and thin integrated circuit microprocessor for wireless power and communication. The microprocessor was thinned down to 50 μm thick and then attached to the contact lens substrate [59]. As developments continue in the area of flexible electronics manufacturing, fully flexible electronic devices will become more prevalent.

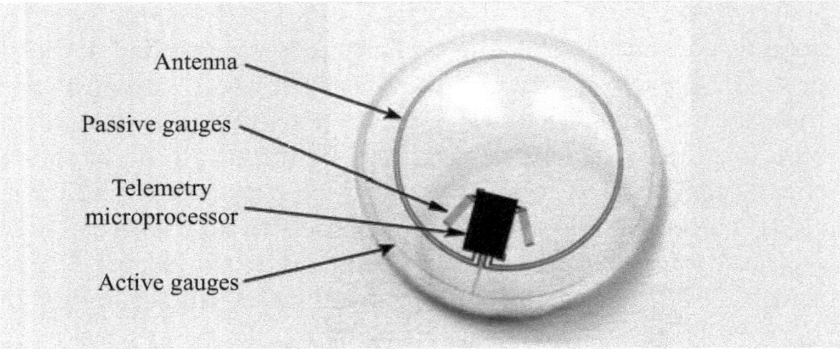

Figure 6.3 Diagram of an intraocular pressure contact lens sensor comprising a strain gauge, antenna, and thin integrated circuit microprocessor. Reprinted from Ref. [59], Copyright 2009, with permission from John Wiley and Sons.

6.3 Flexible Nanocarbon-Based Electronics

Due to the superior electrical, thermal, and mechanical characteristics of nanocarbons, low-dimensional carbon structures such as CNTs, carbon nanofibers, and graphene are promising candidates for flexible electronic devices. Recently, conductive films made from nanocarbons have been demonstrated in the development of flexible transistors. While each application has specific requirements regarding the nanocarbon films, they generally need to have good mechanical flexibility [60], low sheet resistance, and often must maintain optical transparency. This section will address several applications of nanocarbons in flexible electronics.

6.3.1 Flexible Transistors

With high intrinsic mobility, semiconducting SWCNTs have a promising application in electronics [61, 62]. Due to their high tensile strength and good flexibility, SWCNTs have begun to be used in flexible electronics [60, 63]. Conventionally, organic semiconductors such as pentacene have been used for printing electronic components such as interconnecting wires [64], but many have shown limited performance characteristics [65, 66]. More recently, SWCNTs have been used to form semiconducting layers in thin-film transistors (TFTs). These thin semiconducting films are generally deposited from a suspension solution of SWCNTs [66–70] or SWCNTs grown by CVD and then transferred to a flexible substrate [65, 71, 72]. Source, drain, and gate electrodes are commonly Au, Pd, or Ag, though a dense CNT network was also demonstrated as a gate electrode to maintain device optical transparency [67]. One of the main characteristics used to compare transistor performance is its carrier mobility. An FET based on a single SWCNT on silicon has achieved >100,000 $cm^2 V^{-1}s^{-1}$ at 50 K.

However, lower performance is observed for transistors fabricated on flexible substrates due to limitations in device resolution and materials integration. Much research is being conducted to improve carrier mobility on flexible transistors.

The charge transport in CNT TFTs experiences ballistic or diffusive transfer depending on the length of the CNT, the mean free path, and the type of contact at the source and drain. When the length of the CNT is less than the mean free path, the charge experiences ballistic transfer, whereas when the length of CNT is more than the mean free path, it experiences diffusive transfer at the contact. Thus, CNT TFTs can be divided into four main types: (1) ohmic-contact ballistic CNT TFT, (2) ohmic-contact diffusive CNT TFT, (3) Schottky-contact ballistic CNT TFT, or (4) Schottky-contact diffusive CNT TFT. Recently, Tang, et al. demonstrated a high charge carrier mobility of ~50 $cm^2 V^{-1}s^{-1}$ and an exceptionally high current density of $J_{on} = I_{ds}/W_{ch}$ = 17.2 μA μm^{-1} for flexible TFTs with low contact resistance. Their reported current density was in the same range of devices fabricated on rigid substrates, which are typically 1–100 μA μm^{-1} [73, 74]. The high current density was attributed to the use of an oxide-functionalized surface molecule, which was used to bind a high density of semiconducting SWCNTs (100 tubes per μm^2). Further, they demonstrated flexible CNT complementary logic gates such as CMOS inverters, NAND and NOR gates [70]. CNT-based TFTs are not only limited to single-layer devices, advances have been made to fabricate multilayer devices with up to two stacks. Zhao et al. fabricated a multilayer CNT TFT to perform logical operation and low-frequency oscillation (Fig. 6.4) [70].

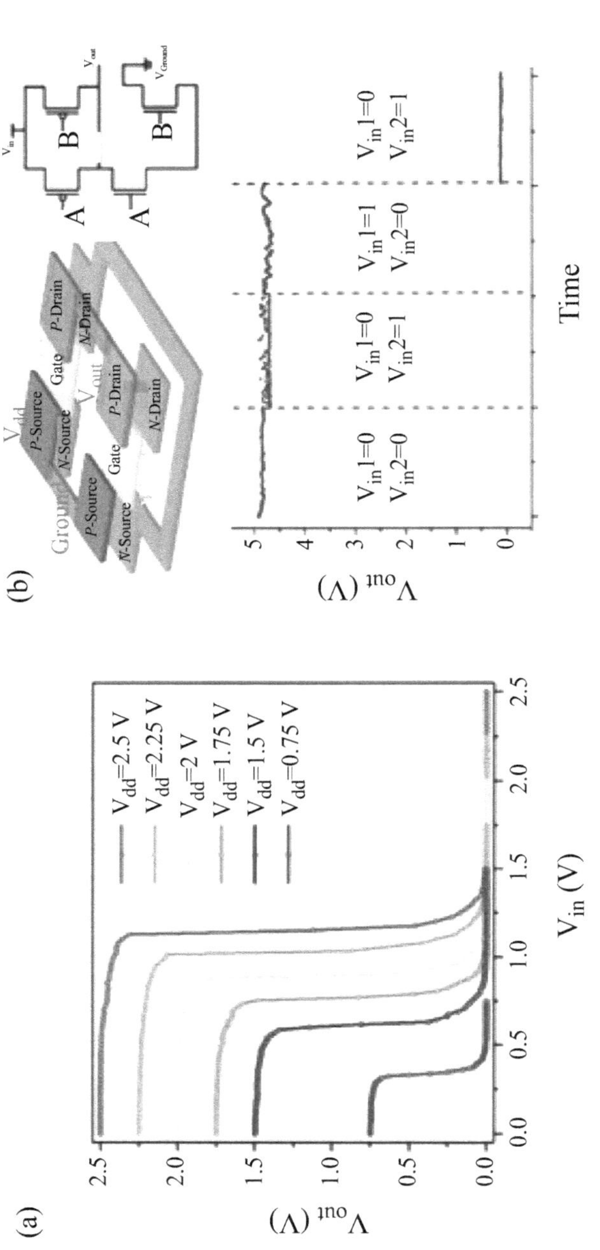
(a)
V_{dd}=2.5 V
V_{dd}=2.25 V
V_{dd}=2 V
V_{dd}=1.75 V
V_{dd}=1.5 V
V_{dd}=0.75 V
V_{out} (V)
V_{in} (V)
(b)
V_{dd}
Ground
P-Source
N-Source
Gate
P-Drain
N-Drain
V_{out}
A
B
V_{in}
V_{out}
V_{Ground}
$V_{in}1=0$ $V_{in}2=0$
$V_{in}1=0$ $V_{in}2=1$
$V_{in}1=1$ $V_{in}2=0$
$V_{in}1=0$ $V_{in}2=1$
V_{out} (V)
Time

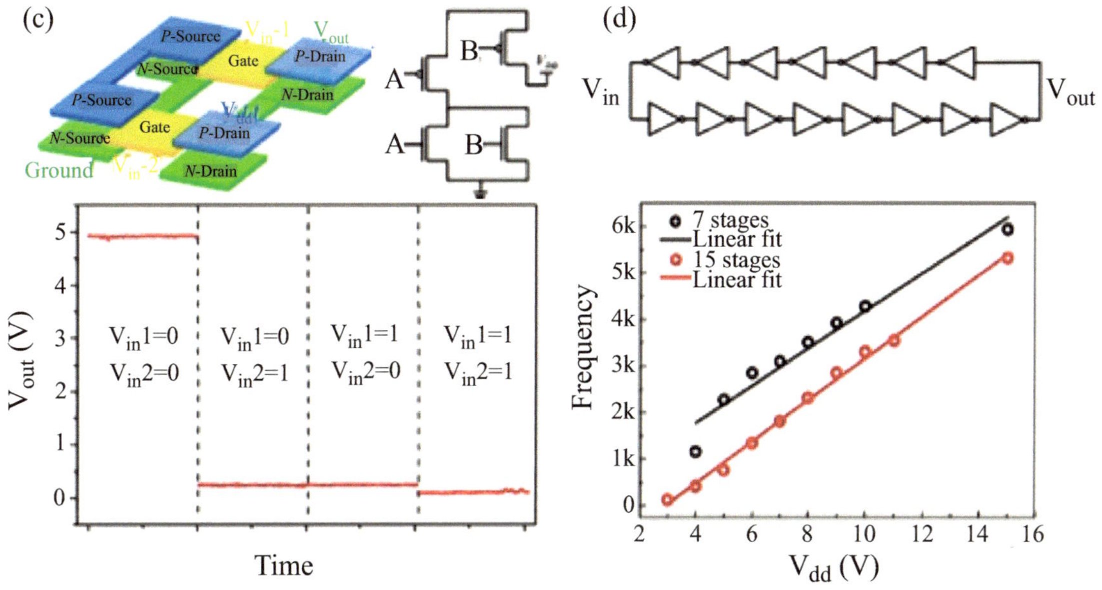

Figure 6.4 Electrical characteristics of flexible CNT TFTs. (a) Transfer curve from V_{dd} = 0.75 V to 2.5 V, showing a high I_{on}/I_{off} ratio above 10^6. (b) Schematic diagram and pictorial layout of CMOS NAND gate with output characteristic of a flexible CNT TFT NAND gate. The input signals are the voltage inputs for A and B, where logics '1' and '0' represent voltages V_{DD} = + 5 V and V_{SS} = 0 V, respectively. (c) Schematic diagram and pictorial layout of CMOS NOR gate with output characteristics of a flexible CNT TFT NOR gate. (c) Schematic diagram of a 15-stage ring oscillator with comparison of 7-stage (black) and 15-stage linear (red) oscillator. Reprinted with permission from Ref. [70], Copyright 2016, American Chemical Society.

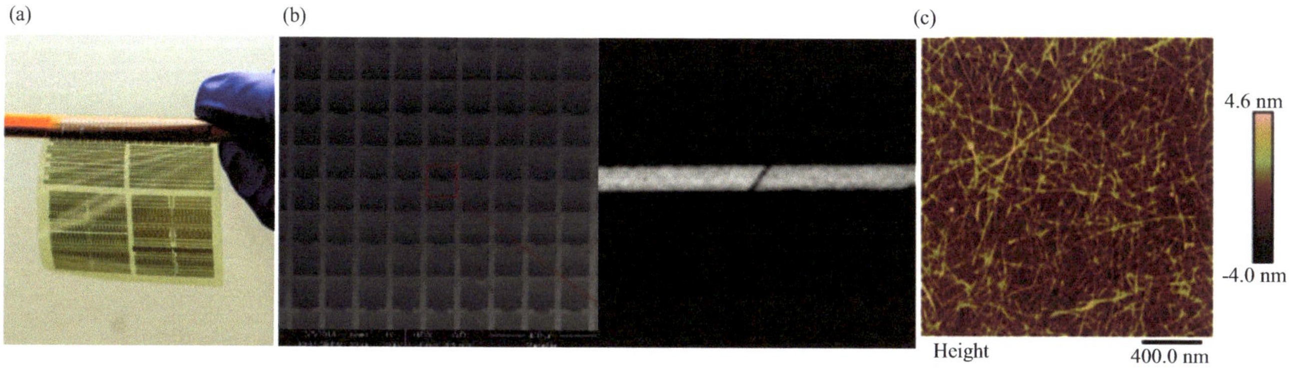

Figure 6.5 CNT TFT integrated circuits on 10 μm polyimide. (a) Photograph of flexible integrated circuits. (b) Scanning electron micrograph of short-channel device matrix used in the circuit. (c) Atomic force microscope image of the deposited CNT film [65].

Even higher mobility was achieved by Cao et al. with an impressive 80 $cm^2\ V^{-1}s^{-1}$ carrier mobility through an SWCNT TFT on a PI substrate [65]. Briefly, the flexible SWCNT integrated circuits use an architecture similar to those demonstrated on silicon. Gate SWCNTs are first grown on silicon by CVD and then subsequently transferred by spin coating polyamic acid, transferring to a polyurethane-coated PI substrate through lamination and then peeling the polymers off the silicon wafer. Au source, drain, and gate electrodes were fabricated on top. Most importantly, this approach was used to create up to ~100 SWCNT TFTs yielding medium-scale integrated circuits with performance uniformity of ~20% for normalized on-state current. This achievement illustrates a crucial step toward large-scale circuit integration. Nevertheless, the CNFETs are prone to process variations even though they are highly purified, which is a problem for large-scale integration. Lately, Ting Lei et al. fabricated mixed signal circuit (operating voltage 2 V) using ultra-pure semiconducting CNT to minimize the variations, as shown in Fig. 6.5 [65]. The deposited CNT film used had a mobility of 23.4 $cm^2\ V^{-1}s^{-1}$ and a density amid 35~40 CNTs/μm.

6.4 Nanocarbon-Based Flexible Sensors

Flexible nanocarbon sensors have been developed for many applications, including motion tracking, strain measurement, detection of small molecules and gases, pH and conductivity, among others. These innovations have been applied to various fields. The most notable applications for these sensors relate to physical health monitoring and point-of-use manufacturing. It should also be noted that there is a broad overlap between these two fields.

Flexible nanocarbon sensors for physical health monitoring offer a host of interesting applications, especially in the domain of wearable technology. Flexible nanocarbon sensors have been integrated into smart contacts, electronic skins, and flex sensors for facial movement. The use of flexible wearables has the potential to revolutionize healthcare by facilitating continuous health monitoring for improved treatment. Such health monitoring can increase personalized data collection allowing for more accurate diagnosis and decrease the number of office visits, thus improving the patient's overall quality of life.

The second application for these sensors is on-site point-of-use manufacturing. This aspect of flexible sensors is particularly interesting for medical and chemical testing in isolated regions where laboratory access is limited, such as remote villages, military field use, and even long-duration space travel expeditions. Under such conditions, the required assays and measurements may vary significantly over time. On-demand manufacturing has the potential for the creation of a multipurpose, mobile printing laboratory for production and use of sensors on an as-needed basis. This can potentially reduce the number of items (i.e., disposable sensors) needed for such expeditions, replacing them with flexible substrates and inks, which can be used in a variety of scenarios.

Because there is a vast overlap between these applications, we have categorized flexible electrochemical sensors based on their sensing mechanisms. These categories are as follows:

- Electroanalytical sensors, which measure a change in voltage or current resulting from a redox reaction.
- Ion-selective sensors, which selectively measure ion concentrations in a sample medium yielding a voltage signal.
- Chemiresistors, which yield a resistance change in the presence of specific chemical species, typically gases.

6.4.1 Electroanalytical Sensors

Electroanalytical sensors measure a change in current resulting from an electrochemical redox reaction at a working electrode surface. The reaction can be driven at the working electrode by applying a voltage between it and a second "counter" electrode, yielding a measurable, predictable current [75]. Often the sensor is operated in a three-electrode configuration where the applied voltage is referenced to a third "reference" electrode to relate the value to a thermodynamic, electrochemical potential. Flexible electrochemical sensors are typically fabricated to include all the necessary electrodes [75]. Glassy carbon and graphite are common electrode materials for electrochemical sensors because they have a wide electrochemical window, are biocompatible, and electrocatalytic toward many biologically relevant redox reactions [76]. Additionally, carbon materials are easy to chemically modify or functionalize. However, these carbon materials are difficult to pattern by lithography. Graphitic inks are commercially available for screen printing, but these inks typically are not suitable for inkjet printing as the large particle sizes often lead to clogging. Nanocarbons, on the other hand, can be patterned by common lithographic techniques and are suitable for inkjet printing, as discussed in Section 6.2, and are consequently one of the most suitable materials for carbon-based working electrodes for flexible sensors.

The use of carbon nanomaterials for electrochemical purposes is fundamentally different from other detection methods. For example, the performance of pristine carbon surfaces is very poor; oxygen defects on carbon materials improve electrode performance [76]. Consequently, the activation of carbon nanomaterials by oxygen plasma or electrochemical etching is often employed during the fabrication process [77, 78]. Alternatively, defect-rich materials, such as reduced graphene oxide, can be used to simplify fabrication [79].

Perhaps the most well-known electrochemical sensor is the blood glucose meter. Commercial blood glucose meters contain an enzyme called glucose oxidase (GOx), which oxidizes glucose to gluconolactone. A by-product of this reaction is H_2O_2, which can be detected by direct electrochemistry [80]. Because this strategy has proven robust, glucose detection is a commonly used proof-of-concept sensing strategy for novel electrode materials. Recent reports have shown in vitro measurements of glucose using this common sensing strategy for silk cloth functionalized with Pt-decorated MWCNTs [81] and graphene [82], SWCNTs functionalized with glucose oxidase patterned on PET [83], and electrochemically activated SWCNTs on PEN [77].

This type of sensor has also been adapted to wearable systems for continuous glucose monitoring. These wearable monitors are advantageous because they can be wirelessly interfaced with an external insulin pump, allowing incremental dosage in real time [84]. Currently, the sensor is integrated into a microdialysis needle with an internal Pt/Ir electrode inserted into the skin [84]. Adapting these wearables to fully flexible electrochemical sensors will be beneficial to patient comfort. Additionally, transitioning from precious metals to nanocarbon materials has the potential to decrease costs as the fabrication methods become more ubiquitous over time. Lee et al. developed a complex, multiplex sensing patch capable of monitoring blood glucose levels as well as monitoring for metrics relevant to diabetic treatment (Fig. 6.6) [79]. This patch was fabricated with graphene on PDMS and included a hybrid graphene/Prussian blue enzymatic glucose meter, along with sensors for pH, humidity, and a strain gauge for detecting tremors induced by hypoglycemia. This suite of sensors was capable of measurements in sweat taken up through a thin membrane and included a micro-needle drug-delivery system.

Despite the proven robustness of traditional glucose-sensing strategies, nanocarbon/inorganic hybrid electrode materials have recently been used to create nonenzymatic glucose assays. These work by direct electrochemical oxidation of glucose rather than the detection of H_2O_2 as the indirect, enzyme-mediated product of glucose oxidation. Many examples of these hybrid devices include a 3D-structured graphene foam with electrodeposited Mn_3O_4 [85], graphene-paper-supported Au with a PtCo alloy [86], printed graphene/CNT/ionic liquid gel with alloyed PtAu [87], inkjet-printed graphene functionalized with Au nanoparticles [88], and laser-scribed graphene with Cu nanoparticles [89]. These flexible, nonenzymatic sensors have shown similar limits of detection to commercial continuous glucose monitors [84, 90–95].

Wang et al. demonstrated a sensor made on AuNP-decorated free-standing graphene paper for selectively detecting *Escherichia coli* in vegetable samples [96]. Here the electrodeposited Au nanoclusters were functionalized with a streptavidin/biotinylated anti-*E. coli* composite allowing for binding of *E. coli* to the electrode surface.

Electrochemical impedance spectroscopy was used to measure the presence of *E. coli* on the electrode surface and clearly showed a linear increase in the charge-transfer resistance with increased target binding yielding a limit of detection of 1.5×10^4 and 1.5×10^3 colony-forming units (cfu) per mL in beef and cucumber samples, respectively. Additionally, the graphene paper structure showed good mechanical durability, with minimal signal loss after 500 bending cycles. This type of sensor shows promise for applications in the food health industry.

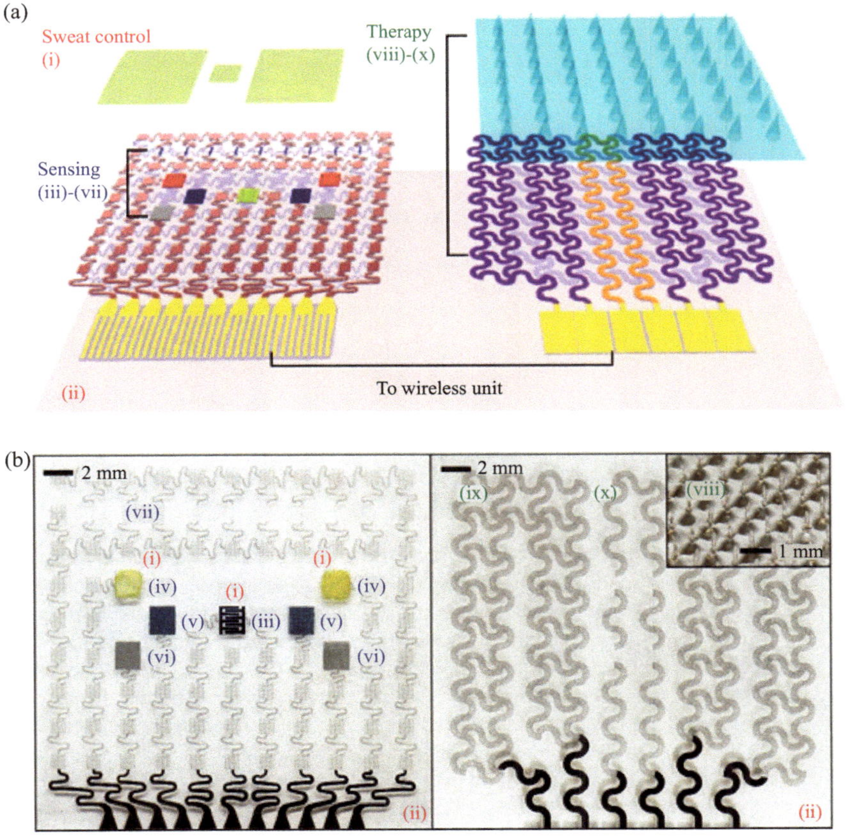

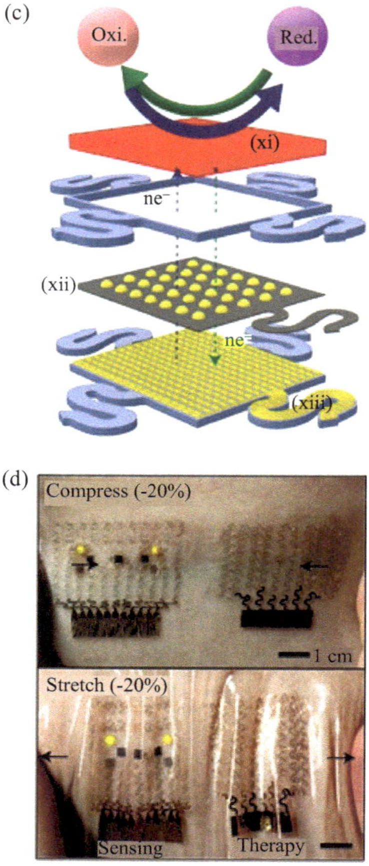

Figure 6.6 (a) Schematic drawings of the diabetes patch, which is composed of the sweat-control (i, ii), sensing (iii–vii), and therapy (viii–x) components. (b) Optical camera image of the electrochemical sensor array (left), therapeutic array (right), and magnified view of the drug-loaded microneedles (inset). (i) Sweat-uptake layer (Nafion); (ii) water-proof film (silicone); (iii) humidity sensor (PEDOT); (iv) glucose sensor (PB); (v) pH sensor (PANi); (vi) counter electrode (Ag/AgCl); (vii) tremor sensor (graphene); (viii) microneedles with drugs (PVP@ PCM); (ix) heater (Au mesh/graphene); and (x) temperature sensor (graphene). (c) Schematic of the GP-hybrid electrochemical unit, which consists of electrochemically active and soft functional materials (xi), gold-doped graphene (xii), and a serpentine Au mesh (xiii), from top to bottom. (d) Optical camera images of the diabetes patch laminated on human skin under mechanical deformations. Reprinted from Ref. [79], Copyright 2016, by permission from Springer Nature.

Another study for the selective detection of DNA was performed by Li and Lee [97]. In this study, they used an MWCNT/PDMS/glass composite. A single-stranded DNA (ss-DNA) probe was electrostatically adsorbed onto the exposed MWCNTs. To establish a baseline signal, methylene blue (MB) was allowed to associate with the ss-DNA, enhancing the electrochemical readout using differential pulse voltammetry. Addition of the ss-DNA target binds to the complimentary ss-DNA probe forming a double-stranded DNA (ds-DNA) sequence. The resulting ds-DNA does not have the same affinity toward the MWCNT surface and consequently leaves the surface along with the associated MB. This results in a decrease in the MB signal. Non-complimentary ss-DNA associates to the surface with additional MB, leading to an increase in signal. Using this mechanism, they were able to show a selective sensor with an LOD of 130 pM for a 24-base-pair probe and target ss-DNA. The simple preparation of this sensor has potential applications in point-of-care medical detection.

An interesting study from the Huang group demonstrated a CNT-based transparent electrode for electrochemical and optical monitoring of live cells [98]. The electrode was prepared using PEDOT-coated SWCNTs, which were transferred onto a PDMS stamp. Once prepared, electrodes were characterized by monitoring an $Fe(CN)_6^3$ redox reaction after bending and stretching and demonstrated high signal stability after 50–100 deformation cycles. To demonstrate the applicability of this sensor, human umbilical vein endothelial cells (HUVECs) were cultured on the electrode. The cells were stimulated using L-arginine to initiate an enzymatic reaction with nitric oxide synthase (NOS), yielding nitric oxide (NO), which can be detected electrochemically. This method was used to test the cultured cells for viability under both normal and applied mechanical stress (stretching the substrate 20% during the measurement) conditions. Figure 6.7a shows successful stimulation of the HUVECs using L-arginine (blue) as well as the control cases where there are no HUVECs present (black trace) and where NO is not produced due to the blocking of NOS by an inhibitor. Figure 6.7b shows the effect of mechanical strain on the cells (note the difference in signal scale). This study demonstrates the utility of a mechanically robust, biocompatible transparent electrode for fundamental cell studies.

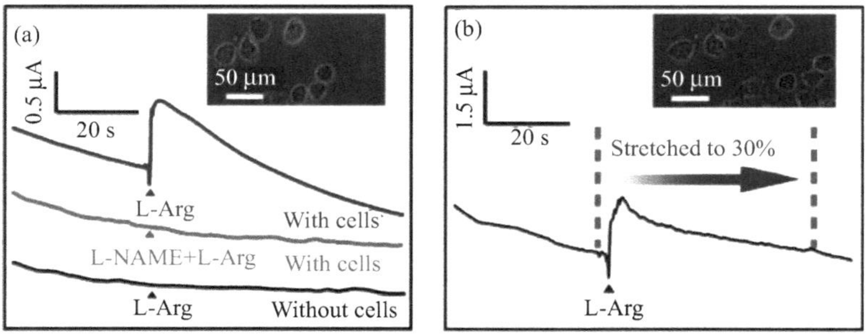

Figure 6.7 Monitoring NO release from HUVECs by SWCNTs/PEDOT/PDMS sensor (a) without stretching and (b) with stretching. The insets are the microphotographs showing the state of HUVECs in (a) and (b), respectively. Reprinted with permission from Ref. [98], Copyright 2017, American Chemical Society.

6.4.2 Ion-Selective Sensors

Ion-selective electrodes (ISEs) are often used to selectively measure concentrations of specific ions in a solution. Generally, ISEs consist of an electrode isolated by a membrane or polymer film with a high complexation affinity toward a specific ionic species (i.e., H^+, Cl^-, NO_3^-, etc.) [75, 99]. When the analyte ions are present, they associate with the ion-selective membrane, shifting the thermodynamic potential of the surface and causing a measurable change in potential relative to a stable reference electrode. The final readout is measured as a voltage, which can then be calibrated to yield a quantifiable value. The most common example of an ISE is the pH meter, which uses a porous glass membrane to detect H^+ along with an internal Ag/AgCl reference electrode [75]. Because the functionality of ISEs primarily relies on the chemistry between the target ion and the ion-selective membrane [99, 100], ISEs are easy to adapt to most electrode materials and configurations. Consequently, several flexible nanocarbon-based ISEs have been developed for targets, including NO_3^- and NH_4^+ [101], K^+ [102, 103], Ca^{2+} [103, 104], Cd^{2+} [105], and H^+ [103], on substrates ranging from laser-induced graphene, to printed CNT and graphene electrodes, and free-standing graphene paper.

Notably, Parrilla et al. expanded on this by demonstrating a wearable, screen printed CNT ISE for the detection of Na^+ and K^+ on polyurethane as pictured in Fig. 6.8 [106]. Figure 6.8a shows a schematic representation of the electrode layout and device design. The device consists of leads made from Ag/AgCl ink. The pads of two of the electrodes are printed with MWCNT ink followed by either a K^+ or Na^+ ion-selective membrane to serve as ISEs. The third electrode pad is unmodified Ag/AgCl, which serves as a reference electrode. Finally, the leads were coated in a layer of Ecoflex polymer leaving the ISEs, reference electrodes, and their respective contact pads exposed. Figure 6.8b shows an example of a fabricated device. This report demonstrated the suitability of polyurethane as a flexible substrate material by evaluating the performance of the ISEs after several cycles of bending, stretching, and immersion in water. The authors proposed that these sensors may later be incorporated into athletic textiles to monitor hydration needs during exertion.

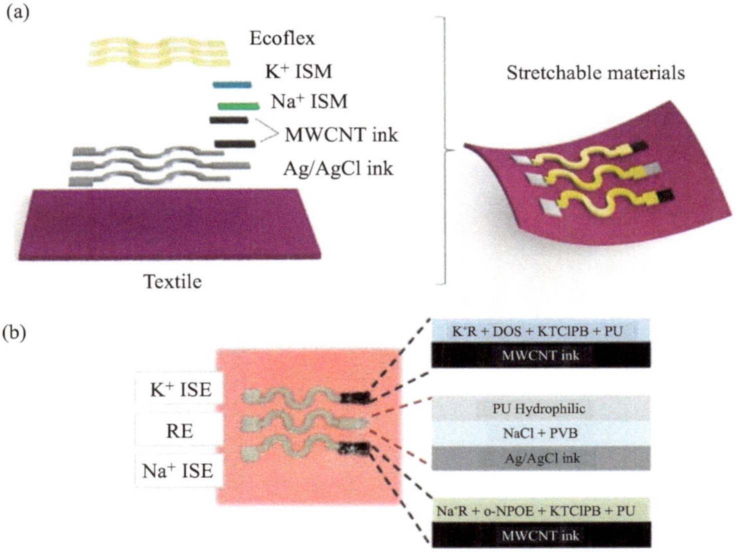

Figure 6.8 Wearable, screen printed ISE. (a) Schematic representation of the tailor-made stretchable materials and manufacturing process. (b) Image depicting the wearable sensor based on textile and ion-selective membranes (ISM) composition. Reprinted from Ref. [106], Copyright 2016, with permission from John Wiley and Sons.

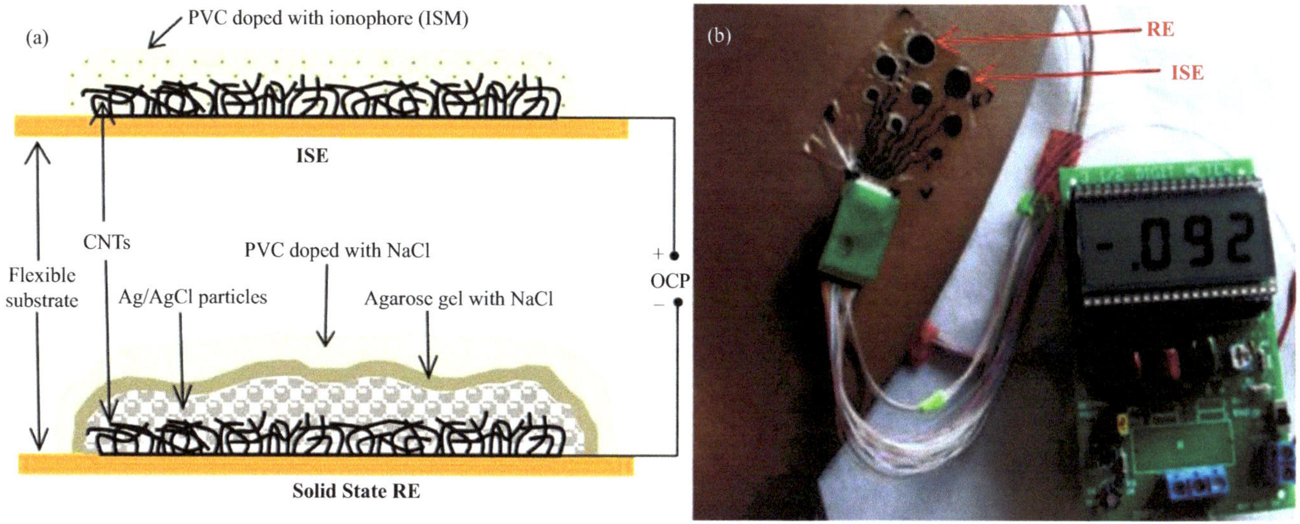

Figure 6.9 Skin-wearable ISE. (a) Schematic representation of the flexible CNT-based ISE and RE. (b) A standalone wearable sensor for in vivo measurements. The portable device includes a high-input impedance voltmeter connected to CNT-based solid-state ISEs and REs, fabricated on a temporary tattoo substrate. Reprinted with permission from Ref. [107], Copyright 2017, American Chemical Society.

Later, Roy et al. developed an ISE for the detection of Na^+ on a temporary tattoo substrate utilizing a CNT electrode coated with an ionophore-doped PVC film, as illustrated in Fig. 6.9a [107]. This device yielded an LOD of 3.16 μM. As shown in Fig. 6.9b, this device has multiple electrodes on the sensor. Each of these electrodes can potentially be modified with different ionophores for simultaneous detection of multiple target ions.

Because of the wide range of ion-selective membrane materials and ionophores, ISEs have a broad potential. However, this potential is limited to ionic species, which can be useful for heavy metal detection using different ionophores. This factor is also quite limiting. While these sensors are useful for biomedical applications, they will be most impactful when integrated into complex sensing suites for simultaneous, multiplexed detection.

6.4.3 Chemiresistors

With their inherently high surface area, nanocarbons have been demonstrated as highly sensitive to their environment and have thus been evaluated as a sensing material for gases and vapors. Gases adsorb to the surface of the nanocarbons, which manifest in a change in electron transport through the material [108]. This has often been measured as a change in resistance, and such a device is called a chemiresistor. Flexible CNT-based chemiresistors have been applied to gas sensing for human health safety and environmental monitoring. Previously, pristine SWCNT chemiresistive sensors were fabricated on PI substrates for the detection of NH_3 [109]. To improve adhesion to cellulose substrates, COOH-functionalized SWCNTs were drop cast onto filter paper to create a humidity sensor [110]. Surface modification has also been employed to improve sensitivity and selectivity toward specific gases. COOH-modified SWCNTs were shown to improve sensitivity, down to 1 ppm, toward CO detection on a flexible PET substrate [111]. Further surface modification by addition of materials has also been performed. Zhang et al. electrodeposited polyaniline onto the surface of SWCNTs and demonstrated improved NH_3 sensitivity by 60% when compared to pristine SWCNTs [112]. Flexible CNT-based chemiresistors have also been modified with nanoparticles to improve selectivity. Recently, a flexible hydrogen chemiresistor was fabricated by electrodepositing

Pd nanoparticles onto SWCNT surfaces and is shown in Fig. 6.10 [113]. In addition to achieving a 500 ppm limit of detection for H_2, the authors demonstrated sensor response reproducibility to over 2000 bending cycles.

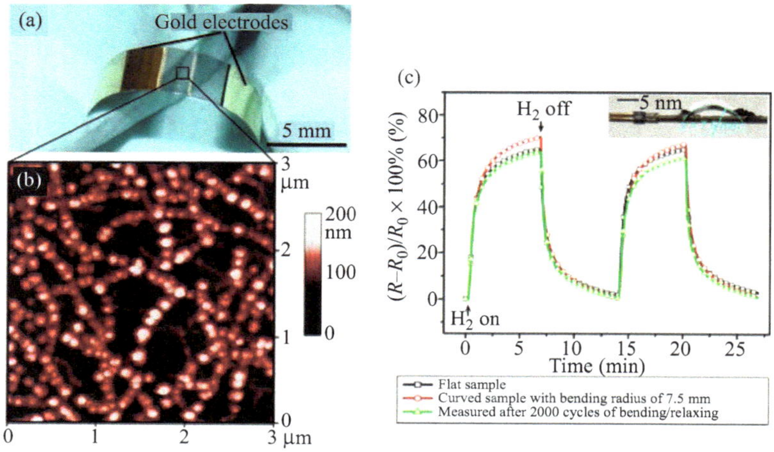

Figure 6.10 Pd-coated SWCNT chemiresistor for H_2 detection. (a) Photograph of flexible chemiresistor comprising Pd-nanoparticle-modified SWCNTs [50]. (b) Atomic force microscope height topograph of Pd-nanoparticle-modified SWCNTs. (c) Sensor response to H_2 upon being bent and after 2000 bending cycles. Reprinted from Ref. [113], Copyright 2007, with the permission of AIP Publishing.

Flexible sensors have the advantage over rigid sensors in that they are suitable to integrate into clothing or become wearable devices. Since clothing will bend as the wearer moves, it is critical that the sensor's performance does not change with that movement. Han et al. coated a yarn thread with SWCNT ink and then wove the functional thread into a cotton textile and observed an 8 ppm NH_3 detection limit [114]. This woven chemiresistor maintained a reproducible output, while the textile was bent 90° and then brought back to 0°. Yun et al. demonstrated an electronic textile gas sensor based on reduced graphene-oxide-coated yarn woven into a textile. Sensor durability was evaluated through several detergent washes and used to detect NO_2 with selectivity against acetone, ethanol, ethylene, and CO_2 [115].

In addition to sensing gas adsorption, SWCNTs are sensitive to UV radiation, as UV radiation induces the dissociation of O_2 from the SWCNT surface, as shown in Fig. 6.11 [116]. Cotton threads were coated with SWCNT ink, bundled into a wristband, and then rinsed to remove any poorly adhered SWCNTs. Sensor responses to 254 nm, 365 nm, visible, and infrared wavelengths were characterized, and it was concluded that the SWCNT chemiresistor wristband could distinguish UV light from visible and infrared. These wearable chemiresistors thus show great promise for personalized monitoring.

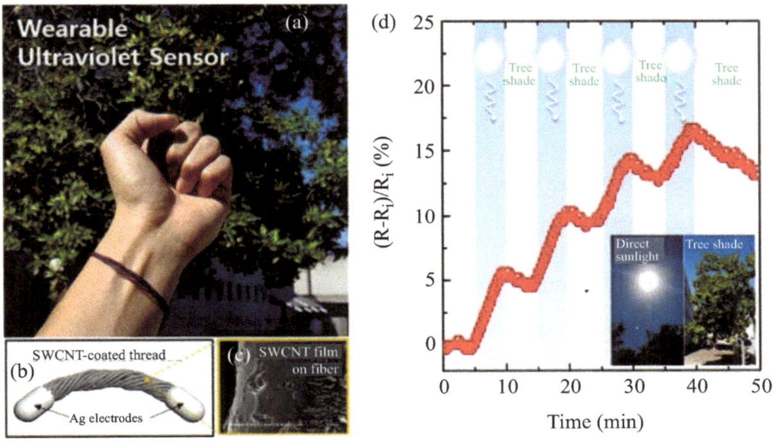

Figure 6.11 Wearable UV wristband sensor. (a) Photograph of wristband sensor. (b) Schematic of SWCNT-coated thread. (c) Scanning electron microscope image of SWCNT-coated thread. (d) Sensor response to direct sunlight. Reprinted with permission from Ref. [116], Copyright 2018, American Chemical Society.

6.5 Conclusion

Nanocarbons have been demonstrated as suitable materials for integration into a wide range of flexible electronic and sensor components and devices. Printed devices using nanomaterial inks have the ability to drastically reduce fabrication time and cost and allow for rapid prototyping and high device tunability. Advances in printed integrated circuits and ultra-thin flexible silicon will continue to aid in the development of fully flexible electronic devices.

Finally, advantages of FHE in rapidly prototyping and point-of-use allow for wearable devices for customized electronics and sensing. In conclusion, nanocarbon-based flexible electronics and sensors show great promise to truly transform the electronics industry.

References

1. R. Das, X. He, and K. Ghaffarzadeh, "Flexible, printed and organic electronics 2019–2029: Forecasts, players and opportunities." [Online]. Available: www.idtechex.com/research/reports/flexible-printed-and-organic-electronics-2019-2029-forecasts-players-and-opportunities-000639.asp. [Accessed: 17-Apr-2019].
2. V. Zardetto, T. M. Brown, A. Reale, and A. Di Carlo, "Substrates for flexible electronics: A practical investigation on the electrical, film flexibility, optical, temperature, and solvent resistance properties," *J. Polym. Sci. Part B Polym. Phys.*, **49**, 9, pp. 638–648, 2011.
3. A. C. Siegel, S. T. Phillips, M. D. Dickey, N. Lu, Z. Suo, and G. M. Whitesides, "Foldable printed circuit boards on paper substrates," *Adv. Funct. Mater.*, **20**, 1, pp. 28–35, 2010.
4. A. W. Martinez, S. T. Phillips, M. J. Butte, and G. M. Whitesides, "Patterned paper as a platform for inexpensive, low-volume, portable bioassays," *Angew. Chemie Int. Ed.*, **46**, 8, pp. 1318–1320, 2007.
5. H. Liu and R. M. Crooks, "Three-dimensional paper microfluidic devices assembled using the principles of origami," *J. Am. Chem. Soc.*, **133**, 44, pp. 17564–17566, 2011.
6. D.-H. Kim, N. Lu, R. Ma, Y.-S. Kim, R.-H. Kim, S. Wang, J. Wu, S. M. Won, H. Tao, A. Islam, K. J. Yu, T.-I Kim, R. Chowdhury, M. Ying, L. Xu, M. Li, H.-J. Chung, H. Keum, M. McCormick, P. Liu, Y.-W. Zhang, F. G. Omenetto, Y. Huang, T. Coleman, and J. A. Rogers, "Epidermal electronics," *Science*, **333**, 6044, pp. 838–843, 2011.
7. Y. Y. Huang and E. M. Terentjev, "Dispersion of carbon nanotubes: Mixing, sonication, stabilization, and composite properties," *Polymers*, **4**, 1, 2012.
8. L. Jiang, L. Gao, and J. Sun, "Production of aqueous colloidal dispersions of carbon nanotubes," *J. Colloid Interface Sci.*, **260**, 1, pp. 89–94, 2003.
9. J. Liu, A. G. Rinzler, H. Dai, J. H. Hafner, R. K. Bradley, P. J. Boul, A. Lu, T. Iverson, K. Shelimov, C. B. Huffman, F. Rodriguez-Macias, Y. S. Shon, T. R. Lee, D. T. Colbert, and R. E. Smalley, "Fullerene pipes," *Science*, **280**, 5367, pp. 1253–1256, 1998.

10. V. Datsyuk, M. Kalyva, K. Papagelis, J. Parthenios, D. Tasis, A. Siokou, I. Kallitsis, and C. Galiotis, "Chemical oxidation of multiwalled carbon nanotubes," *Carbon N. Y.*, **46**, 6, pp. 833–840, 2008.

11. K. D. Ausman, R. Piner, O. Lourie, R. S. Ruoff, and M. Korobov, "Organic solvent dispersions of single-walled carbon nanotubes: Toward solutions of pristine nanotubes," *J. Phys. Chem. B*, **104**, 38, pp. 8911–8915, 2000.

12. A. Ciesielski and P. Samorì, "Graphene via sonication assisted liquid-phase exfoliation," *Chem. Soc. Rev.*, **43**, 1, pp. 381–398, 2014.

13. Y. Hernandez, V. Nicolosi, M. Lotya, F. M. Blighe, Z. Sun, S. De, I. T. McGovern, B. Holland, M. Byrne, Y. K. Gun'Ko, J. J. Boland, P. Niraj, G. Duesberg, S. Krishnamurthy, R. Goodhue, J. Hutchison, V. Scardaci, A. C. Ferrari, and J. N. Coleman, "High-yield production of graphene by liquid-phase exfoliation of graphite," *Nat. Nanotechnol.*, **3**, p. 563, 2008.

14. J. I. Paredes, S. Villar-Rodil, A. Martínez-Alonso, and J. M. D. Tascón, "Graphene oxide dispersions in organic solvents," *Langmuir*, **24**, 19, pp. 10560–10564, 2008.

15. X. Gong, J. Liu, S. Baskaran, R. D. Voise, and J. S. Young, "Surfactant-assisted processing of carbon nanotube/polymer composites," *Chem. Mater.*, **12**, 4, pp. 1049–1052, 2000.

16. V. C. Moore, M. S. Strano, E. H. Haroz, R. H. Hauge, R. E. Smalley, J. Schmidt, and Y. Talmon, "Individually suspended single-walled carbon nanotubes in various surfactants," *Nano Lett.*, **3**, 10, pp. 1379–1382, 2003.

17. R. Tortorich and J.-W. Choi, "Inkjet printing of carbon nanotubes," *Nanomaterials*, **3**, 3, pp. 453–468, 2013.

18. T. Fujigaya and N. Nakashima, "Methodology for homogeneous dispersion of single-walled carbon nanotubes by physical modification," *Polym. J.*, **40**, pp. 577–589, 2008.

19. L. Maillaud, P. Poulin, M. Pasquali, and C. Zakri, "Effect of the rheological properties of carbon nanotube dispersions on the processing and properties of transparent conductive electrodes," *Langmuir*, **31**, 21, pp. 5928–5934, 2015.

20. W. Yang and C. Wang, "Graphene and the related conductive inks for flexible electronics," *J. Mater. Chem. C*, **4**, 30, pp. 7193–7207, 2016.

21. A. Larmagnac, S. Eggenberger, H. Janossy, and J. Vörös, "Stretchable electronics based on Ag-PDMS composites," *Sci. Rep.*, **4**, p. 7254, 2014.

22. W. J. Hyun, E. B. Secor, M. C. Hersam, C. D. Frisbie, and L. F. Francis, "High-resolution patterning of graphene by screen printing with a

silicon stencil for highly flexible printed electronics," *Adv. Mater.*, **27**, 1, pp. 109–115, 2015.

23. E. P. Randviir, D. A. C. Brownson, J. P. Metters, R. O. Kadara, and C. E. Banks, "The fabrication, characterisation and electrochemical investigation of screen-printed graphene electrodes," *Phys. Chem. Chem. Phys.*, **16**, 10, pp. 4598–4611, 2014.
24. D. S. Kinnamon, S. Krishnan, S. Brosler, E. Sun, and S. Prasad, "Screen printed graphene oxide textile biosensor for applications in inexpensive and wearable point-of-exposure detection of influenza for at-risk populations," *J. Electrochem. Soc.*, **165**, 8, pp. B3084–B3090, 2018.
25. Y. Lin, F. Lu, and J. Wang, "Disposable carbon nanotube modified screen-printed biosensor for amperometric detection of organophosphorus pesticides and nerve agents," *Electroanalysis*, **16**, 1–2, pp. 145–149, 2004.
26. P. Fanjul-Bolado, P. Queipo, P. J. Lamas-Ardisana, and A. Costa-García, "Manufacture and evaluation of carbon nanotube modified screen-printed electrodes as electrochemical tools," *Talanta*, **74**, 3, pp. 427–433, 2007.
27. J. Li, F. Rossignol, and J. Macdonald, "Inkjet printing for biosensor fabrication: Combining chemistry and technology for advanced manufacturing," *Lab Chip*, **15**, 12, pp. 2538–2558, 2015.
28. G. D. Martin, S. D. Hoath, and I. M. Hutchings, "Inkjet printing: The physics of manipulating liquid jets and drops," *J. Phys. Conf. Ser.*, **105**, p. 12001, 2008.
29. K. Kordás, T. Mustonen, G. Tóth, H. Jantunen, M. Lajunen, C. Soldano, S. Talapatra, S. Kar, R. Vajtai, and P. M. Ajayan, "Inkjet printing of electrically conductive patterns of carbon nanotubes," *Small*, **2**, 8–9, pp. 1021–1025, 2006.
30. M. in het Panhuis, A. Heurtematte, W. R. Small, and V. N. Paunov, "Inkjet printed water sensitive transparent films from natural gum–carbon nanotube composites," *Soft Matter*, **3**, 7, pp. 840–843, 2007.
31. P. Chen, H. Chen, J. Qiu, and C. Zhou, "Inkjet printing of single-walled carbon nanotube/RuO_2 nanowire supercapacitors on cloth fabrics and flexible substrates," *Nano Res.*, **3**, 8, pp. 594–603, 2010.
32. O.-S. Kwon, H. Kim, H. Ko, J. Lee, B. Lee, C.-H. Jung, J.-H. Choi, and K. Shin, "Fabrication and characterization of inkjet-printed carbon nanotube electrode patterns on paper," *Carbon N. Y.*, **58**, pp. 116–127, 2013.

33. H. Okimoto, T. Takenobu, K. Yanagi, Y. Miyata, H. Shimotani, H. Kataura, and Y. Iwasa, "Tunable carbon nanotube thin-film transistors produced exclusively via inkjet printing," *Adv. Mater.*, **22**, 36, pp. 3981–3986, 2010.

34. J.-W. Song, J. Kim, Y.-H. Yoon, B.-S. Choi, J.-H. Kim, and C.-S. Han, "Inkjet printing of single-walled carbon nanotubes and electrical characterization of the line pattern," *Nanotechnology*, **19**, 9, p. 95702, 2008.

35. S. Matsuzaki, Y. Nobusa, R. Shimizu, K. Yanagi, H. Kataura, and T. Takenobu, "Continuous electron doping of single-walled carbon nanotube films using inkjet technique," *Jpn. J. Appl. Phys.*, **51**, p. 06FD18, 2012.

36. J. H. Yun, H. Chang-Soo, J. Kim, J. W. Song, D. H. Shin, and Y. G. Park, "No Title," in *Proc. IEEE Int. Conf. Nano/Micro Eng. Mol. Sys.*, 2008, pp. 506–509.

37. J.-W. Song, Y.-S. Kim, Y.-H. Yoon, E.-S. Lee, C.-S. Han, Y. Cho, D. Kim, J. Kim, N. Lee, Y.-G. Ko, H.-T. Jung, and S.-H. Kim, "The production of transparent carbon nanotube field emitters using inkjet printing," *Phys. E Low-dimensional Syst. Nanostructures*, **41**, 8, pp. 1513–1516, 2009.

38. E. Gracia-Espino, G. Sala, F. Pino, N. Halonen, J. Luomahaara, J. Mäklin, G. Tóth, K. Kordás, H. Jantunen, M. Terrones, P. Helistö, H. Seppä, P. M. Ajayan, and R. Vajtai, "Electrical transport and field-effect transistors using inkjet-printed SWCNT films having different functional side groups," *ACS Nano*, **4**, 6, pp. 3318–3324, 2010.

39. J. Li, F. Ye, S. Vaziri, M. Muhammed, M. C. Lemme, and M. Östling, "Efficient inkjet printing of graphene," *Adv. Mater.*, **25**, 29, pp. 3985–3992, 2013.

40. E. B. Secor, P. L. Prabhumirashi, K. Puntambekar, M. L. Geier, and M. C. Hersam, "Inkjet printing of high conductivity, flexible graphene patterns," *J. Phys. Chem. Lett.*, **4**, 8, pp. 1347–1351, 2013.

41. X. Liu, D. Liu, J.-H. Lee, Q. Zheng, X. Du, X. Zhang, H. Xu, Z. Wang, Y. Wu, X. Shen, J. Cui, Y.-W. Mai, and J.-K. Kim, "Spider-web-inspired stretchable graphene woven fabric for highly sensitive, transparent, wearable strain sensors," *ACS Appl. Mater. Interfaces*, **11**, 2, pp. 2282–2294, 2019.

42. E. T. Alonso, D. P. Rodrigues, M. Khetani, D.-W. Shin, A. D. Sanctis, H. Joulie, I. de Schrijver, A. Baldycheva, H. Alves, A. I. S. Neves, S. Russo, and M. F. Craciun, "Graphene electronic fibres with touch-sensing and

light-emitting functionalities for smart textiles," *npj Flex. Electron.*, **2**, 1, p. 25, 2018.

43. T. Purkait, G. Singh, D. Kumar, M. Singh, and R. S. Dey, "High-performance flexible supercapacitors based on electrochemically tailored three-dimensional reduced graphene oxide networks," *Sci. Rep.*, **8**, 1, p. 640, 2018.

44. K. Jiang, Q. Li, and S. Fan, "Spinning continuous carbon nanotube yarns," *Nature*, **419**, 6909, p. 801, 2002.

45. M. Zhang, K. R. Atkinson, and R. H. Baughman, "Multifunctional carbon nanotube yarns by downsizing an ancient technology," *Science*, **306**, 5700, pp. 1358–1361, 2004.

46. T. Carey, S. Cacovich, G. Divitini, J. Ren, A. Mansouri, J. M. Kim, C. Wang, C. Ducati, R. Sordan, and F. Torrisi, "Fully inkjet-printed two-dimensional material field-effect heterojunctions for wearable and textile electronics," *Nat. Commun.*, **8**, 1, p. 1202, 2017.

47. G. Tong, Z. Jia, and J. Chang, "Flexible hybrid electronics: Review and challenges," in *2018 IEEE International Symposium on Circuits and Systems (ISCAS)*, 2018, pp. 1–5.

48. Y. Khan, M. Garg, Q. Gui, M. Schadt, A. Gaikwad, D. Han, N. A. D. Yamamoto, P. Hart, R. Welte, W. Wilson, S. Czarnecki, M, Poliks, Z. Jin, K. Ghose, F. Egitto, J. Turner, and A. C. Arias, "Flexible hybrid electronics: Direct interfacing of soft and hard electronics for wearable health monitoring," **26**, 47, pp. 8764–8775, 2016.

49. L. Yang, R. Zhang, D. Staiculescu, C. P. Wong, and M. M. Tentzeris, "A novel conformal RFID-enabled module utilizing inkjet-printed antennas and carbon nanotubes for gas-detection applications," *IEEE Antennas Wirel. Propag. Lett.*, **8**, pp. 653–656, 2009.

50. R. Vyas, V. Lakafosis, H. Lee, G. Shaker, L. Yang, G. Orecchini, A. Traille, M. M. Tentzeris, and L. Roselli, "Inkjet printed, self powered, wireless sensors for environmental, gas, and authentication-based sensing," *IEEE Sens. J.*, **11**, 12, pp. 3139–3152, 2011.

51. A. Vena, L. Sydänheimo, M. M. Tentzeris, and L. Ukkonen, "A fully inkjet-printed wireless and chipless sensor for CO_2 and temperature detection," *IEEE Sens. J.*, **15**, 1, pp. 89–99, 2015.

52. J. Kim, M. Kim, M.-S. Lee, K. Kim, S. Ji, Y.-T. Kim, J. Park, K. Na, K.-H. Bae, H. K. Kim, F. Bien, C. Y. Lee, and J.-U. Park, "Wearable smart sensor systems integrated on soft contact lenses for wireless ocular diagnostics," *Nat. Commun.*, **8**, p. 14997, 2017.

53. M. S. Mannoor, H. Tao, J. D. Clayton, A. Sengupta, D. L. Kaplan, R. R. Naik, N. Verma, F. G. Omenetto, and M. C. McAlpine, "Graphene-based wireless bacteria detection on tooth enamel," *Nat. Commun.*, **3**, p. 763, 2012.

54. S. Gupta, W. T. Navaraj, L. Lorenzelli, and R. Dahiya, "Ultra-thin chips for high-performance flexible electronics," *npj Flex. Electron.*, **2**, 8, 2018.

55. F. Schmitt and M. Zernak, "Thin wafer manufacturing and handling using low cost carriers," in *Ultra-thin Chip Technology and Applications*, Springer, New York, NY, 2011, pp. 21–32.

56. C. Landesberger, S. Scherbaum, and K. Bock, "Ultra-thin wafer fabrication through dicing-by-thinning," in *Ultra-thin Chip Technology and Applications*, Springer, New York, NY, 2011, pp. 33–44.

57. M. Zimmermann, J. N. Burghartz, W. Appel, and C. Harendt, "Fabrication of ultra-thin chips using silicon wafer with buried cavities," in *Ultra-thin Chip Technology and Applications*, Springer, New York, NY, 2011, pp. 69–80.

58. A. Kugler, M. Koyuncu, and A. Zimmerman, "Chip embedding in laminates," in *Ultra-thin Chip Technology and Applications*," Springer, New York, NY, 2011, pp. 159–166.

59. M. Leonardi, E. M. Pitchon, A. Bertsch, P. Renaud, and A. Mermoud, "Wireless contact lens sensor for intraocular pressure monitoring: Assessment on enucleated pig eyes," *Acta Ophthalmol.*, **87**, 4, pp. 433–437, 2009.

60. M. R. Falvo, G. J. Clary, R. M. Taylor 2nd, V. Chi, F. P. Brooks Jr, S. Washburn, and R. Superfine, "Bending and buckling of carbon nanotubes under large strain," *Nature*, **389**, 6651, pp. 582–584, 1997.

61. X. Zhou, J.-Y. Park, S. Huang, J. Liu, and P. L. McEuen, "Band structure, phonon scattering, and the performance limit of single-walled carbon nanotube transistors," *Phys. Rev. Lett.*, **95**, 14, p. 146805, 2005.

62. S. J. Kang, C. Kocabas, T. Ozel, M. Shim, N. Pimparkar, M. A. Alam, S. V. Rotkin, and J. A. Rogers, "High-performance electronics using dense, perfectly aligned arrays of single-walled carbon nanotubes," *Nat. Nanotechnol.*, **2**, p. 230, 2007.

63. N. Rouhi, D. Jain, and P. J. Burke, "High-performance semiconducting nanotube inks: Progress and prospects," *ACS Nano*, **5**, 11, pp. 8471–8487, 2011.

64. A. Dodabalapur, "Organic and polymer transistors for electronics," *Mater. Today*, **9**, 4, pp. 24–30, 2006.

65. T. Lei, L.-L. Shao, Y.-Q. Zheng, G. Pitner, G. Fang, C. Zhu, S. Li, R. Beausoleil, H.-S. P. Wong, T.-C. Huang, K.-T. Cheng, and Z. Bao, "Low-voltage high-performance flexible digital and analog circuits based on ultrahigh-purity semiconducting carbon nanotubes," *Nat. Commun.*, **10**, 1, 2161, 2019.

66. K. Bradley, J.-C. P. Gabriel, and G. Grüner, "Flexible nanotube electronics," *Nano Lett.*, **3**, 10, pp. 1353–1355, 2003.

67. E. Artukovic, M. Kaempgen, D. S. Hecht, S. Roth, and G. Grüner, "Transparent and flexible carbon nanotube transistors," *Nano Lett.*, **5**, 4, pp. 757–760, 2005.

68. P. H. Lau, K. Takei, C. Wang, Y. Ju, J. Kim, Z. Yu, T. Takahashi, G. Cho, and A. Javey, "Fully printed, high performance carbon nanotube thin-film transistors on flexible substrates," *Nano Lett.*, **13**, 8, pp. 3864–3869, 2013.

69. S. H. Jin, J. Shin, I.-T. Cho, S. Y. Han, D. J. Lee, C. H. Lee, J.-H. Lee, and J. A. Rogers, "Solution-processed single-walled carbon nanotube field effect transistors and bootstrapped inverters for disintegratable, transient electronics," *Appl. Phys. Lett.*, **105**, 1, p. 13506, 2014.

70. Y. Zhao, Q. Li, X. Xiao, G. Li, Y. Jin, K. Jiang, J. Wang, and S. Fan, "Three-dimensional flexible complementary metal–oxide–semiconductor logic circuits based on two-layer stacks of single-walled carbon nanotube networks," *ACS Nano*, **10**, 2, pp. 2193–2202, 2016.

71. C. Kocabas, S.-H. Hur, A. Gaur, M. A. Meitl, M. Shim, and J. A. Rogers, "Guided growth of large-scale, horizontally aligned arrays of single-walled carbon nanotubes and their use in thin-film transistors," *Small*, **1**, 11, pp. 1110–1116, 2005.

72. D. Sun, M. Y. Timmermans, Y. Tian, A. G. Nasibulin, E. I. Kauppinen, S. Kishimoto, T. Mizutani, and Y. Ohno, "Flexible high-performance carbon nanotube integrated circuits," *Nat. Nanotechnol.*, **6**, p. 156, 2011.

73. Q. Cao, S. Han, G. S. Tulevski, Y. Zhu, D. D. Lu, and W. Haensch, "Arrays of single-walled carbon nanotubes with full surface coverage for high-performance electronics," *Nat. Nanotechnol.*, **8**, p. 180, 2013.

74. G. J. Brady, A. J. Way, N. S. Safron, H. T. Evensen, P. Gopalan, and M. S. Arnold, "Quasi-ballistic carbon nanotube array transistors with current density exceeding Si and GaAs," *Sci. Adv.*, **2**, 9, p. e1601240, 2016.

75. A. J. Bard and L. R. Faulkner, *Electrochemical Methods: Fundamentals and Applications*, 2nd ed. New York: John Wiley & Sons, Inc., 2001.

76. R. L. McCreery, "Advanced carbon electrode materials for molecular electrochemistry," *Chem. Rev.*, **108**, 7, pp. 2646–2687, 2008.

77. K. Nishimura, T. Ushiyama, N. X. Viet, M. Inaba, S. Kishimoto, and Y. Ohno, "Enhancement of the electron transfer rate in carbon nanotube flexible electrochemical sensors by surface functionalization," *Electrochim. Acta*, **295**, pp. 157–163, 2019.

78. J. Y. Lee, E. J. Park, C. J. Lee, S. W. Kim, J. J. Pak, and N. K. Min, "Flexible electrochemical biosensors based on O_2 plasma functionalized MWCNT," *Thin Solid Films*, **517**, 14, pp. 3883–3887, 2009.

79. H. Lee, T. K. Choi, Y. B. Lee, H. R. Cho, R. Ghaffari, L. Wang, H. J. Choi, T. D. Chung, N. Lu, T. Hyeon, S. H. Choi, and D.-H. Kim, "A graphene-based electrochemical device with thermoresponsive microneedles for diabetes monitoring and therapy," *Nat. Nanotechnol.*, **11**, p. 566, 2016.

80. C. M. Girardin, C. Huot, M. Gonthier, and E. Delvin, "Continuous glucose monitoring: A review of biochemical perspectives and clinical use in type 1 diabetes," *Clin. Biochem.*, **42**, 3, pp. 136–142, 2009.

81. C. Chen, R. Ran, Z. Yang, R. Lv, W. Shen, F. Kang, and Z.-H. Huang, "An efficient flexible electrochemical glucose sensor based on carbon nanotubes/carbonized silk fabrics decorated with Pt microspheres," *Sens. Actuat. B Chem.*, **256**, pp. 63–70, 2018.

82. B. Liang, L. Fang, Y. Hu, G. Yang, Q. Zhu, and X. Ye, "Fabrication and application of flexible graphene silk composite film electrodes decorated with spiky Pt nanospheres," *Nanoscale*, **6**, 8, pp. 4264–4274, 2014.

83. X. H. Pham, M.-P. N. Bui, C. A. Li, K. N. Han, J. H. Kim, H. Won, and G. H. Seong, "Electrochemical characterization of a single-walled carbon nanotube electrode for detection of glucose," *Anal. Chim. Acta*, **671**, 1–2, pp. 36–40, 2010.

84. C. Chen, X.-L. Zhao, Z.-H. Li, Z.-G. Zhu, S.-H. Qian, and A. J. Flewitt, "Current and emerging technology for continuous glucose monitoring," *Sensors (Basel)*, **17**, 1, p. 182, 2017.

85. P. Si, X. C. Dong, P. Chen, and D. H. Kim, "A hierarchically structured composite of Mn_3O_4/3D graphene foam for flexible nonenzymatic biosensors," *J. Mater. Chem. B*, **1**, 1, pp. 110–115, 2013.

86. F. Xiao, Y. Li, H. Gao, S. Ge, and H. Duan, "Growth of coral-like PtAu–MnO_2 binary nanocomposites on free-standing graphene paper for flexible nonenzymatic glucose sensors," *Biosens. Bioelectron.*, **41**, 1, pp. 417–423, 2013.

87. W. He, Y. Sun, J. Xi, A. A. M. Abdurhman, J. Ren, and H. Duan, "Printing graphene-carbon nanotube-ionic liquid gel on graphene paper: Towards flexible electrodes with efficient loading of PtAu alloy nanoparticles for electrochemical sensing of blood glucose," *Anal. Chim. Acta*, **903**, pp. 61–68, 2016.

88. Z. Pu, R. Wang, J. Wu, H. Yu, K. Xu, and D. Li, "A flexible electrochemical glucose sensor with composite nanostructured surface of the working electrode," *Sens. Actuat. B Chem.*, **230**, pp. 801–809, 2016.

89. S. Lin, W. Feng, X. Miao, X. Zhang, S. Chen, Y. Chen, W. Wang, and Y. Zhang, "A flexible and highly sensitive nonenzymatic glucose sensor based on DVD-laser scribed graphene substrate," *Biosens. Bioelectron.*, **110**, pp. 89–96, 2018.

90. A. Heras, F. Vulcano, J. Garoz-Ruiz, N. Porcelli, F. Terzi, A. Colina, R. Seeber, and C. Zanardi, "A flexible platform of electrochemically functionalized carbon nanotubes for NADH sensors," *Sensors*, **19**, 3, p. 518, 2019.

91. W. Al-Graiti, Z. Yue, J. Foroughi, X.-F. Huang, G. Wallace, R. Baughman, and J. Chen, "Probe sensor using nanostructured multi-walled carbon nanotube yarn for selective and sensitive detection of dopamine," *Sensors*, **17**, 4, p. 884, 2017.

92. J. Xu, W. Su, Z. Li, W. Liu, S. Liu, and X. Ding, "A modularized and flexible sensor based on MWCNT/PDMS composite film for on-site electrochemical analysis," *J. Electroanal. Chem.*, **806**, pp. 68–74, 2017.

93. J. Liu, Z. He, J. Xue, and T. T. Yang Tan, "A metal-catalyst free, flexible and free-standing chitosan/vacuum-stripped graphene/polypyrrole three-dimensional electrode interface for high performance dopamine sensing," *J. Mater. Chem. B*, **2**, 17, pp. 2478–2482, 2014.

94. N. X. Viet, S. Kishimoto, and Y. Ohno, "Highly uniform, flexible microelectrodes based on clean single-walled carbon nanotube thin film with high electrochemical activity," *ACS Appl. Mater. Interfaces*, **11**, pp. 6389–6395, 2019.

95. J. W. Oh, J. Heo, and T. H. Kim, "An electrochemically modulated single-walled carbon nanotube network for the development of a transparent flexible sensor for dopamine," *Sens. Actuat. B Chem.*, **267**, pp. 438–447, 2018.

96. Y. Wang, J. Ping, Z. Ye, J. Wu, and Y. Ying, "Impedimetric immunosensor based on gold nanoparticles modified graphene paper for label-free detection of *Escherichia coli* O157: H7," *Biosens. Bioelectron.*, **49**, pp. 492–498, 2013.

97. J. Li and E.-C. Lee, "Carbon nanotube/polymer composite electrodes for flexible, attachable electrochemical DNA sensors," *Biosens. Bioelectron.*, **71**, pp. 414–419, 2015.

98. Z.-H. Jin, Y.-L. Liu, J.-J. Chen, S.-L. Cai, J.-Q. Xu, and W.-H. Huang, "Conductive polymer-coated carbon nanotubes to construct stretchable and transparent electrochemical sensors," *Anal. Chem.*, **89**, 3, pp. 2032–2038, 2017.

99. E. Bakker and E. Pretsch, "The new wave of ion-selective," *Anal. Chem.*, **74**, 15, pp. 420A–426A, 2002.

100. E. Zdrachek and E. Bakker, "Potentiometric sensing," *Anal. Chem.*, **91**, 1, pp. 2–26, 2019.

101. N. T. Garland, E. S. McLamore, N. D. Cavallaro, D. Mendivelso-Perez, E. A. Smith, D. Jing, and J. C. Claussen, "Flexible laser-induced graphene for nitrogen sensing in soil," *ACS Appl. Mater. Interfaces*, **10**, 45, pp. 39124–39133, 2018.

102. Q. He, S. R. Das, N. T. Garland, D. Jing, J. A. Hondred, A. A. Cargill, S. Ding, C. Karunakaran, and J. C. Claussen, "Enabling inkjet printed graphene for ion selective electrodes with postprint thermal annealing," *ACS Appl. Mater. Interfaces*, **9**, 14, pp. 12719–12727, 2017.

103. J. Ping, Y. Wang, K. Fan, W. Tang, J. Wu, and Y. Ying, "High-performance flexible potentiometric sensing devices using free-standing graphene paper," *J. Mater. Chem. B*, **1**, 37, p. 4781, 2013.

104. J. Ping, Y. Wang, Y. Ying, and J. Wu, "Application of electrochemically reduced graphene oxide on screen-printed ion-selective electrode," *Anal. Chem.*, **84**, 7, pp. 3473– 3479, 2012.

105. C. Jiang, X. Li, Y. Yao, Y. Ying, and J. Ping, "Fully written flexible potentiometric sensor using two-dimensional nanomaterial-based conductive ink," *Anal. Chem.*, **90**, 21, pp. 13088–13095, 2018.

106. M. Parrilla, R. Cánovas, I. Jeerapan, F. J. Andrade, and J. Wang, "A textile-based stretchable multi-ion potentiometric sensor," *Adv. Healthc. Mater.*, **5**, 9, pp. 996–1001, 2016.

107. S. Roy, M. David-Pur, and Y. Hanein, "Carbon nanotube-based ion selective sensors for wearable applications," *ACS Appl. Mater. Interfaces*, **9**, 40, pp. 35169–35177, 2017.

108. S. Peng and K. Cho, "Chemical control of nanotube electronics," *Nanotechnology*, **11**, 2, pp. 57–60, 2000.

109. D.-I. Moon, B. Kim, R. Peterson, K. Badokas, M.-L. Seol, D. G. Senesky, J.-W. Han, and M. Meyyappan, "A single input multiple output (SIMO)

variation-tolerant nanosensor," *ACS Sensors*, **3**, 9, pp. 1782–1788, 2018.

110. J.-W. Han, B. Kim, J. Li, and M. Meyyappan, "Carbon nanotube based humidity sensor on cellulose paper," *J. Phys. Chem. C*, **116**, 41, pp. 22094–22097, 2012.

111. D. Fu, H. Lim, Y. Shi, X. Dong, S. G. Mhaisalkar, Y. Chen, S. Moochhala, and L.-J. Li, "Differentiation of gas molecules using flexible and all-carbon nanotube devices," *J. Phys. Chem. C*, **112**, 3, pp. 650–653, 2008.

112. T. Zhang, M. B. Nix, B.-Y. Yoo, M. A. Deshusses, and N. V. Myung, "Electrochemically functionalized single-walled carbon nanotube gas sensor," *Electroanalysis*, **18**, 12, pp. 1153–1158, 2006.

113. Y. Sun and H. H. Wang, "Electrodeposition of Pd nanoparticles on single-walled carbon nanotubes for flexible hydrogen sensors," *Appl. Phys. Lett.*, **90**, 21, pp. 213107/1–213107/3, 2007.

114. J.-W. Han, B. Kim, J. Li, and M. Meyyappan, "A carbon nanotube based ammonia sensor on cotton textile," *Appl. Phys. Lett.*, **102**, 19, p. 193104, 2013.

115. Y. Ju Yun, W. G. Hong, N.-J. Choi, B. Hoon Kim, Y. Jun, and H.-K. Lee, "Ultrasensitive and highly selective graphene-based single yarn for use in wearable gas sensor," *Sci. Rep.*, **5**, p. 10904, 2015.

116. S. J. Kim, D.-I. Moon, M.-L. Seol, B. Kim, J.-W. Han, and M. Meyyappan, "Wearable UV sensor based on carbon nanotube-coated cotton thread," *ACS Appl. Mater. Interfaces*, **10**, 46, pp. 40198–40202, 2018.

Chapter 7

Electrospun Carbon Nanofibers for Energy Conversion and Storage

Seok-Hwan Park[a] **and Jun Li**[b]
[a]*Department of Chemical Engineering, Pohang University of Science and Technology, Pohang 37673, South Korea*
[b]*Department of Chemistry, Kansas State University, Manhattan, Kansas 66506, USA*
shpark1@postech.ac.kr, junli@ksu.edu

7.1 Introduction

As environmental pollution has become a serious problem, developing energy conversion and storage devices to support the utilization of renewable energy technologies is increasingly important. Lithium-ion batteries (LIBs) and solar cells (SCs) are representative examples. Better LIBs with high energy density, high power density, long cycle life, and improved safety are in high demands. Electrode materials are the key to define the LIB performance. Carbon materials have been a key component in today's LIBs due to their chemical stability, excellent electrical conductivities, and large surface areas. Among them, carbon nanofibers (CNFs) have attracted particular at-

Nanocarbon Electronics
Edited by Changjian Zhou, Min Zhang, and Cary Y. Yang

ISBN 978-981-4877-11-4 (Hardcover), 978-1-003-04308-9 (eBook)
www.jennystanford.com

tentions. Initially, carbon fibers were developed as one of the important industrial materials using various carbon precursors, including polyacrylonitrile (PAN) through a melt spinning process. PAN was later commonly used as a polymeric precursor in electrospinning processes, which can be converted into CNFs through oxidation stabilization and carbonization processes.

In general, there are two methods of producing CNFs: vapor growth and spinning. Methods to make CNFs in vapor phase were studied in the 1970s and 1980s [1, 2]. Vapor growth is a method to obtain CNFs (and later carbon nanotubes) by catalytic decomposition of specific hydrocarbons using metal particles (catalyst), but it is difficult to mass produce, particularly for membrane assemblies with well-defined configurations such as alignment, porosity, and purity. Therefore, in terms of mass production of self-standing membranes, the spinning method of producing CNFs by spinning and heat treatment of polymeric precursor fibers has become the most common method. While many materials with carbon backbones can be used as carbon precursors, CNFs are typically made of three polymer precursors: PAN, cellulose, and pitch, which provide good yields.

CNFs prepared by electrospinning and vapor-phase growth show very different characteristics linked to the processing conditions. An important feature of the CNFs produced by electrospinning is that they are made into webs as they are collected by rotating a cylindrical drum while the polymer fibers are spinning. When used as an electrode material of LIBs, electrospun CNFs can be directly used as a self-standing paper-like membrane electrode without incorporating a binder. In contrast, CNFs produced by the vapor-phase growth method are in the form of powders, so a binder is essential when used as an thin-film electrode material.

In addition, the structure and properties of CNFs prepared by electrospinning can be modified through structure and properties through the addition of additives into the polymeric cursors during electrospinning. Also, since the web-like structure is made of a stacking of 1D fibers, it is easy to obtain a high porosity (over 90%) even if the membrane thickness is increased. The high porosity also allows deposition of other active energy materials inside the membrane to form unique hybrid materials using electrophoretic deposition (EPD) or electrochemical deposition.

7.2 Materials Preparation

7.2.1 Electrospinning Techniques

Electrospinning is based on applying a high voltage to a polymer solution to release the polymer in the form of fibers using the interaction between the charged polymer and the external electric field. Electrospinning is the most efficient method of producing carbon fibers of nanoscale diameter from tens of nanometers to hundreds of nanometers among spinning technologies. The electrospinning technology is also useful for the future energy material industry because it can be scaled up. Electrospinning is much more effective in fabricating CNFs with smaller diameters and higher electrical conductivities than conventional spinning techniques using only mechanical forces and can produce CNFs at much lower cost than those produced by chemical vapor deposition (CVD).

An electrospinning equipment generally includes a syringe pump for regulating the release rate of polymer solution, a syringe needle for narrowing down the diameter of the released fibers, a rotating cylindrical collector for collecting the fibers, and a high voltage supply for forming a strong electric field between the syringe needle and the collector. When a high voltage is applied to the polymer solution, electrostatic force is generated in the polymer. The balance between the electrostatic force and the surface tension of the polymer solution makes the surface of the polymer solution hemispherical or conical, forming the so-called "Taylor cone." When the electric field between the syringe needle and the collector becomes strong enough, the electrostatic force overcomes the surface tension and the polymer begins to be ejected from the syringe needle (Fig. 7.1) [3]. After ejection, the ejected polymer solution moves from the needle to the collector, and the electrostatic repulsion force in the charged polymer solution becomes strong due to the evaporation of the solvent, so that the polymer solution is separated into a plurality of small polymer fibers. When the polymer fibers arrive at the collector, the solvent evaporates further and solidifies into a completely fibrous form. The actual mechanism of the electrospinning process is very complex because it depends on the property of the polymer solution and environmental factors. Several theoretical models have been proposed to facilitate the interpretation of the mechanism. The first model is the leaky dielectric model that was developed by Taylor in the 1960s [4]. The leakage dielectric model can be applied when the electrical conductivity of the polymer solution is limited and the onset of ejection can be predicted. Then a variety of models were developed to predict the steps of jet thinning at the stage of collecting fibers in a cylindrical collector and of jet coagulation for the formation of nanofibers and were used to predict a part of the electrospinning process, but the final diameter and orientation of the fibers could not be predicted quantitatively. Developing a theoretical model that fully explains the phenomena that occur during electrospinning is still the key to develop fibers of various structures, including core–shell structures with fiber alignment and diameter control.

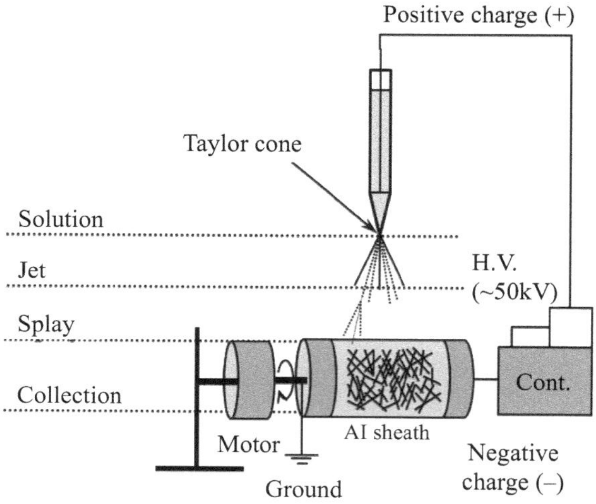

Figure 7.1 Schematic diagram of an electrospinning device.

To produce CNFs through electrospinning, a suitable precursor solution that can be electrospun has to be identified first. The most widely used polymer for manufacturing CNFs is polyacrylonitrile (PAN), which can be dissolved in dimethylformamide (DMF) and used as a precursor solution. The carbon backbone of the PAN polymer and the acrylonitrile functional group are transformed into a hexagonal molecular structure including nitrogen through a low-temperature heat-treatment process at 280°C. Due to the hexagonal molecular structure, the PAN polymer nanofibers are carbonized into CNFs at temperature above 800°C (Fig. 7.2). The most important feature of CNFs produced by electrospinning is that it is obtained in the form of a web, so no binder or support is required to use them as a membrane electrode. Since the CNFs can self-support, the process can be simplified and the price can be lowered when used as an electrode material. Further, because there is no need to use a binder or a substrate, this leads to an improvement in the mass fraction and energy density of the active materials, which has a great advantage in energy storage devices. In addition, it is possible to change the properties of the CNFs by adding various additives to the precursor solution for specific applications, as well as to control the diameter, surface area, porosity, and structure of the fiber by changing the process conditions such as composition of the precursor solution and the temperature profile of heat treatments.

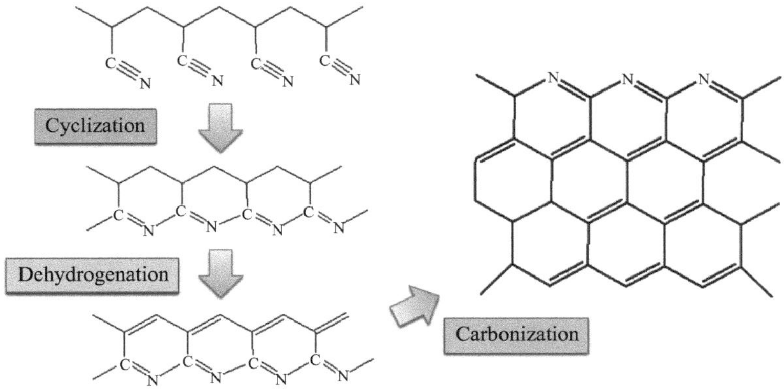

Figure 7.2 General chemical mechanism for PAN stabilization and carbonization.

7.2.1.1 Polymer blends

Since electrospinning is a technique of making a fiber using a polymer precursor solution, it is possible to incorporate other additives for electrospinning. Normally, in order to improve the properties of the polymer fiber, other polymers are mixed in the precursor (for example, a hydrophilic polymer is mixed with a hydrophobic polymer to impart hydrophilicity to the fiber). In the case of CNFs, a polymer precursor solution can be prepared by mixing a polymer having a high yield to carbon when a temperature is applied, and a polymer having a low yield, which produces porous CNFs. The structural characteristics of the pore are determined by the difference in the carbon yield of the polymers.

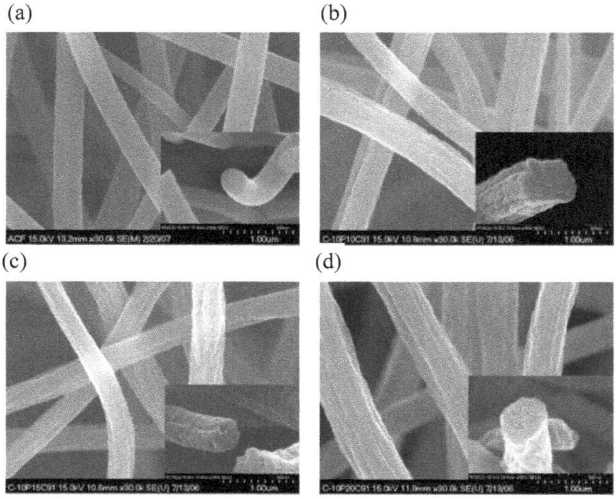

Figure 7.3 SEM images of (a) carbon nanofibers with pure PAN (CP), (b) carbon nanofibers with 10% CA (10CP), (c) carbon nanofibers with 15% CA (15CP), and (d) carbon nanofibers with 20% CA (20CP) [5].

Figure 7.3 shows SEM images of CNFs obtained by heat treatment of polymer nanofibers prepared by mixing a PAN polymer and a cellulose acetate (CA) polymer solution during electrospinning. Fig. 7.3a is a CNF made of 100% PAN polymer, and Fig. 7.3d is a CNF made by mixing PAN and CA in the ratio of 8:2. The surface of the CA-free CNF is smooth, while the surface of the CNF with 20% CA contains many pores. These pores are due to the difference in the carbon yield of PAN polymer and CA polymer at high temperature. In the TGA graph, the PAN polymer has a yield of 35.1% at 800°C, while the CA polymer is only 11.5% (Fig. 7.4). Depending on which polymer is blended with the PAN polymer, the shape of the fiber and the formation of the pores are different, which requires further investigation of the mechanism involved in polymer interactions during the thermal decomposition. The resulting pores greatly

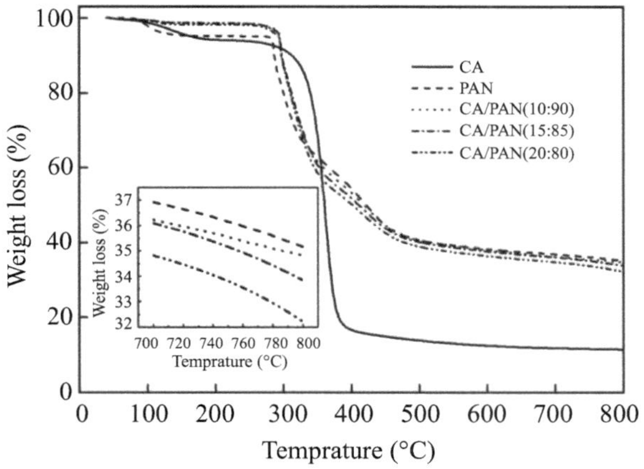

Figure 7.4 Profiles of TGA of electrospun CA, PAN, and three CA/PAN fibers [5].

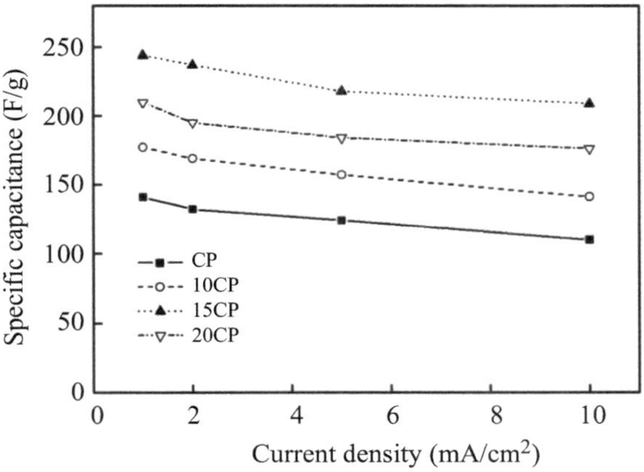

Figure 7.5 Specific capacitances of CP, 10CP, 15CP, and 20CP as a function of various current densities [5].

expand the interfacial reaction area when CNFs are used as the electrode material. Figure 7.5 shows that CA-free CNFs have a specific capacitance of 113 F/g, but the CNFs mixed with CA have higher specific capacitances than that of CA-free CNFs. In particular, the specific capacitance of carbon nanofibers made by adding 15% of CA is the highest at 213 F/g. This is because the specific surface area of the carbon nanofibers is greatly increased by mixing the appropriate amount of polymer, which has the effect of lowering the reaction resistance at the interface. Moreover, CA has an effect of improving the electrical conductivity of carbon nanofibers due to the acetate functional groups, making CNFs more suitable for electrode materials. The polymer blend is one of the ways to modify the properties of the CNFs for specific applications by using a suitable polymer.

7.2.1.2 Carbon nanofibers containing nanoparticles

As a method of imparting a porous structure to electrospun CNFs, there is a method of electrospinning by the addition of nanoparticles to a polymer precursor solution in addition to the polymer blends method. This method can be used in various cases. Silica nanoparticles can be added to the CNFs to provide hydrophilicity to the CNFs. Electrospinning with CNTs can increase the electrical conductivity of the CNFs. In addition, silica nanoparticles can be easily etched with NaOH solutions, opening another way to make porous CNFs.

Figure 7.6a illustrates how to make CNFs via electrospinning by adding MWCNTs and silica nanoparticles. Figure 7.6c is an SEM image of mesoporous activated CNFs (Meso-ACNF) prepared by electrospinning through the addition of silica NPs in the precursor and then removing them with NaOH etching after electrospinning. Figure 7.6d shows mesoporous MWCNT-embedded CNFs (CNT/Meso-ACNF) prepared by electrospinning of the precursor containing both MWCNTs and silica NPs, followed by removal of silica NPs by NaOH etching. In Figs. 7.6a and c, removing the silica NPs positioned inside the CNFs results in well-defined pores throughout the CNFs. It should be noted that the CNT/Meso-ACNF sample not only improves the electrical conductivity by the MWCNTs but also forms larger pore size than the Meso-ACNF sample (Table 7.1) [6]. This means that the addition of MWCNTs facilitates better mass transport of the electrolyte into the CNFs. This property can be used as an electrode material of a cell in which an electrolyte is a liquid and shows that the addition of NPs to a polymer precursor solution can effectively change the structure, morphology, and porosity of CNFs.

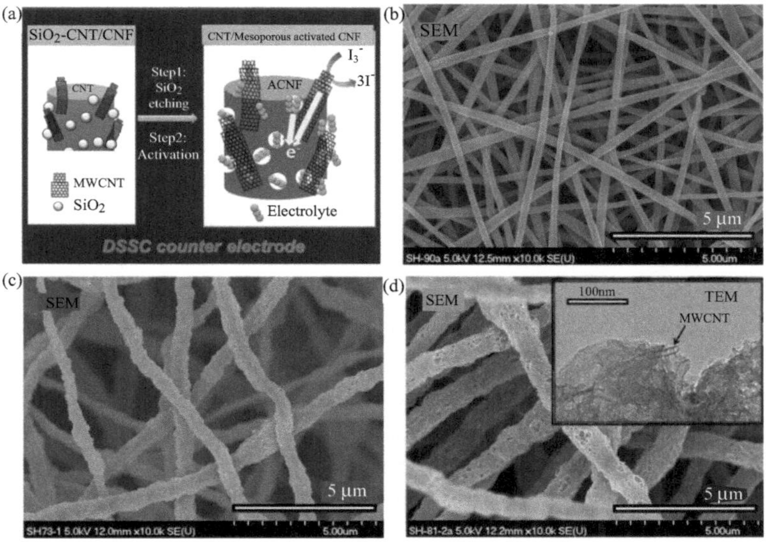

Figure 7.6 The sample preparing model and catalytic activity for (a) CNT/Meso-ACNF; SEM and TEM (inset) images for (b) CNT/ACNF, (c) Meso-ACNF, and (d) CNT/Meso-ACNF [6].

Table 7.1 Pore characteristics, conductivity, and elemental composition of CNT/ACNF, Meso-ACNF, and CNT/Meso-ACNF

	Elemental composition (atomic%)[a]				Electrical conductivity	BET surface area	Total pore volume	Mesopore avg. dia.[b]
	C	O	Si	Na	(S/cm)	(m^2/g)	(cm^3/g)	(mm)
CNT/ACNF	94.0	6.0	—	—	1.0×10^{-1}	626.8	0.32	10.83
Meso-ACNF	69.3	24.8	5.0	0.9	9.6×10^{-3}	249.3	0.65	19.72
CNT/Meso-ACNF	90.6	8.0	1.4	—	1.9×10^{-1}	562.1	1.16	27.32

[a]Determined by EDX results
[b]Calculated with 4 V/A of BJH method

7.2.1.3 Concentric electrospinning

Concentric electrospinning uses two coaxial nozzles (inner and outer nozzles). In this way, core–shell polymer nanofibers can be made, and the properties of the core–shell polymer nanofibers depend on which polymer is electrospun through each nozzle. For example, if a hydrophilic polymer is electrospun through the inner nozzle and a hydrophobic polymer is electrospun through an outer nozzle, a core–shell polymer nanofiber having both hydrophilic and hydrophobic properties can be produced.

CNFs prepared by electrospinning have a diameter of about 250 nm, which is larger than CNFs produced by CVD. Because of its larger diameter, it has a small specific surface area. Concentric electrospinning is introduced to overcome this disadvantage. The core–shell polymer nanofiber comprising the pyrolytic core and the carbonizable shell can be obtained when the pyrolytic polymer is electrospun through the inner nozzle and the carbonizable polymer is electrospun through the outer nozzle. When the core–shell polymer nanofiber is carbonized through heat treatment, the thermally decomposable polymer is removed, while the shell is converted to carbon, making it possible to produce a hollow carbon nanofiber.

In addition to concentric electrospinning, there are also fiber template methods and emulsion electrospinning methods for making hollow CNFs. The fiber template method is a technique of using a fiber as a template. The surface of the fiber template can be coated with carbon by CVD, hydrothermal synthesis, or plasma-coating method. Hollow CNF can be formed by removing the fiber template. The emulsion electrospinning method utilizes the property of two immiscible solutions. When they are mixed, a relatively small amount of solution is present in the suspension in the form of droplets. These droplets move through the syringe needle during electrospinning and form a core phase by agglomerating. The emulsion electrospinning method seems simpler than the fiber template method since it does not need to remove the template in later steps. But the core is not stable, and the process is difficult to control. Compared to these two methods, concentric electrospinning is much more reliable in the production of core–shell CNFs using two nozzles to feed stable precursors separately. One of the important requirements in concentric electrospinning is that the two polymer solutions must be immiscible. Poly(methyl methacrylate) (PMMA) and PAN polymers, which are typically used for making hollow CNFs, are known to be immiscible with each other presumably because PMMA is an amorphous polymer and PAN is a crystalline polymer.

Figure 7.7 shows the method of concentric electrospinning using PAN and PMMA polymers. Hollow CNFs were prepared by electrospinning a PMMA/DMF solution on the inner nozzle and a PAN/DMF solution on the outer nozzle. Figures 7.8b and 7.8d are SEM and TEM images of hollow CNFs. Essentially all fibers have the nearly ideal hollow structure. Since both the outer and inner surfaces can be used, the hollow activated CNFs (HACNF) have a specific surface area about twice that of the solid activated CNFs (ACNF) (Table 7.2) [7].

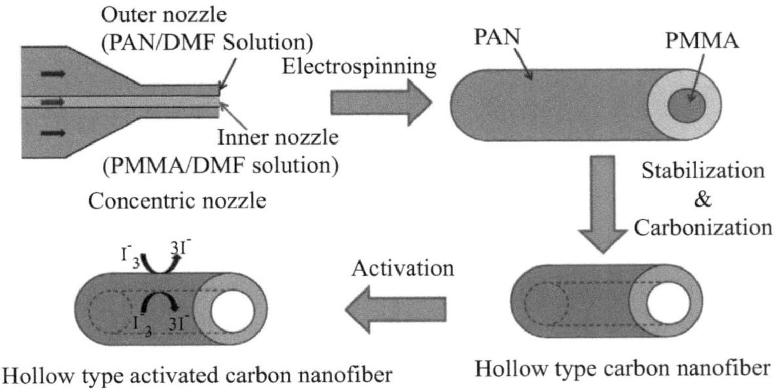

Figure 7.7 The fabrication process of hollow-type activated carbon nanofibers and their use as counter electrode in dye-sensitized solar cell for I_3^- reduction into I^- [7].

Table 7.2 Pore characteristics of ACNF and HACNF

	Micropore surface area (m^2g^{-1})	Mesopore surface area (m^2g^{-1})	Total surface area (m^2g^{-1})	Total pore volume (cm^3g^{-1})	APD[a] (m)	Mesopore avg. dia.[b] (nm)
ACNF	588.3	32	620.3	0.31	2	8.88
HACNF	939.7	97.9	1037.5	0.70	2.7	17.40

[a]Average pore diameter
[b]Calculated with 4 V/A of BJH method

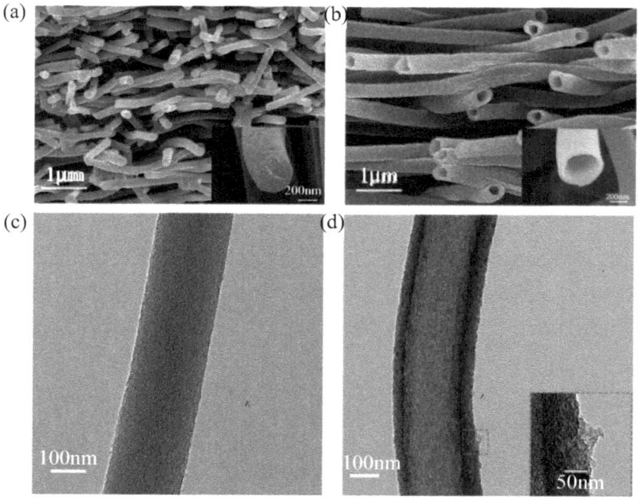

Figure 7.8 SEM images for (a) ACNF, and (b) HACNF; TEM images for (c) ACNF, and (d) HACNF [7].

7.2.2 Electrochemical Deposition Techniques

The electrospun CNFs form a highly conductive continuous membrane, which can be directly used as an electrode for further deposition of active materials for energy applications, particularly metal oxides. These include three methods: (1) direct electrochemical deposition, (2) electrolysis induced precipitation, and (3) electrophoretic deposition.

Most energy materials, particularly those used for batteries and supercapacitors, are high-valence-state metal oxides (such as MnO_2 [8] and V_2O_5 [9]), which are stable and insoluble in aqueous solutions. These can be readily deposited onto the CNF electrode by anodic oxidation of soluble low-valence-state species (such as salt solutions containing Mn^{2+} and VO^{2+} ions) at high positive potentials. This is most straightforward, and the deposition quantity can be controlled by measuring the accumulated charge generated by the deposition.

Alternatively, when CNFs are used as a cathode at sufficiently negative electrode potential, it induces water electrolysis to produce hydrogen and release OH^- ions. Accompanying the reaction, the pH value near the electrode increases, leading to the formation of insoluble metal hydroxides (such as $Mn(OH)_2$) at the electrode surface area [10, 11]. The 3D porous CNF membrane provides a large surface area, which enhances the reaction rate of water electrolysis, thus making the precipitation more facile. Interestingly, the oxidation potential is highly pH dependent. As the pH value at the CNF electrode surface is increased due to water electrolysis, Mn^{2+} or $Mn(OH)_2(s)$ can be electrochemically oxidized into more stable and insoluble $MnO_2(s)$ or $Mn_3O_4(s)$, respectively, even though the potential is at negative values [8]. These processes are irreversible, and thus the metals oxides will sustain after the pH restores the same as the bulk solution after the deposition potential is turned off. In some other studies, the low-valence metal hydroxides can be subjected to thermal treatments in the air to convert them to high-valence metal oxides. Furthermore, hydrogen gas bubbles generated at the CNF electrode surface during the process block a part of the electrode surface and result in very interesting 3D porous honeycomb structures [10, 11]. Such hierarchical nanoporous structure is beneficial to ion transport, which is a critical factor for LIB performance, as will be discussed later. The metal oxides, including CuO, NiO, Mn_3O_4, and Co_3O_4, coaxially coated on electrospun CNFs, which will be discussed later, were deposited by this method by holding CNF membranes at a constant cathodic potential of 10 V in ethanol solutions [12–15].

The EPD method is based on electrostatic interaction with the particles in solution using an electric field. The desired metal oxide NPs can be first dispersed in solvent to a sol solutions. These NPs are typically positively charged. When an electric field is generated by applying a voltage between the CNF cathode and an anodic counter electrode, the positively charged NPs move toward the CNF cathode and deposit on the surface as a porous film. Thermal treatment may be applied later to improve the quality of the film. The EPD process, in principle, mainly manipulates the preformed metal oxide NPs by an electric field and does not change their valence states. However, after reaching the surface, redox reaction may occur through the other two mechanisms. As a result, it could be a combined process. One advantage of EPD is that other organic or inorganic materials can be mixed with the metal oxide NPs during deposition to form a composite film through this low-temperature process.

In most studies, the metal oxide depositions were done by applying a constant voltage or current over certain time to generate the desired amount deposits. However, the deposition may not be uniform across the 3D CNF membrane due to the mass transport. The outer surface of the membrane may have more deposition than the internal surface. This problem can be reduced by applying a pulsed deposition method, in which a short period (for example 10 s) of deposition at the desired potential is applied and then followed by a longer resting time (for example 20 to 60 s) at the potential that no deposition occurs [8, 9]. This allows the solution to re-establish equilibrium through thermal diffusion. Thus, more uniform electrodeposition can be achieved.

7.3 Energy Conversion Applications

7.3.1 Dye-Sensitized Solar Cells

Dye-sensitized solar cells (DSSCs) have attracted extensive attention as a potential technology to solve energy problems because of lower production cost than silicon solar cells and the advantage of generating energy with visible light. However, since commercial DSSCs use expensive platinum (Pt) as the counter electrode material, extensive research has been carried out to investigate other materials to replace Pt and lower the production cost. The requirement is that the alternative materials need to have a catalytic property capable of reducing triiodide ion ($I^{-}{}_{3}$) to I^{-}. Carbon materials, metal oxide, and conductive polymer have been identified as materials having such catalytic property. Among them, carbon materials are the most promising materials due to their low cost, excellent chemical stability, high electrical conductivity, and large specific surface area.

When the CNFs prepared by electrospinning are directly used as a counter electrode in DSSCs, the power conversion efficiency is not high owing to their low specific surface area. Therefore, additional techniques to improve the structure of CNFs are needed. Figure 7.9 represents a method of manufacturing mesoporous hollow CNFs (Meso-HACNF) with a very high specific surface area by using concentric electrospinning and polymer blend technology simultaneously. Hollow CNFs have about twice the specific surface area as solid carbon nanofibers because they can utilize both the inner and outer shell surface areas. Meso-HACNFs have an even higher specific surface area. Mesopores can be formed in the hollow CNF shell by applying a polymer blend technique to the shell (Fig. 7.10). This further enhances the reduction reaction of triiodide ions. Table 7.3 shows that as higher PMMA content is added to the shell, the specific surface area and the pore size increase. However, if the content of PMMA is too high, the pore size becomes too large and this leads to a decrease in specific surface area. In addition, the hollow structure tends to be slightly distorted due to the thermal decomposition of PMMA. Therefore, it is important to adjust the content of PMMA in order to obtain a high efficiency and to make a structure optimized for the reduction reaction of triiodide ions.

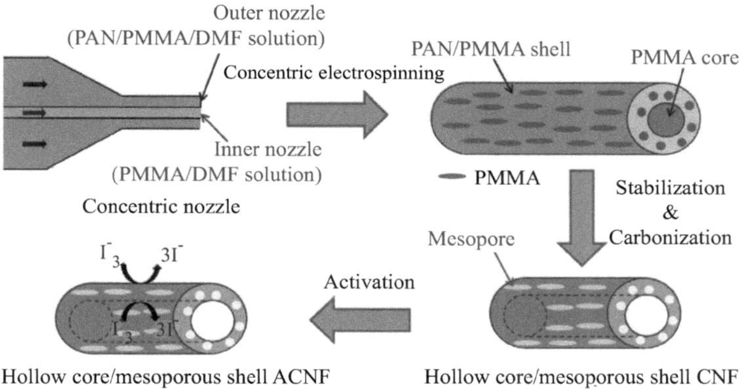

Figure 7.9 The fabrication process of activated carbon nanofiber with hollow core and mesoporous shell structure (Meso-HACNF), and its application as counter electrodes in DSSCs for I_3^- reduction [16].

Table 7.3 Pore characteristics of HACNF, Meso-HACNF1, Meso-HACNF2

	Micropore surface area (m^2g^{-1})	**Mesopore surface area (m^2g^{-1})**	**Total surface area (m^2g^{-1})**	**Total pore volume (cm^3g^{-1})**	**APD[a] (nm)**	**Mesopore avg. dia.[b] (nm)**
HACNF	940	98	1038	0.70	2.7	17.4
Meso-HACNF1	1039	151	1191	0.99	3.3	17.9
Meso-HACNF2	867	145	1012	0.89	3.5	18.6

[a]Average pore diameter
[b]Calculated with 4 V/A of the BJH method

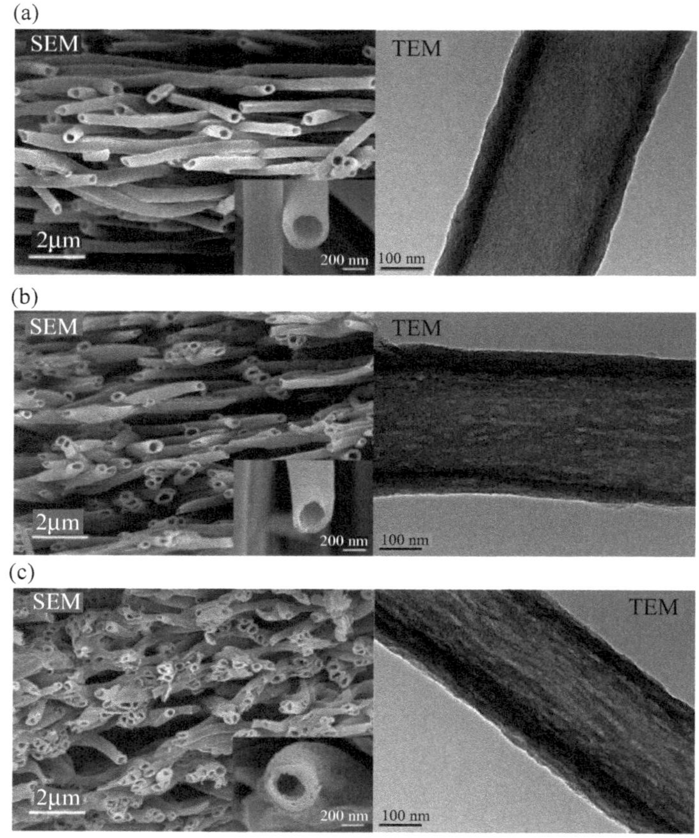

Figure 7.10 The cross-sectional SEM and TEM images for (a) HACNF (PMMA-free in the shell), (b) Meso-HACNF1 (PAN : PMMA = 9:1), and (c) Meso-HACNF2 (PAN : PMMA = 8:2) [16].

Figure 7.11 is the electrochemical impedance spectroscopy (EIS) curve of various CNFs. The charge transfer resistance (R_{ct}) indicates the catalytic activity of reducing the triiodide ion, which is closely related to the semicircle in the high-frequency region. The R_{ct} value of Meso-HACNF1 is about half of that of the Pt electrode, indicating that the Meso-HACNF1 electrode has a higher catalytic activity than Pt due to the high specific surface area. Therefore, Meso-HACNF1 shows a higher fill factor (FF) than that of Pt (Table 7.4). However, even though Meso-HACNF1 has a higher FF value, the energy conversion efficiency (η) is slightly lower than that of Pt. This is because the surface of Pt electrode is smoothly coated on the FTO glass and behaves like a mirror so that the unabsorbed light passing through the photoanode is reflected back at the Pt counter electrode. Thus, the efficiency of the solar cell is enhanced. On the other hand, counter electrodes made of black CNFs strongly absorb the visible light instead of reflecting it back. Thus, the Pt counter electrode shows a higher maximum current value and energy conversion efficiency than the CNF counter electrode (Fig. 7.12 and Table 7.4) [16]. However, it is significant that CNFs can achieve a relatively high energy conversion efficiency of at least 7%, making it possible to replace Pt in sustainable DSSCs for large-scale applications.

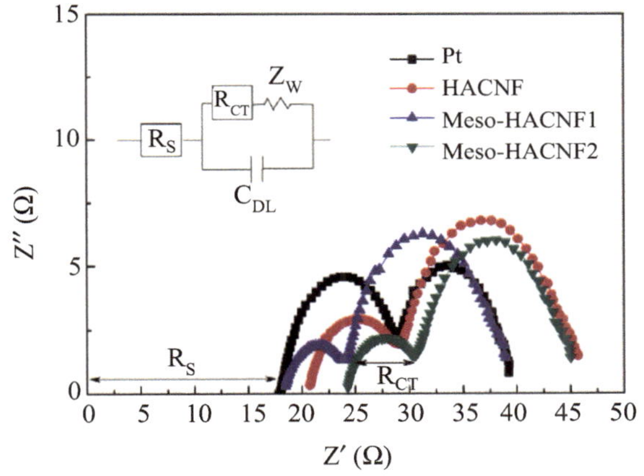

Figure 7.11 EIS spectra of HACNF, Meso-HACNF1, Meso-HACNF2, and Pt counter electrode [16].

Table 7.4 Photoelectrochemical parameters of DSSCs for HACNF, Meso-HACNF1, Meso-HACNF2, and Pt counter electrodes

	J_{sc} (mA cm^{-2})	V_{oc} (V)	FF (%)	η (%)	R_s (Ωcm^2)	R_{ct} (Ωcm^2)
Pt (reference)	17.6	0.72	60	7.69	18.0	10.9
HACNF	15.8	0.72	61	6.96	20.7	8.2
Meso-HACNF1	15.4	0.73	64	7.21	18.4	5.4
Meso-HACNF2	15.3	0.74	61	6.91	24.2	6.3

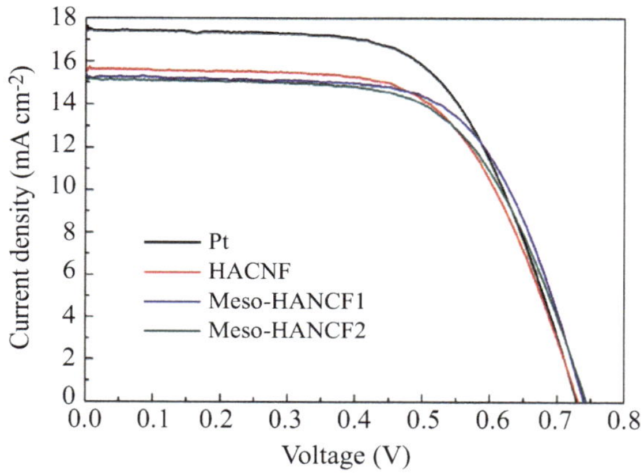

Figure 7.12 *I–V* curves of HACNF, Meso-HACNF1, Meso-HACNF2, and Pt counter electrode [16].

7.3.2 Polymer Electrolyte Membrane Fuel Cells

Polymer electrolyte membrane fuel cells (PEMFCs) have been recognized as a promising power source because of their high conversion efficiency and low-temperature operation. PEMFCs use platinum as a catalyst for ionizing hydrogen, but since platinum is expensive, it is used to be dispersed on a catalyst support in order to utilize platinum efficiently. A commonly used catalyst support for PEMFC is Vulcan XC-72R. The total specific surface area of XC-72R and electrospun CNFs (e-CNF) are approximately similar, at 235 $m^2\ g^{-1}$ and 307 $m^2\ g^{-1}$, respectively. But there is a big difference in the pore size distribution. In details, the pore size distribution of XC-72R is composed of micropore surface area (S_{micro}) of 156 $m^2\ g^{-1}$ and mesopore surface area (S_{meso}) of 79 $m^2\ g^{-1}$ (Table 7.5) [17]. The mesopore surface area is almost 30% of the overall specific surface area. In addition, the average pore diameter (APD) and mesopore volume (V_{meso}) of XC-72R are 10.92 nm and 0.578 $cm^3\ g^{-1}$, respectively. The deep pores are very disadvantageous for the formation of three-phase boundaries with Nafion ionomer (4–200 nm diameter) because the platinum nanoparticles buried deeply in the pores have decreased efficiency (Fig. 7.13). On the contrary, electrospun CNFs consist of S_{mico} of 292 $m^2\ g^{-1}$ and S_{meso} of

15 m^2 g^{-1}, so S_{meso} occupies a small fraction of 5% of the total specific surface area. Furthermore, micropores of the electrospun CNFs are very shallow, which is very advantageous for forming a three-phase boundary with the Nafion ionomer (Fig. 7.13). On top of that, since CNFs have 1D morphology, it can form hierarchical membranes with high porosity to allow efficient diffusion of hydrogen gas compared to XC-72R.

Table 7.5 Surface characterization of XC-72R and e-CNF

Sample	SSA (m^2 g^{-1})	V_{micro} (cm^3 g^{-1})	V_{meso} (cm^3 g^{-1})	V_{tot} (cm^3 g^{-1})	S_{micro} (m^2 g^{-1})	S_{meso} (m^2 g^{-1})	APD (nm)	Conductivity (S cm^{-1})
XC-72R	235	0.062	0.578	0.640	156	79	10.92	0.35
e-CNF	307	0.128	0.054	0.182	292	15	2.36	0.69
Pt/XC-72R	194	0.049	0.550	0.599	132	62	14.28	—
Pt/e-CNF	302	0.125	0.053	0.178	288	14	2.31	—

SSA: specific surface area; V_{micro}: micropore volume; V_{meso}: mesopore volume; V_{tot}: total pore volume; S_{micro}: micropore surface area; S_{meso}: mesopore surface area; APD: average pore diameter

Figure 7.14 shows TEM images of XC-72R and electrospun carbon nanofibers loaded with platinum NPs. Clearly, platinum NPs are uniformly dispersed on the surface of both XC-72R and electrospun CNFs. The average particle sizes of platinum on XC-72R and CNFs are 3.6 nm and 2.9 nm, respectively. The platinum NPs dispersed in CNFs are slightly smaller than those on XC-72R. This is likely because XC-72R has deep and wide pores formed on its surface, while CNFs have shallow micropores formed on the surface. The shallow micropores are effective for platinum dispersion as well as for forming three-phase boundaries. Therefore, the electrospun CNFs have a pore size distribution favorable to the dispersion of platinum than XC-72R.

The cyclic voltammograms show how efficiently the dispersed platinum NPs on the surface are participating in the reaction (Fig. 7.15). The Pt utilization (%) is obtained by calculation using the adsorption/desorption charge of hydrogen ions, the electrochemical surface area (ESA), and the chemical specific surface area (CSA). As shown in Table 7.6 [17], the Pt utilization of Pt/XC-72R is 35.2%, whereas Pt utilization for Pt/e-CNF is 69.1%, which is almost twice of the former. This is attributed to the 1D morphology of electrospun CNFs and the formation of shallow micropores on the surface, which helps to disperse Pt and formed better three-phase boundary with Nafion ionomer.

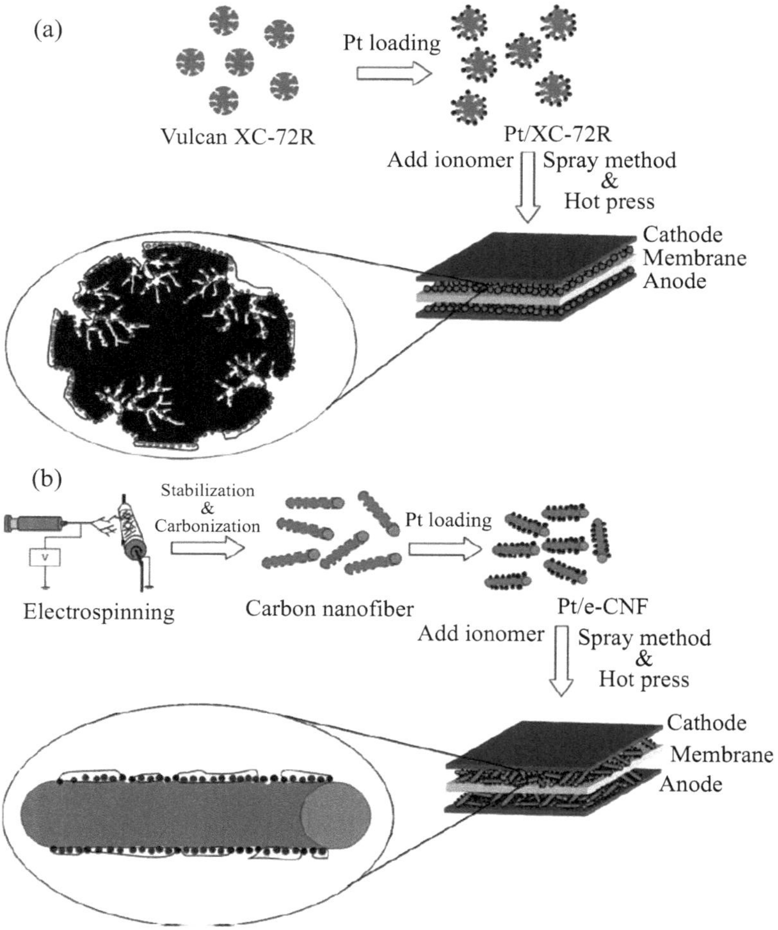

Figure 7.13 Schematic diagrams of (a) Pt/XC-72R and (b) Pt/e-CNF [17].

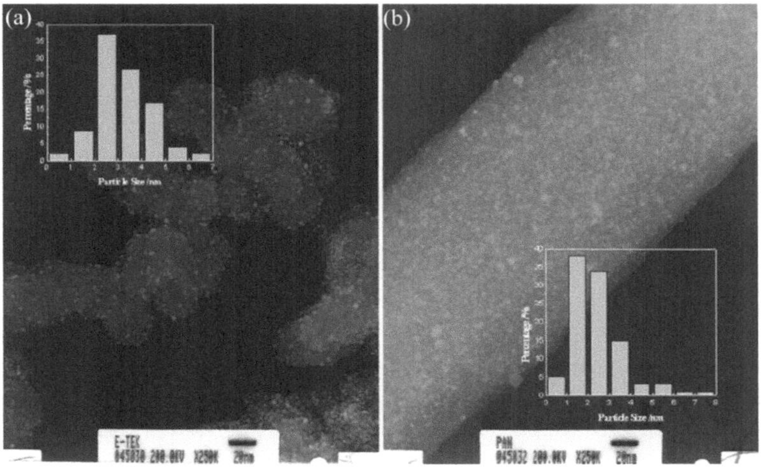

Figure 7.14 TEM images of (a) Pt/XC-72R and (b) Pt/e-CNF [17].

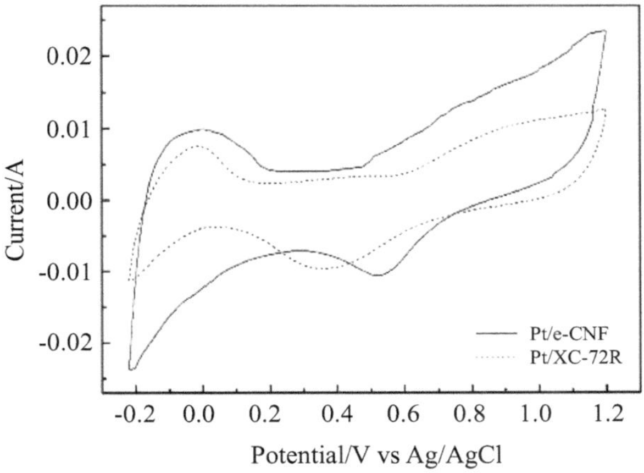

Figure 7.15 Cyclic voltammograms of Pt/XC-72R and Pt/e-CNF at scan rate of 10 mV s^{-1} in 1 M H_2SO_4 [17].

Table 7.6 Pt particle size, ESA, CSA, and Pt utilization in XC-72R and e-CNF

	d_{TEM} (nm)	d_{XRD} (nm)	ESA (m^2g^{-1} Pt)	CSA (m^2g^{-1} Pt)	Pt utilization (%)
Pt/XC-72R	3.6	3.2	30.8	87.6	35.2
Pt/e-CNF	2.9	2.6	74.5	107.8	69.1

d_{TEM}: Pt particle size by TEM; d_{XRD}: Pt particle size by XRD; ESA: electrochemical surface area; CSA: chemical surface area

7.4 Energy Storage Applications

7.4.1 Metal-Embedded Carbon Nanofibers for LIBs

For LIBs, volume changes should be considered when using anode materials such as metal, which have a large volume change during Li intercalation extraction. Generally, when the volume of the active material changes greatly, the gap between the active material and the conductive material widens during charge–discharge cycles, and the electric contact site is rapidly deteriorated.

A way to stably use an active material with a large volume change as an anode for LIB is to embed a metal in CNFs using electrospinning. This method is slightly different from the method of electrospinning by adding NPs to a polymer precursor solution. If the metal is added to the polymer precursor solution in the form of preformed NPs, it is difficult to disperse them uniformly throughout the fibers. Furthermore, it is not easy to form alloys with other metals. Therefore, strategic approaches are needed to overcome these issues. First, a metal precursor that is appropriate for embedding the metal NPs into the nanofiber needs to be identified. The metal precursor determines the shape and size of the final metal NPs. Second, it is necessary to determine the polymer that serves as a dispersant so that the metal NPs can be evenly distributed throughout the CNFs. The dispersing polymers are important because they can turn metals into metal oxides and cause segregation of metal NPs on the edges of CNFs. Third, the proper carbonization temperature needs to be determined. As the temperature rises, metal precursors and polymers react and are pyrolyzed to form metal NPs. Especially, since the property and size of the metal NPs and the morphology of the CNFs vary with the temperature, an appropriate temperature balancing the carbonization and metal NP formation is critical for this application.

Figure 7.16 represents the process of making CNFs embedded with cobalt–tin alloys. In this study, cobalt acetate and tin acetate were used as metal precursors, and polyvinylpyrrolidone (PVP) and PAN polymers were used as a dispersant and carbon precursor, respectively. It was expected that the carbonyl group of PVP could effectively disperse cobalt ions and tin ions. Especially, cobalt was anticipated to improve stability of tin by acting as a framework when used as an anode material for LIBs. Figure 7.17 provides the SEM and TEM images of Co–Sn alloy embedded CNFs carbonized at various temperatures. As the temperature rises, the size of the Co–Sn alloy NPs increases. In particular, when the carbonization temperature is raised up to 900°C, decomposition of the CNFs becomes significant due to the catalytic effect of Co–Sn (Fig. 7.17c). The XRD graph (Fig. 7.18) shows that SnO_2 only exists when carbonized at 700°C. When carbonized at 800°C, it starts to form Co–Sn alloy, and when carbonized at 900°C, $CoSn_2$ becomes the largest proportion. SnO_2 is known as an anode material having a high specific capacity for LIBs, but its charge–discharge cycle stability is poor due to a large volume change. $CoSn_2$ has a two-step reaction mechanism involving irreversible reaction during charge and discharge cycles and thus also has poor charge–discharge cycle stability. CoSn has a high specific capacity and good charge–discharge cycle stability because cobalt acts as a framework to improve the stability of Sn during charge and discharge. As a result, Co–Sn alloy embedded CNFs carbonized at 800°C have the highest specific capacity and good charge–discharge cycle stability due to the formation of abundant Co–Sn alloy (Fig. 7.19).

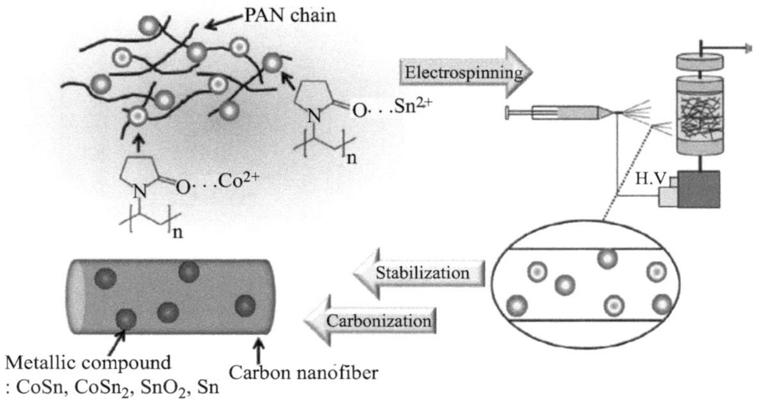

Figure 7.16 The fabrication process of Co–Sn/CNFs composites [18].

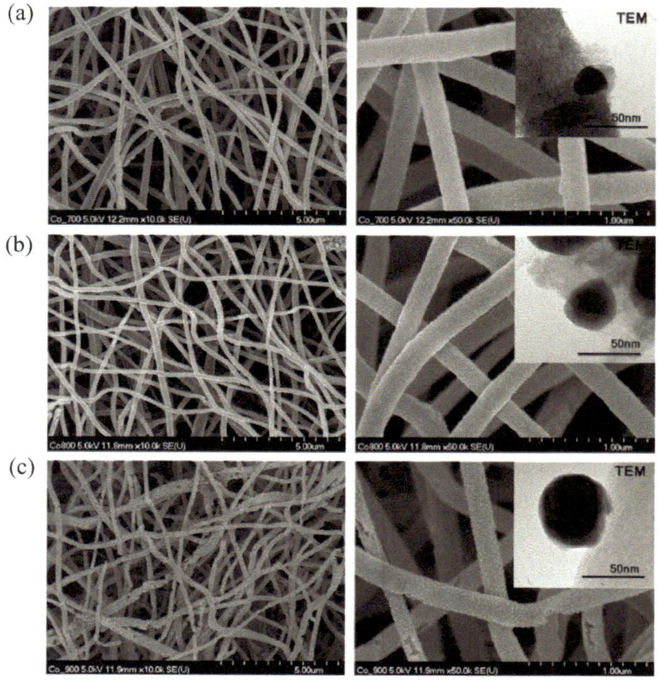

Figure 7.17 SEM and TEM images of Co–Sn/CNFs with various temperatures: (a) 700°C, (b) 800°C, and (c) 900°C [18].

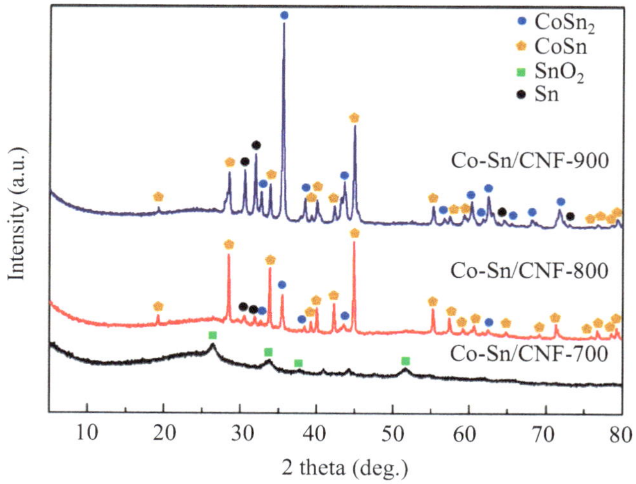

Figure 7.18 XRD patterns of Co–Sn/CNFs with various temperatures [18].

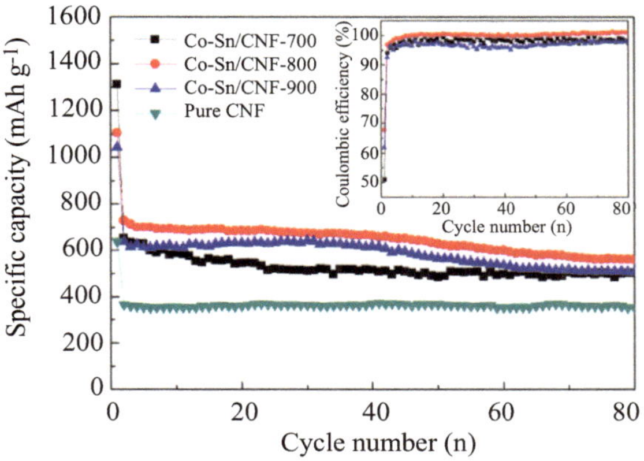

Figure 7.19 Cycle performance of pure-CNF-800 and Co–Sn/CNFs with various temperatures at 0.3 C (161 mA/g) in 1 M $LiPF_6$/EC/DMC [18].

7.4.2 Core–Shell CNFs for LIBs

The more efficient anode electrode for LIBs than the metal-embedded CNFs described above is the core–shell structure built on CNF cores. Metal-embedded CNFs are inefficient because metal NPs are deeply embedded inside CNFs and not easily accessed by Li^+ ions. In contrast, the lithium-active materials prepared as thin shells on CNFs are more ready to react.

Core–shell CNFs containing a CNF core and a metal oxide shell can be easily prepared using the EPD method. Generally, sputtering is used as a physical deposition method for coating metal oxides on CNFs. However, this method is difficult to form uniform deposits inside the CNF film. It is also difficult to scale-up due to the high cost and slow throughput. On the other hand, the EPD method, particularly with pulsed procedures, is able to uniformly coat almost all surfaces of CNFs over the entire layer. In addition, scale-up is possible using simple equipment. For example, the EPD method described above can be used, in which a CNF web is dipped in a precursor solution and coated by applying an electric current. A large electrode can be manufactured defined only by the size of the used CNF webs. For reference, the size of the CNF web produced by the electrospinning equipment in laboratory is equal or greater than the size of an A4 paper. But it can be much larger in future industrial equipment. More importantly, the core–shell structure is the optimal structure to mitigate the large volume change of metal oxide that occurs during charge–discharge cycles. Figure 7.20 shows the charge–discharge process of the CNF–NiO core–shell structure. When the lithium ion is intercalated into the NiO shell during charging, NiO NPs expand, causing the outer surface of the shell to extend in the radial direction while the inner shell surface remains strongly interfaced with the CNF surface. In addition, since the CNFs have an elastic property, they provide a buffering effect to the NiO shell so that neither the fibers break nor the shell delaminates from the core as the NiO shell repeatedly expands and shrinks during the charge–discharge cycles [15]. This is why it is stable despite the large volume change of the NiO shell. In addition, NiO is converted into Ni particles according to the conversion mechanism ($NiO + 2Li^+ + 2e^- \leftrightarrow 4Li_2O + Ni$) during charging. This reaction occurs mainly around the CNF cores, which supply the needed electrons. As a result, Ni particles are distributed near the fiber core. When discharging, Ni particles are easily converted to NiO because the CNF core and connected Ni particles act as an efficient electron pathway. This charge–discharge mechanism is applicable to many metal oxides besides NiO.

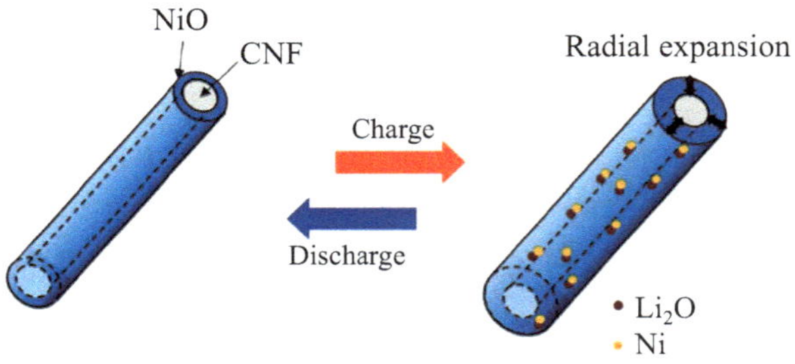

Figure 7.20 Schematic illustration of NiO/CNF during charge–discharge processes [15].

Figure 7.21 represents the change in EIS curves by Nyquist plots for Co_3O_4/CNF, CuO/CNF, NiO/CNF, and Mn_3O_4/CNF electrodes. The internal resistance (R_Ω) is located in the intercept of the semicircle in the high-frequency region of the real axis. All four electrodes with and without CNF cores have very similar R_Ω values likely because it is dominated by the electrolyte resistance. However, the charge transfer resistance (R_{ct}) is dramatically reduced in the core–shell samples. The R_{ct} value is approximately represented by the size of the semicircle of the EIS. Clearly, the R_{ct} value of all electrodes having core–shell structure is lower than that of the electrode with just metal oxide particles. There are two main reasons. First, the shells are coated with metal oxide nanoparticles of 5 to 10 nm through the EPD method. This leads to a sufficient space for the diffusion of lithium ions in the metal oxide shell, thereby improving the reaction efficiency of the entire metal oxide shell. Second, the CNF core improves electrical conductivity. The electron pathways by the CNF core contribute to the structural stability of the metal oxide shell and the reaction rate during charge–discharge processes. Therefore, because of these advantages, the CNF–metal-oxide core–shell structures have substantially higher specific capacity and the improved charge–discharge stability than corresponding metal oxides alone (Fig. 7.22).

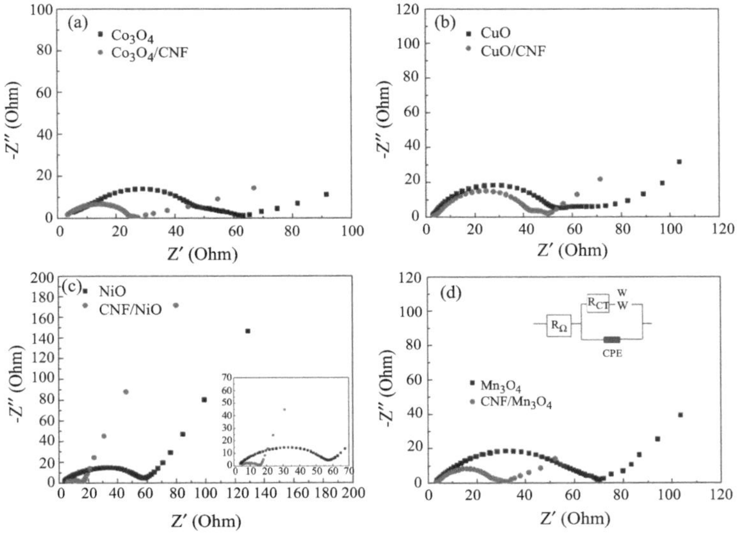

Figure 7.21 AC impedance spectra of (a) Co_3O_4/CNF, (b) CuO/CNF, (c) NiO/CNF, and (d) Mn_3O_4/CNF in 1 M $LiPF_6$/EC/DMC [12–15].

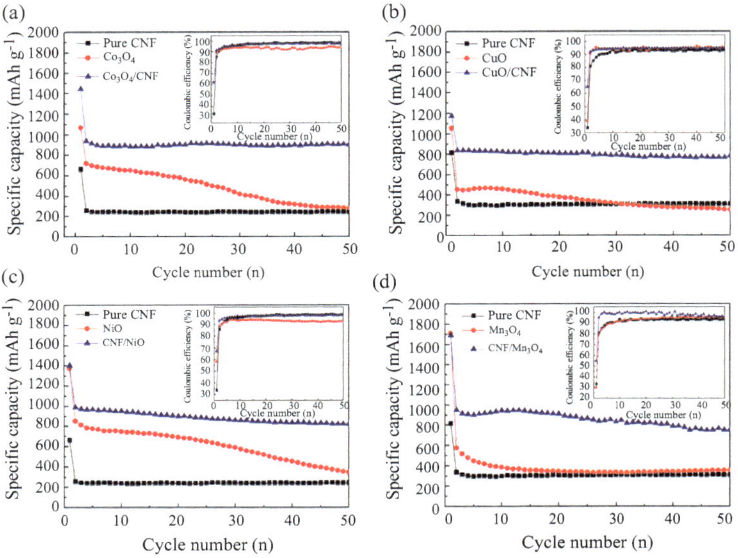

Figure 7.22 Cycle performance of (a) Co_3O_4/CNF, (b) CuO/CNF, (c) NiO/CNF, and (d) Mn_3O_4/CNF in 1 M $LiPF_6$/EC/DMC [12–15].

7.5 Summary and Outlook

In this chapter, we have discussed the basic theory of electrospinning, the techniques of electrospinning, the types and characteristics of electrospun carbon nanofibers, and applications for energy conversion and storage. As the importance of renewable energy continues to grow and the demand for mass production of electrode materials increases, the importance of electrospun CNFs, which can be mass produced, is increasing. In addition, since electrospinning can be applied creatively to other technologies, CNFs with new structures and novel properties are expected to be developed and used in the future.

References

1. N. M. Rodriguez, *Journal of Materials Research*, **8** (1993), 3233–3250.
2. G. G. Tibbetts, M. L. Lake, K. L. Strong, and B. P. Rice, *Composites Science and Technology*, **67** (2007), 1709–1718.

3. A. L. Yarin, S. Koombhongse, and D. H. Reneker, *Journal of Applied Physics*, **90** (2001), 4836–4846.
4. G. I. Taylor, *Proceedings of the Royal Society of London A*, **313** (1969), 453–475.
5. Y. W. Ju, S. H. Park, H. R. Jung, and W. J. Lee, *Journal of the Electrochemical Society*, **156** (2009), A489–A494.
6. S. H. Park, H. R. Jung, B. K. Kim, and W. J. Lee, *Journal of Photochemistry and Photobiology A: Chemistry*, **246** (2012), 45–49.
7. S. H. Park, H. R. Jung, and W. J. Lee, *Electrochimica Acta*, **102** (2013), 423–428.
8. J. Liu, J. Essner, and J. Li, *Chemistry of Materials*, **22** (2010), 5022–5030.
9. E. Brown, S.-H. Park, A. Elangovan, Y. Yuan, J. Kim, X. S. Sun, X. Zhang, G. Wang, and J. Li, *Electrochimica Acta*, **269** (2018), 144–154.
10. D. Liu, B. B. Garcia, Q. Zhang, Q. Guo, Y. Zhang, S. Sepehri, and G. Cao, *Advanced Functional Materials*, **19** (2009), 1015–1023.
11. D. Liu, Q. Zhang, P. Xiao, B. B. Garcia, Q. Guo, R. Champion, and G. Cao, *Chemistry of Materials*, **20** (2008), 1376–1380.
12. S. H. Park and W. F. Lee, *Carbon*, **89** (2015), 197–207.
13. S. H. Park and W. J. Lee, *Scientific Reports*, **5** (2015).
14. S. H. Park and W. J. Lee, *Journal of Power Sources*, **281** (2015), 301–309.
15. S. H. Park and W. J. Lee, *RSC Advances*, **5** (2015), 23548–23555.
16. S. H. Park, B. K. Kim, and W. J. Lee, *Journal of Power Sources*, **239** (2013), 122–127.
17. J. H. Park, Y. W. Ju, S. H. Park, H. R. Jung, K. S. Yang, and W. J. Lee, *Journal of Applied Electrochemistry*, **39** (2009), 1229–1236.
18. H. R. Jung, S. H. Park, and W. J. Lee, *Materials Chemistry and Physics*, **135** (2012), 340–347.

Chapter 8

Nanocarbon Electronics Prospects

Changjian Zhou[a] and Cary Y. Yang[b]
[a]*School of Microelectronics, South China University of Technology, Guangzhou 510641, China*
[b]*Center for Nanostructures, Santa Clara University, Santa Clara, California 95053, USA*
zhoucj86@gmail.com

8.1 Market Trends of Nanocarbons

The global carbon nanotube (CNT) market reached US$4.55 billion in 2018 and is estimated to reach US$9.84 billion by 2023 [1]. Figure 8.1 shows the revenues of various sectors of the CNT-based end-user industry. Electronics and energy are the two largest sectors, where multi-walled CNTs (MWCNTs) are the most widely adopted materials. In general, their light weight, high electrical and thermal conductivities, large surface-to-volume ratio, and relatively low cost are the underlying factors for adopting MWCNTs as the filler materials for composites and coatings. In terms of electronics, which is the focus of this book, the market mainly consists of conductive electrodes, conductive heating films, nano inks, sensors, and displays. CNT field-effect transistors (CNFETs) and CNFET-based integrated

Nanocarbon Electronics
Edited by Changjian Zhou, Min Zhang, and Cary Y. Yang

ISBN 978-981-4877-11-4 (Hardcover), 978-1-003-04308-9 (eBook)
www.jennystanford.com

circuits (ICs), which would require high-purity semiconducting SWCNTs, are still confined to the domain of the research laboratory, and various challenges have to be overcome before the traditional silicon-dominated semiconductor industry seriously considers the introduction of CNT in its half-trillion-plus US dollar business. Nevertheless, among the various application fields, CNT electronics is expected to grow at the highest rate, suggesting the significant appeal of CNT for future-generation electronics applications.

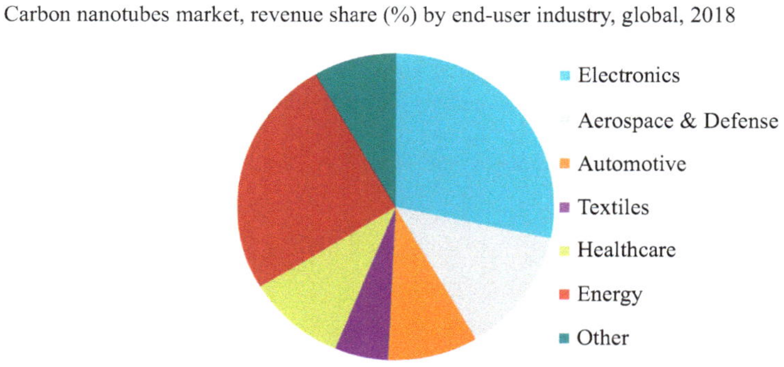

Figure 8.1 CNT market, revenues, and application fields (Source: Mordor Intelligence).

Graphene had a global market size of US$15–50 million in 2015 to a projected value of US$200–2000 million around 2025 [2]. While the scientific definition of graphene is a one-atom-thick honeycomb structure, the graphene industry covers a family of graphene-related materials, including monolayer graphene, multilayer graphene flakes, graphene oxide, reduced graphene oxide, and graphene composites. Thus, the graphene market is assessed based on the predicted revenue and compound annual growth rate (CAGR) [2] for each sector, as shown in Fig. 8.2. Energy and composites possess the largest market value as well as a large CAGR. Compared with graphene composites, which have already received end-user recognition in various applications such as sport facility and automotive coating, it is much more difficult for graphene to break into the semiconductor industry in active or passive electronic devices. Most market predictions show that graphene-based electronics and optoelectronics are expected to have a relatively high growth rate of 20–30%, but a small revenue by 2025 [2]. At the product

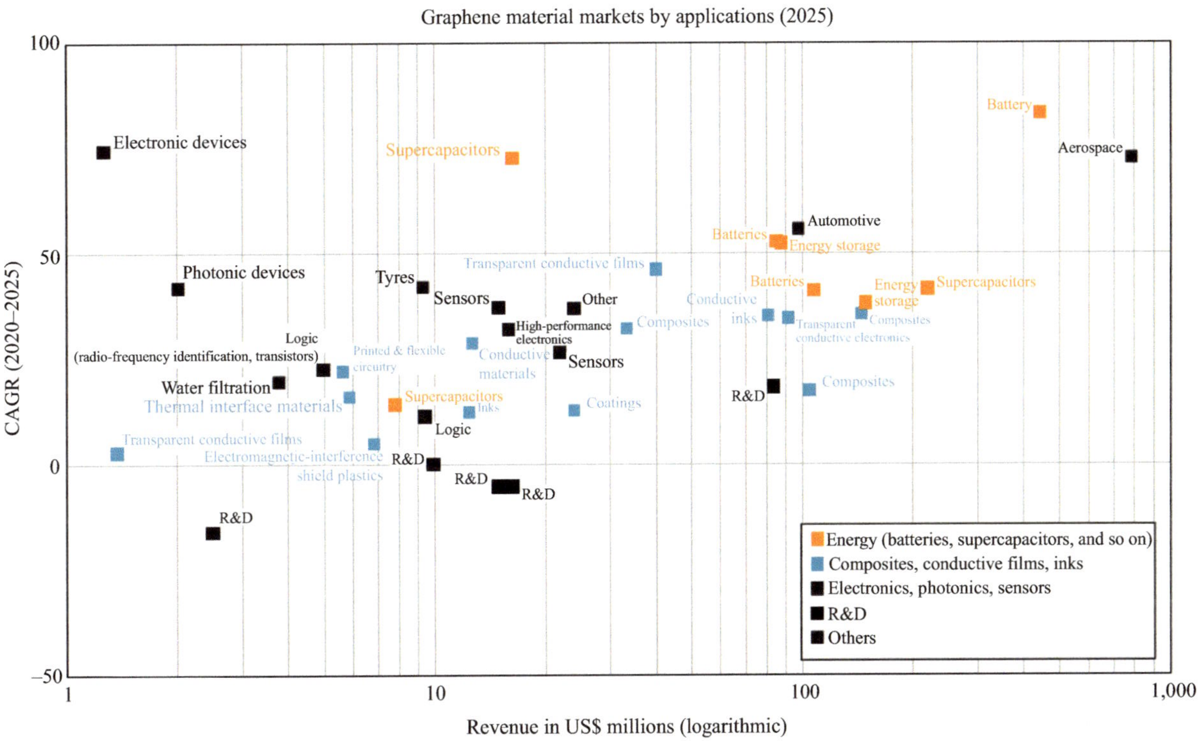

Figure 8.2 Estimated graphene market by revenue and CAGR. Reprinted from Ref. [2], Copyright 2019, by permission from Springer Nature.

level, commercially successful applications are mostly in the energy sector (battery and supercapacitors), composites, and electrodes. In terms of electronics and optoelectronics, successful prototype devices include photodetectors in the visible to infrared range and Hall sensors, but it might take 5–10 years before these devices attain adequate commercial success even in niche application fields. Unlike CNT, the lack of a sizable bandgap in graphene has precluded graphene from logic circuit developments. On the other hand, the high-frequency RF circuits have a smaller market size and well-established compound semiconductor industry. Therefore, there is a limited market opportunity for graphene-based electronic devices. It has been predicted that composites and inks will enter the market in the short term (<5 years), and graphene-based electrodes will be commercialized in supercapacitors, batteries, and solar cells in 5–10 years, while applications in optoelectronics and sensors, including medical devices, might emerge in the next 5–15 years [2].

8.2 Current Challenges and Future Directions

8.2.1 CNT-Based Nanocarbon Electronics

Even if the most stringent requirement of single-chirality semiconducting SWCNTs is lowered to a target that is easier to achieve such as high-purity semiconducting SWCNT network with similar bandgaps, a mature technology is still needed to be developed for optimal device performances, including high on/off current ratio, high on current, and low subthreshold slope. Obtaining pure semiconducting SWCNTs for electron devices such as FETs remains the most challenging task to achieve mass production. In Chapter 2, we learned that common vapor–liquid–solid growth mechanism usually results in SWCNT networks with random chirality distributions. Over the past three decades, researchers have made great progress using the following approaches to obtain controllable CNT synthesis in terms of electronic type (semiconducting or metallic), as well as spatial orientation and diameter control [3].

An approach that is usually adopted to prepare high-purity semiconducting SWCNT films relies on post-growth sorting and placement methods. The as-grown CNT is processed in the solution form by density gradient ultracentrifugation (DGU), or agarose-gel-based chromatograph separation to obtain a CNT solution with semiconducting-enriched SWCNTs. The semiconducting percentage can reach 99.99% with ultrahigh speed DGU, although the cost is very high. The high-purity semiconducting SWCNTs are then transferred to the targeting substrate by dielectrophoresis or laminar Langmuir–Blodgett methods. High packing density can be achieved with this method on various targeting substrates, including flexible materials. Many high-performance CNT devices and small-scale circuits are reported based on the solution-based high-purity semiconducting CNTs [4, 5]. The high-purity CNT solution is already commercialized [6], but the market is limited to research labs due to the high cost. Another problem is the high defect density introduced because of the harsh treatment steps. The resultant CNT length is very small, which makes it challenging to build large-scale ICs.

Another approach is to directly grow type-selected horizontally aligned SWCNTs at the desired location, which entails the in situ or post-growth removal/etching of the metallic SWCNTs or direct growth of semiconducting CNTs by engineering catalyst nanoparticles. The electrical breakdown approach by Joule heating was adopted to improve the on/off current ratio of CNFETs by selectively burning the metallic CNTs out. This method was used to build the first CNT computer [7]. For future ultra-scaled CNT devices whose channel consists of CNTs with high packing density, the burning process is not likely to be performed reliably as the heat generation would destroy surrounding CNTs. On the other hand, gas phase plasma using H_2/CH_4 selectively etches the metallic CNTs while retaining semiconducting ones due to the high reactivity of the metallic CNTs. Similarly, light irradiation preferentially oxidizes metallic ones, and microwave radiation selectively removes metallic ones due to its high dielectric constants (higher density of free charges). In situ selective etching during the CVD synthesis can also be performed based on the same principle. The radicals produced in the CVD chamber can selectively react with the metallic CNTs and the CNTs with small diameters. This approach has the potential to be scaled up to the wafer level, which can be promising for future IC development.

Through the choice of optimal substrate, it is also possible to directly synthesize semiconducting SWCNTs on specific crystalline substrates. Aligned CNT arrays can be grown on crystalline substrates such as quartz and sapphire due to van der Waals interaction between the substrates and grown CNTs. Catalyst nanoparticles with uniform size and crystallized structures can be used to synthesize CNTs with a narrow distribution of diameters and electronic type. A high purity of 92% single-chirality (12, 6) SWCNTs were obtained based on the bimetallic compound nanoparticles W_6Co_7 [8]. The treatment of nanoparticles under high vacuum/NH_3 annealing prior to the growth for narrowing the particle size distribution is beneficial to narrow the diameter distribution.

Currently, we are still far from the "holy grail" of CNT synthesis, that is, single-chirality SWCNT for a target application. One promising approach called CNT cloning from a CNT seed with the same chirality provides the solution toward this target [9]. Open-ended SWCNT segments are used as the seeds for following CNT growth, and the grown CNTs can maintain the same chirality of the small segments. If the seeds are presorted to contain a single-chirality CNT, the resultant CNT preserves the single-chirality characteristics without any catalyst for elongating the CNT segments. Combining all existing techniques such as the arc-discharge method to obtain high-quality CNT raw material, sorting technique to separate metallic and semiconducting CNTs to obtain high-purity semiconducting CNT solution, and dielectrophoresis to deposit aligned semiconducting CNT arrays with a high packing density, which can be used as the seeds for cloning CNT growth without catalyst, it is optimistic to reach the final goal for practical large-scale CNT IC fabrication.

Once the materials challenge is resolved, the next step would be polarity control to obtain the CMOS building block if we follow the roadmap of silicon-based ICs. Fortunately, polarity control through contact metal is quite mature to obtain both p-type and n-type CNT transistors. For example, high-performance p-type and n-type CNFETs are demonstrated with Pd and Sc forming respective ohmic contacts. With a self-aligned top-gate structure, symmetric p-type and n-type CNFETs with a scaled channel length down to 5 nm have been demonstrated with performance superior to their silicon counterparts with the same dimensions [10]. Compared to current silicon technology node, one advantage of adopting CNT transistors in the sub-10 nm regime is the lesser complexity of the device structure, as one can simply adopt the planar structure to control the ultrathin CNT channel instead of using the FinFET device structure. In ultra-scaled FETs, the velocity saturation effect dominates the saturation characteristics of the transistor. As the carriers in CNT have a much higher saturation velocity and much smaller effective mass than that of silicon, it is expected that CNT could provide a higher on current and more energy efficiency than silicon-based FETs.

For circuit fabrication, uniformity over a large area such as a 300 mm diameter wafer is also a daunting challenge. The growth of ultra-long CNT along one direction on the wafer surface would enable large-scale IC design with standard layout strategy. Recently, Zhu [11] demonstrated that the atomic assembly rate of the semiconducting CNT is 10 times higher than that of the metallic CNT. Although the initially grown CNTs contain both semiconducting and metallic CNTs, more and more metallic CNTs decay as the CNT length increases, leaving 99.99% semiconducting CNTs when the length exceeds a certain value. On the other hand, the packing density of the CNTs is substantially lower as more nanoparticles lose their catalytic activity during the elongating growth process [11]. Another problem is the packing density requirement. To compete with its silicon counterpart, the packing density should be higher than 125/μm [12].

In general, the main challenge for the foreseeable future is obtaining well-aligned, all-semiconducting SWCNT films with high uniformity, high array density, and low defect density on large wafers. Until now, we have not discussed interconnects in CNT ICs. The currently demonstrated CNT ICs consist of a relatively small number of CNFETs, and the layout design and the interconnect structure are simple compared to those for advanced silicon ICs with more than 10 layers of interconnects. For future complex CNT ICs, novel interconnects such as CNT and/or graphene-based interconnects discussed in Chapter 3 should be seriously considered to exploit fully the merits of CNFETs. Such development would lead to an all-carbon IC in the decades ahead.

8.2.2 Graphene-Based Nanocarbon Electronics

Fundamental research on graphene reveals major scientific discoveries such as Dirac fermions, integer quantum Hall effect, fractional quantum Hall effect, unconventional superconductivity in twisted bilayer graphene, to name just a few [13]. For technology developments, the properties of interests are high mobility and atomic thickness for electronic applications, strong intrinsic mechanical strength and flexibility for flexible electronics and sensor applications, and photoresponse in a wide spectrum, high optical absorption per layer combined with high carrier mobility for optoelectronic applications, lightweight, high surface-to-volume ratio, and high electrical/thermal conductivity for composites and energy applications.

For electronic applications, CVD graphene on copper is a mature process to yield single-layer graphene [14]. Despite the success of the CVD growth process, the resulting graphene is mostly polycrystalline with a domain size in the micrometer to millimeter range. Carrier mobility is not the hurdle for these applications even with the non-ideal polycrystalline graphene flakes. Scaling up is limited by the carbon phase transition rate on top of the copper catalytic surface. Cost could be even lowered if the growth process precludes high vacuum and high-temperature annealing steps. To apply the as-grown graphene materials, a post-growth process is required to transfer the graphene to the target substrate. Roll-to-roll transfer process is ideal for large-scale production, but this method also has the problem of introducing residues and cracks [14]. Dry transfer or semi-dry transfer, during which the graphene is not in direct contact with the polymer, maybe a better choice but needs to be scaled up in the future.

One recent study revealed that surface contamination is mainly due to the high-temperature process itself but not the transfer process [15]. Super-clean graphene was obtained based on a modified CVD process in which the Cu foil–Cu foam stacked structure is used as the growth substrate. The copper foam on top of the foil suppressed the carbon supply rate and the formation of amorphous carbon. The large surface area of the upper copper foam helps to continuously provide sufficient supply of Cu vapor to catalyze the decomposed carbon species. Meter-scale super-clean graphene with a lower contact resistance and larger carrier mobilities was obtained with this method, which is very promising for high-performance electronic and optoelectronic applications [15].

Graphene was perhaps overhyped in terms of applications following its discovery. As we stated at the beginning of this chapter, graphene for applications in the marketplace is quite different from the well-defined one in research labs and reported in scientific journals. The low quality and variety of graphene-based materials and the lack of standard on grade assessment of graphene have made it difficult to judge the true integrity of graphene in end-user products. Kauling [16] has critically reviewed graphene products in the market and found that many of these products had little relationship with true graphene but were more like graphitic microplatelets. The lack of stringent standards on graphene products is an impediment to robust development of electronic applications based on graphene nanomaterials [16]. To tackle this problem, a third-party inspection and assessment process is needed to provide end-product developers with valuable and transparent characterizations of graphene materials in the marketplace.

It requires enormous efforts to exploit the unique properties of graphene in practical applications. The existing graphene-related products include sports equipment such as motorcycle helmet, and coating for heat dissipation and electrostatic discharge in CMOS-integrated linear array sensors and earphones [17]. The atomic-scale thickness and ultrahigh mechanical strength enable its application as transparent electrodes. The unique electronic properties of graphene can be maintained even at a 5% strain, which are appealing for flexible electronics. Graphene has already been adopted in touchscreens for mobile phones, traditionally dominated by the brittle and expensive indium tin oxide (ITO) [18]. Similar applications include electrodes for solar cells, wearable electronics, and RF identifications.

The lack of sizable bandgaps has precluded graphene FETs, but the unique linear energy-momentum dispersion relation of graphene enables a strong photoresponse within a wide range of the spectrum ranging from ultraviolet to infrared frequency, which is beneficial for optoelectronic applications [19]. Various applications such as photodetectors, modulators, optical communications, and polarizers have been demonstrated with even better performance than their Si counterparts. The scaling up or further commercialization of these graphene-based optoelectronics will be mainly limited by the graphene quality, uniformity, and clean transfer process over a large area. A CMOS-compatible integration process protocol should be established for incorporating the transferred graphene for device fabrication based on the current mature semiconductor industry.

Other than directly using graphene for electronic products, emerging applications of graphene as the growth template to form high-quality 3D crystalline materials seem promising. Recent demonstrations include high-quality AlN growth on graphene for ultraviolet LED applications [20]. The weak van der Waals interaction between graphene and the 3D material enables epitaxial growth without stringent lattice match requirement, and the grown crystalline materials can be easily exfoliated and transferred to other substrates for device development.

8.2.3 From Dream to Reality

As the two most studied nanocarbons, CNT and graphene have unquestionably unique and superior properties. However, while taking advantage of these properties to meet market requirements, a thorough scientific understanding and innovative utilization of existing and future technology tools and cost–benefit analysis must also be considered. It is unclear at this time which is and will be the key driver in realizing large-scale manufacturing of nanocarbon electronic products with sizable markets. Perhaps it could be a combination of scientific breakthrough, advanced manufacturing technology development, and cost reduction that will eventually make it all possible. If and when that scenario comes to past, it would be conceivable that the successful and timely developments of CNT ICs, all-carbon interconnects, and graphene sensors and energy conversion/storage devices can pave the way for their integration into technology developments in artificial intelligence and related fields.

References

1. https://www.marketsandmarkets.com.
2. T. Reiss, K. Hjelt, and A. C. Ferrari. 2019. Graphene is on track to deliver on its promises, *Nature Nanotechnology*, **14**, 907–910.

3. I. Ibrahim, T. Gemming, W. M. Weber, T. Mikolajick, Z. Liu, and M. H. Rümmeli. 2016. Current progress in the chemical vapor deposition of type-selected horizontally aligned single-walled carbon nanotubes, *ACS Nano*, **10**, 7248–7266.
4. S.-J. Han, J. Tang, B. Kumar, A. Falk, D. Farmer, G. Tulevski, K. Jenkins, A. Afzali, S. Oida, J. Ott, J. Hannon, and W. Haensch. 2017. High-speed logic integrated circuits with solution-processed self-assembled carbon nanotubes, *Nature Nanotechnology*, **12**, 861–865.
5. L. Liu, L. Ding, D. Zhong, J. Han, S. Wang, Q. Meng, C. Qiu, X. Zhang, L.-M. Peng, and Z. Zhang. 2019. Carbon nanotube complementary gigahertz integrated circuits and their applications on wireless sensor interface systems, *ACS Nano*, **13**, 2526–2535.
6. http://nanointegris.com.
7. M. M. Shulaker, G. Hills, N. Patil, H. Wei, H.-Y. Chen, H. S. P. Wong, and S. Mitra. 2013. Carbon nanotube computer, *Nature*, **501**, 526–530.
8. F. Yang, X. Wang, J. Si, X. Zhao, K. Qi, C. Jin, Z. Zhang, M. Li, D. Zhang, J. Yang, Z. Zhang, Z. Xu, L.-M. Peng, X. Bai, and Y. Li. 2017. Water-assisted preparation of high-purity semiconducting (14,4) carbon nanotubes, *ACS Nano*, **11**, 186–193.
9. Y. Yao, C. Feng, J. Zhang, and Z. Liu. 2009. "Cloning" of single-walled carbon nanotubes via open-end growth mechanism, *Nano Letters*, **9**, 1673–1677.
10. C. Qiu, Z. Zhang, M. Xiao, Y. Yang, D. Zhong, and L.-M. Peng. 2017. Scaling carbon nanotube complementary transistors to 5-nm gate lengths, *Science*, **355**, 271.
11. Z. Zhu, N. Wei, W. Cheng, B. Shen, S. Sun, J. Gao, Q. Wen, R. Zhang, J. Xu, Y. Wang, and F. Wei. 2019. Rate-selected growth of ultrapure semiconducting carbon nanotube arrays, *Nature Communications*, **10**, 4467.
12. A. D. Franklin. 2013. The road to carbon nanotube transistors, *Nature*, **498**, 443–444.
13. F. Pulizzi, O. Bubnova, S. Milana, D. Schilter, D. Abergel, and A. Moscatelli. 2019. Graphene in the making, *Nature Nanotechnology*, **14**, 914–918.
14. W. Kong, H. Kum, S.-H. Bae, J. Shim, H. Kim, L. Kong, Y. Meng, K. Wang, C. Kim, and J. Kim. 2019. Path towards graphene commercialization from lab to market, *Nature Nanotechnology*, **14**, 927–938.
15. L. Lin, J. Zhang, H. Su, J. Li, L. Sun, Z. Wang, F. Xu, C. Liu, S. Lopatin, Y. Zhu, K. Jia, S. Chen, D. Rui, J. Sun, R. Xue, P. Gao, N. Kang, Y. Han, H. Q. Xu, Y.

Cao, K. S. Novoselov, Z. Tian, B. Ren, H. Peng, and Z. Liu. 2019. Towards super-clean graphene, *Nature Communications*, **10**, 1912.

16. A. P. Kauling, A. T. Seefeldt, D. P. Pisoni, R. C. Pradeep, R. Bentini, R. V. B. Oliveira, K. S. Novoselov, and A. H. Castro Neto. 2018. The worldwide graphene flake production, *Advanced Materials*, **30**, 1803784.
17. http://graphene-flagship.eu.
18. https://graphene-flagship.eu/news/Pages/Future-Phone.aspx.
19. F. Bonaccorso, Z. Sun, T. Hasan, and A. C. Ferrari. 2010. Graphene photonics and optoelectronics, *Nature Photonics*, **4**, 611.
20. Z. Chen, X. Zhang, Z. Dou, T. Wei, Z. Liu, Y. Qi, H. Ci, Y. Wang, Y. Li, H. Chang, J. Yan, S. Yang, Y. Zhang, J. Wang, P. Gao, J. Li, and Z. Liu. 2018. High-brightness blue light-emitting diodes enabled by a directly grown graphene buffer layer, *Advanced Materials*, **30**, 1801608.

Index